MW00514558

Programmable Digital
Signal Processors

Signal Processing and Communications

Programmable Digital Signal Processors

Architecture, Programming, and Applications

edited by

Yu Hen Hu

University of Wisconsin–Madison
Madison, Wisconsin

MARCEL DEKKER, INC.

NEW YORK • BASEL

ISBN: 0-8247-0647-1

This book is printed on acid-free paper.

Headquarters
Marcel Dekker, Inc.
270 Madison Avenue, New York, NY 10016
tel: 212-696-9000; fax: 212-685-4540

Eastern Hemisphere Distribution
Marcel Dekker AG
Hutgasse 4, Postfach 812, CH-4001 Basel, Switzerland
tel: 41-61-261-8482; fax: 41-61-261-8896

World Wide Web
http://www.dekker.com

The publisher offers discounts on this book when ordered in bulk quantities. For more information, write to Special Sales/Professional Marketing at the headquarters address above.

Current printing (last digit):
10 9 8 7 6 5 4 3 2 1

PRINTED IN THE UNITED STATES OF AMERICA

Series Introduction

Over the past 50 years, digital signal processing has evolved as a major engineering discipline. The fields of signal processing have grown from the origin of fast Fourier transform and digital filter design to statistical spectral analysis and array processing, and image, audio, and multimedia processing, and shaped developments in high-performance VLSI signal processor design. Indeed, there are few fields that enjoy so many applications—signal processing is everywhere in our lives.

When one uses a cellular phone, the voice is compressed, coded, and modulated using signal processing techniques. As a cruise missile winds along hillsides searching for the target, the signal processor is busy processing the images taken along the way. When we are watching a movie in HDTV, millions of audio and video data are being sent to our homes and received with unbelievable fidelity. When scientists compare DNA samples, fast pattern recognition techniques are being used. On and on, one can see the impact of signal processing in almost every engineering and scientific discipline.

Because of the immense importance of signal processing and the fast-growing demands of business and industry, this series on signal processing serves to report up-to-date developments and advances in the field. The topics of interest include but are not limited to the following:

- Signal theory and analysis
- Statistical signal processing
- Speech and audio processing
- Image and video processing

- Multimedia signal processing and technology
- Signal processing for communications
- Signal processing architectures and VLSI design

I hope this series will provide the interested audience with high-quality, state-of-the-art signal processing literature through research monographs, edited books, and rigorously written textbooks by experts in their fields.

K. J. Ray Liu

Preface

Since their inception in the late 1970s, programmable digital signal processors (PDSPs) have gradually expanded into applications such as multimedia signal processing, communications, and industrial control. PDSPs have always played a dual role: on the one hand, they are programmable microprocessors; on the other hand, they are designed specifically for digital signal processing (DSP) applications. Hence they often contain special instructions and special architecture supports so as to execute computation-intensive DSP algorithms more efficiently. This book addresses various programming issues of PDSPs and features the contributions of some of the leading experts in the field.

In Chapter 1, Kittitornkun and Hu offer an overview of the various aspects of PDSPs. Chapter 2, by Managuli and Kim, gives a comprehensive discussion of programming methods for very-long-instruction-word (VLIW) PDSP architectures; in particular, they focus on mapping DSP algorithms to best match the underlying VLIW architectures. In Chapter 3, Lee and Fiskiran describe native signal processing (a technique to enhance the performance of multimedia signal processing by general-purpose microprocessors) and compare various formats for multimedia extension (MMX) instruction. Chapter 4, by Tessier and Burleson, presents a survey of academic research and commercial development in reconfigurable computing for DSP systems over the past 15 years.

The next three chapters focus on issues in software development. In Chapter 5, Wu and Wolf examine the pros and cons of various options for implementing video signal processing applications. Chapter 6, by Yu and Hu, details a methodology for optimal compiler linear code generation. In Chapter 7, Chen et al. offer practical advice on proper design of multimedia algorithms using MMX instruction sets.

Chapter 8, by Bhattacharyya, addresses the relationship between hardware synthesis and software design, focusing particularly on automated mapping of high-level specifications of DSP applications onto programmable DSPs. In Chapter 9, Catthoor et al. discuss critical, yet often overlooked, issues of storage system architecture and memory management.

I would like to express my appreciation to the authors of each chapter for their dedication to this project and for their outstanding scholarly work. Thanks also go to chapter reviewers James C. Abel, Jack Jean, Konstantinos Konstantinides, Grant Martin, Miodrag Potkonjak, and Frederic Rousseau. Throughout this project, B. J. Clark, acquisitions editor, and Ray K. J. Liu, series editor, have provided strong encouragement and assistance. I thank them for their support and trust. I would also like to express my gratitude to Michael Deters, production editor, for his cooperation and patience.

Yu Hen Hu

Contents

Contributors

Shuvra S. Bhattacharyya   Department of Electrical and Computer Engineering and Institute for Advanced Computer Studies, University of Maryland at College Park, College Park, Maryland

Wayne Burleson   Department of Electrical and Computer Engineering, University of Massachusetts, Amherst, Massachusetts

Francky Catthoor   Design Technology for Integrated Information, IMEC, Leuven, Belgium

Yen-Kuang Chen   Microprocessor Research Laboratories, Intel Corporation, Santa Clara, California

Koen Danckaert   Design Technology for Integrated Information, IMEC, Leuven, Belgium

A. Murat Fiskiran   Department of Electrical Engineering, Princeton University, Princeton, New Jersey

Yu Hen Hu   Department of Electrical and Computer Engineering, University of Wisconsin–Madison, Madison, Wisconsin

Yongmin Kim   Department of Bioengineering, University of Washington, Seattle, Washington

Surin Kittitornkun   Department of Electrical and Computer Engineering, University of Wisconsin–Madison, Madison, Wisconsin

Chidamber Kulkarni   Design Technology for Integrated Information, IMEC, Leuven, Belgium

Ruby B. Lee   Department of Electrical Engineering, Princeton University, Princeton, New Jersey

Ravi A. Managuli   Department of Bioengineering, University of Washington, Seattle, Washington

Thierry Omnès   Design Technology for Integrated Information, IMEC, Leuven, Belgium

Birju Shah   Microprocessor Research Laboratories, Intel Corporation, Santa Clara, California

Russell Tessier   Department of Electrical and Computer Engineering, University of Massachusetts, Amherst, Massachusetts

Wayne Wolf   Department of Electrical Engineering, Princeton University, Princeton, New Jersey

Zhao Wu   Department of Electrical Engineering, Princeton University, Princeton, New Jersey

Jim K. H. Yu*   Department of Electrical and Computer Engineering, University of Wisconsin–Madison, Madison, Wisconsin

Nicholas Yu   Microprocessor Research Laboratories, Intel Corporation, Santa Clara, California

---

* *Current affiliation*: Tivoli Systems, Austin, Texas

1

Programmable Digital Signal Processors: A Survey

Surin Kittitornkun and Yu Hen Hu
University of Wisconsin–Madison, Madison, Wisconsin

1   INTRODUCTION

Programmable digital signal processors (PDSPs) are general-purpose micropro-cessors designed specifically for digital signal processing (DSP) applications. They contain special instructions and special architecture supports so as to exe-cute computation-intensive DSP algorithms more efficiently.

Programmable digital signal processors are designed mainly for embedded DSP applications. As such, the user may never realize the existence of a PDSP in an information appliance. Important applications of PDSPs include modem, hard drive controller, cellular phone data pump, set-top box, and so forth.

The categorization of PDSPs falls between the general-purpose micropro-cessor and the custom-designed, dedicated chip set. The former have the advan-tage of ease of programming and development. However, they often suffer from disappointing performance for DSP applications due to overheads incurred in both the architecture and the instruction set. Dedicated chip sets, on the other hand, lack the flexibility of programming. The time-to-market delay due to chip development may be longer than the program coding of programmable devices.

1.1   A Brief Historical Scan of PDSP Development

1.1.1   The 1980s to the 1990s

A number of PDSPs appeared in the commercial market in the early 1980s. Around 1980, Intel introduced the Intel2920, featuring on-chip A/D (analog-to-digital) and D/A (digital-to-analog) converters. Nonetheless, it had no hardware multiplier, and it was difficult to load program parameters into the chip due to the lack of digital interface. At almost the same time, NEC introduced the NEC MPD7720. It is equipped with a hardware multiplier and is among the first to adopt the Harvard architecture with physically separate on-chip data memory and program memory. Texas Instruments introduced the TMS320C10 in 1982. Similar to the MPD7720, the 'C10 adopts the Harvard architecture and has a hardware multiplier. Furthermore, the 'C10 is the first PDSP that can execute instructions from off-chip program memory without performance penalty due to off-chip memory input/output (I/O). This feature brought PDSPs closer to the microprocessor/microcontroller programming model. In addition, the emphasis on development tools and libraries by Texas Instruments led to widespread applications of PDSP. The architectural features of several representative examples of these early PDSP chips are summarized in Table 1.

In these early PDSPs, DSP-specific instructions such as MAC (multiply-and-accumulate), DELAY (delay elements), REPEAT (loop control), and other flow control instructions are devised and included in the instruction set so as to improve both programmability and performance. Moreover, special address generator units with bit-reversal addressing mode support have been incorporated to enable efficient execution of the fast Fourier transform (FFT) algorithm. Due to limitation of chip area and transistor count, the on-chip data and program memories are quite small in these chips. If the program cannot fit into the on-chip memory, a significant performance penalty will incur.

Later, floating-point PDSPs, such as Texas Instruments' TMS320C30 and Motorola's DSP96001 appeared in the market. With fixed-point arithmetic as in early PDSPs, the dynamic range of the intermediate results must be carefully monitored to prevent overflow. Some reports estimated that as much as one-third of the instruction cycles in executing PDSP programs are wasted on checking the overflow condition of intermediate results. A key advantage of a floating-point arithmetic unit is its extensive dynamic range. Later on, some PDSPs also included on-chip DMA (direct memory access) controllers, as well as a dedicated DMA bus that allowed concurrent data I/O at the DMA unit, and signal processing computation in the CPU.

1.1.2   The 1990s to 2000

In this decade, the following trends in PDSPs emerged.

Consolidation of PDSP Market.   Unlike the 1980s, in which numerous PDSP architectures had been developed, the 1990s are noted for a consolidation of the PDSP market. Only very few PDSPs are now available in the market. Notably, Texas Instrument's TMS320Cxx series captured about 70% of the PDSP

Table 1 Summary of Characteristics of Early PDSPs

Model	Manufacturer	Year	On-chip data RAM	On-chip data ROM	On-chip program RAM	Multiplier
A100	Inmos		—	—	—	4, 8, 12, 16
ADSP2100	Analog Device	1986	—	—	—	$16 \times 16 \rightarrow 32$
DSP16	AT&T	1984	512×16	$2K \times 16$	—	$16 \times 16 \rightarrow 32$
DSP32	AT&T		$1K \times 32$	512×32	—	$32 \times 32 \rightarrow 40$
DSP32C	AT&T	1988	$1K \times 32$	$2K \times 32$	—	$32 \times 32 \rightarrow 40$
DSP56000	Motorola	1986	512×24	512×24	512×24	$24 \times 24 \rightarrow 56$
DSP96001	Motorola	1988	$1K \times 32$	$1K \times 32$	544×32	$32 \times 32 \rightarrow 96$
DSSP-VLSI	NTT	1986	512×18	—	$4K \times 18$	(18-bit) 12E6
Intel2920	Intel	1980	40×25	—	192×24	—
LM32900	National		—	—	—	—
MPD7720	NEC	1981	128×16	512×13	512×23	$16 \times 16 \rightarrow 32$
MSM 6992	OKI	1986	256×32	—	$1K \times 32$	(22-bit) 16E6
MSP32	Mitsubishi		256×16	—	$1K \times 16$	$32 \times 16 \rightarrow 32$
MB8764	Fujitsu		256×16	—	$1K \times 24$	—
NEC77230	NEC	1986	$1K \times 32$	$1K \times 32$	$2K \times 32$	$24E8 \rightarrow 47E8$
TS68930	Thomson		256×16	512×16	$1K \times 32$	$16 \times 16 \rightarrow 32$
TMS32010	TI	1982	144×16	—	$1.5K \times 16$	$16 \times 16 \rightarrow 32$
TMS320C25	TI	1986	288×16	—	$4K \times 16$	$16 \times 16 \rightarrow 32$
TMS320C30	TI	1988	$2K \times 32$	—	$4K \times 32$	$32 \times 32 \rightarrow 32E8$
ZR34161 VSP	Zoran		128×32	$1K \times 16$	—	16-Bit vector engine

market toward the end of this decade. Within this family, the traditional TMS320C10/20 series has evolved into TMS320C50 and has become one of the most popular PDSPs. Within this TMS family, TMS320C30 was introduced in 1988 and its floating-point arithmetic unit has attracted a number of scientific applications. Other members in this family that were introduced in the 1990s include TMS320C40, a multiprocessing PDSP, and TMS320C80, another multi-processing PDSP designed for multimedia (video) applications. TMS320C54xx and TMS320C6xx are the recent ones in this family. Another low-cost PDSP that has emerged as a popular choice is Analog Device's SHARC processor. These modern PDSP architectures will be surveyed in later sections of this chapter.

DSP Core Architecture.   As the feature size of the digital integrated circuit continues to shrink, more and more transistors can be packed into a single chip. As such, it is possible to incorporate peripheral (glue) logics and supporting logic components into the same chip in addition to the PDSP. This leads to the notion of the *system on (a) chip* (SoC). In designing a SoC system, an existing PDSP core is incorporated into the overall system design. This design may be represented as a VHDL (very-high-speed integrated circuit hardware description language)/Verilog core, or in a netlist format. A PDSP that is used in this fashion is known as a *processor core* or a *DSP core*.

In the 1990s, many existing popular PDSP designs had been converted into DSP cores so that the designers could design new applications using familiar instruction sets or even existing programs. On the other hand, several new PDSP architectures are being developed and licensed as DSP cores. Examples of these DSP cores, including Carmel, R.E.A.L., StarCore, and V850, will be reviewed in Section 4.

Multimedia PDSPs.   With the development of international multimedia standards such as JPEG image compression (Pennebaker and Mitchell, 1993), MPEG video coding (Mitchell et al., 1997), and MP3 audio, there is an expanding market for low-cost, dedicated multimedia processors. Due to the complexity of these standards, it is difficult to develop a multimedia processor architecture without any programmability. Thus, a family of multimedia enhanced PDSPs—such as MPACT, TriMedia, TMS320C8x, and DDMP (Terada et el., 1999)—have been developed. A key feature of these multimedia PDSPs is that they are equipped with various special multimedia-related function units, for instance, the YUV to RGB (color coordinates) converter, the VLC (variable-length code) entropy encoder/decoder, and the motion estimation unit. In addition, they facilitate direct multimedia signal I/O, bypassing the bottleneck of a slow system bus.

Native Signal Processing with Multimedia Extension Instructions.   By native signal processing (NSP), the signal processing tasks are executed in the general-purpose microprocessor, rather than in a separate coprocessing PDSP. As their speed increases, a number of signal processing operations can be performed without additional hardware or dedicated chip sets. In the mid-1990s, Intel introduced the MMX (MultiMedia eXtension) instruction set to the Pentium series microprocessor. Because modern microprocessors have a long internal word length of 32, 64, or even extended 128 bits, several 8-bit or 16-bit multimedia data samples can be packed into a single internal word to facilitate the so-called *subword parallelism*. By processing several data samples in parallel in a single instruction, better performance can be accomplished while processing especially multimedia streams.

1.1.3   Hardware Programmable Digital Signal Processors: FPGA

An FPGA (field programmable gate array) is a software-configurable hardware device that contains (1) a substantial amount of uncommitted combinational logic; (2) preimplemented flip-flops; and (3) programmable interconnections among the combinational logic, flip-flops, and the chip I/O pins. The downloaded configuration bit stream programs all the functions of the combinational logic, flip-flops, and the interconnections. Although not the most efficient, an FPGA can be used to accelerate DSP applications in several different ways (Knapp, 1995):

1. An FPGA can be used to implement a complete application-specific integrated circuit (ASIC) DSP system. A shortcoming of this approach is that current FPGA technology does not yield the most efficient hardware implementation. However, FPGA implementation has several key advantages: (1) time-to-market is short, (2) upgrade to new architecture is relatively easy, and (3) low-volume production is cost effective.
2. An FPGA can act as a coprocessor to a PDSP to accelerate certain specific DSP functions that cannot be efficiently implemented using conventional architecture.
3. Furthermore, an FPGA can be used as a rapid prototyping system to validate the design of an ASIC and to facilitate efficient, hardware-in-the-loop debugging.

1.2   Common Characteristics of DSP Computation

1.2.1   Real-Time Computation

Programmable digital signal processors are often used to implement real-time applications. For example, in a cellular phone, the speed of speech coding must match that of normal conversation. A typical real-time signal processing application has three special characteristics:

1. The computation cannot be initiated until the input signal samples are received. Hence, the result cannot be precomputed and stored for later use.
2. Results must be obtained before a prespecified deadline. If the deadline is violated, the quality of services will be dramatically degraded and even render the application useless.
3. The program execution often continues for an indefinite duration of time. Hence, the total number of mathematical operations needed to be performed per unit time, known as *throughput*, becomes an important performance indicator.

1.2.2   Data Flow Dominant Computation

Digital signal processor applications involve stream media data types. Thus, instead of supporting complex control flow (e.g., context switch, multithread processing), a PDSP should be designed to streamline data flow manipulation. For example, special hardware must be designed to facilitate efficient input and output of data from PDSP to off-chip memory, to reduce overhead involved in accessing arrays of data in various fashions, and to reduce overhead involved in the execution of multilevel nested DO loops.

1.2.3   Specialized Arithmetic Computation

Digital signal processor applications often require special types of arithmetic operations to make computations more efficient. For example, a convolution operation

$$y(n) = \sum_{k=0}^{K-1} x(k)h(n-k)$$

can be realized using a recursion

$$y(n) = 0; \quad y(n) = y(n) + x(k) * h(n-k), \quad k = 0, 1, 2, \ldots, K-1$$

For each k, a multiplication and an addition (accumulation) are to be performed. This leads to the implementation of MAC instruction in many modern PDSPs:

$$R4 \leftarrow R1 + R2 * R3$$

Modern PDSPs often contain hardware support of the so-called *saturation arithmetic*. In saturation arithmetic, if the result of computation exceeds the dynamic range, it is clamped to either the maximum or the minimum value; that is, $9 + 9 = 15$ ($01001_2 + 01001_2 = 01111_2$) in 2's complement arithmetic. Therefore, for applications in which saturation arithmetic is applicable, there will be no need to check for overflow during the execution. These special instructions are also implemented in hardware. For example, to implement a saturation addition function using 2's complement arithmetic without intrinsic function support, we have the following C code segment:

```
int sadd(int a, int b) {
int result;
    result = a + b;
    if (((a ∧ b) & 0x80000000) == 0) {
        if ((result ∧ a) & 0x80000000) {
            result = (a < 0) ? 0x80000000 : 0x7fffffff;
        }
    }
    return (result);
}
```

However, with a special _sadd intrinsic function support in TMS320C6x (Texas Instruments, 1998c) the same code segment reduces to the single line:

```
result = _sadd(a,b);
```

1.2.4 Execution Control

Many DSP algorithms can be formulated as nested, indefinite Do loops. In order to reduce the overhead incurred in executing multilevel nested loops, a number of special hardware supports are included in PDSPs to streamline the control flow of execution.

1. Zero-overhead hardware loop: A number of PDSPs contain a special REPEAT instruction to support efficient execution of multiple loop nests using dedicated counters to keep track of loop indices.
2. Explicit instruction-level parallelism (ILP): Due to the deterministic data flow of many DSP algorithms, ILP can be exploited at compile time by an optimizing compiler. This led several modern PDSPs to adopt the very long instruction word (VLIW) architecture to efficiently utilize the available ILP.

1.2.5 Low-Power Operation and Embedded System Design

1. The majority of applications of PDSPs are embedded systems, such as a disk drive controller, modem, and cellular phone. Thus, many PDSPs are highly integrated and often contain multiple data I/O function units, timers, and other function units in a single chip packaging.
2. Power consumption is a key concern in the implementation of embed-

ded systems. Thus, PDSPs are often designed to compromise between conflicting requirements of high-speed data computation and low-power consumption (Borkar, 1999). The specialization of certain key functions allows efficient execution of the desired operations using a high degree of parallelism while holding down the power source voltage and overall clock frequency to conserve energy.

1.3   Common Features of PDSPs

1.3.1   Harvard Architecture

A key feature of PDSPs is the adoption of a *Harvard memory architecture* that contains separate program and data memory so as to allow simultaneous instruction fetch and data access. This is different from the conventional *Von Neumann architecture*, where program and data are stored in the same memory space.

1.3.2   Dedicated Address Generator

The address generator allows rapid access of data with complex data arrangement without interfering with the pipelined execution of main ALUs (arithmetic and logic units). This is useful for situations such as two-dimensional (2D) digital filtering and motion estimation. Some address generators may include a bit-reversal address calculation to support the efficient implementation of FFT, and circular buffer addressing for the implementation of infinite impulse response (IIR) digital filters.

1.3.3   High Bandwidth Memory and I/O Controller

To meet the intensive input and output demands of most signal processing applications, several PDSPs have built-in multichannel DMA channels and dedicated DMA buses to handle data I/O without interfering with CPU operations. To maximize data I/O efficiency, some modern PDSPs even include a dedicated video and audio codec (coder/decoder) as well as a high-speed serial/parallel communication port.

1.3.4   Data Parallelism

A number of important DSP applications exhibit a high degree of data parallelism that can be exploited to accelerate the computation. As a result, several parallel processing schemes, SIMD (single instruction, multiple data) and MIMD (multiple instruction, multiple data) architecture have been incorporated in the PDSP. For example, many multimedia-enhanced instruction sets in general-purpose microprocessors (e.g., MMX) employed subword parallelism to speed up the execution. It is basically an SIMD approach. A number of PDSPs also facilitate MIMD implementation by providing multiple interprocessor communication links.

2   APPLICATIONS OF PDSP

In this section, both real-world and prototyping applications of PDSPs are surveyed. These applications are divided into three categories: communication systems, multimedia, and control/data acquisitions.

2.1   Communications Systems

Programmable digital signal processors have been applied to implement various communication systems. Examples include caller ID [using TMS320C2xx (Texas Instruments Europe, 1997)], cordless handset, and many others. For voice communication, an acoustic-echo cancellation based on the normalized least mean square (NLMS) algorithm for hands-free wireless system is reported in (Texas Instruments, 1997). Implemented with a TMS320C54, this system performs both active-channel and double-talk detection. A 40-MHz TMS320C50 fixed-point processor is used to implement a low-bit-rate (1.4 Kbps), real-time vocoder (voice coder) (Yao et al., 1998). The realization also includes both the decoder and the synthesizer. A telephone voice dialer (Pawate and Robinson, 1996) is implemented with a 16-bit fixed-point TMS320C5x PDSP. It is a speaker-independent speech recognition system based on the hidden Markov model algorithm.

Modern PDSPs are also suitable for error correction in digital communication. A special Viterbi shift left (VSL) instruction is implemented on both the Motorola DSP56300 and the DSP56600 PDSPs (Taipale, 1998) to accelerate the Viterbi decoding. Another implementation of the ITU V.32bis Viterbi decoding algorithm using a TMS320C62xx is reported by Yiu (Yiu, 1998). Yet another example is the implementation of the U.S. digital cellular error-correction coding algorithm, including both the tasks of source coding/decoding and ciphering/ deciphering on a TMS320C541 evaluation module (Chishtie, 1994).

Digital baseband signal processing is another important application of PDSPs. A TMS320C25 DSP-based GMSK (Gaussian minimum shift keying) modem for Mobitex packet radio data communication is reported in (Resweber, 1996). In this implementation, transmitted data in packet form is level-shifted and Gaussian-filtered digitally within the modem algorithm so that it is ready for transmitter baseband interface, either via a D/A converter or by direct digital modulation. Received data at either baseband or the intermediate frequency (IF) band from the radio receiver is digitized and processed. Packet synchronization is also handled by the modem, assuring that the next layer sees only valid Mobitex packets.

System prototyping can be accomplished using PDSP due to its low cost and ease of programming. A prototype of reverse channel transmitter/receiver for asymmetric digital subscriber line (ADSL) algorithm (Gottlieb, 1994) is implemented using a floating-point DSP TMS320C40 chip clocked at 40 MHz. The program consisted of three parts: synchronization, training, and decision-directed detection.

Navigation using the Global Positioning System (GPS) has been widely accepted for commercial applications such as electronic direction finding. A software-based GPS receiver architecture using the TMS320C30 processor is described in (Kibe et al., 1996). The 'C30 is in charge of signal processing tasks such as correlation, FFT, digital filtering, decimation, demodulation, and Viterbi decoding in the tracking loop. Further investigation on the benefits of using a PDSP in a GPS receiver with special emphasis on fast acquisition techniques is reported in (Daffara and Vinson, 1998). The GPS L1 band signal is down-converted to IF. After A/D conversion, the signal is processed by a dedicated hardware in conjunction with algorithms (software) on a PDSP. Functions that are fixed and require high-speed processing should be implemented in dedicated hardware. On the contrary, more sophisticated functions that are less time-sensitive can be implemented using PDSPs.

For the defense system application, a linear array of TMS320C30 as the front end and a Transputer™ processor array as the back end for programmable radar signal processing are developed to support the PDDR (Point Defense Demonstration Radar) (Alter et al., 1991). The input signal is sampled at 10 MHz to 16-bit, complex-valued samples. The PDSP front end performs pulse compression, moving target indication (MTI), and constant false alarm (CFA) rate detection.

2.2 Multimedia

2.2.1 Audio Signal Processing

The audible signals cover the frequency range from 20 to 20,000 Hz. PDSP applications to audio signal processing can be divided into three categories according to the qualities and audible range of the signal (Ledger and Tomarakos, 1998): professional audio products, consumer audio products, and computer audio multimedia systems. The DSP algorithms used in particular products are summarized in Table 2.

MP3 (MPEG-I Layer 3 audio coding) has achieved the status of the most popular audio coding algorithm in recent years. The PDSP implementation of MP3 decoder can be found in Robinson et al. (1998). On the other hand, most

Table 2 DSP Algorithms for Audio Applications

Application	DSP algorithms used
Professional audio products	
Digital audio effects processors (reverb, chorus, flanging, vibrato pitch shifting, dyn ran. compression, etc.)	Delay-line modulation/interpolation, digital filtering (comb, FIR, etc.)
Digital mixing consoles level detection, volume control	Filtering, digital amplitude panning
Digital audio tape (DAT)	Compression techniques: MPEG
Electronic music keyboards physical modeling	Wavetable/FM synthesis, sample playback
Graphic and parametric equalizers	Digital FIR/IIR filters
Multichannel digital audio recorders	ADPCM, AC-3
Room equalization	Filtering
Speaker equalization	Filtering
Consumer audio products	
CD-I	ADPCM, AC-3, MPEG
CD players and recorders	PCM
Digital amplifiers/speakers	Digital filtering
Digital audio broadcasting equipment	AC-3, MPEG, and so forth
Digital graphic equalizers	Digital filtering
Digital versatile disk (DVD) players	AC-3, MPEG, and so forth
Home theater systems (surround-sound receivers/tuners)	AC-3, Dolby ProLogic, THX DTS, MPEG, hall/auditorium effects
Karaoke	MPEG, audio effects algorithms
Satellite (DBS) broadcasting	AC-3, MPEG
Satellite receiver systems	AC-3
Computer audio multimedia systems	
Sound card	ADPCM, AC-3, MP3, MIDI, etc.
Special-purpose headsets	3D positioning (HRTFs)

synthesized sounds such as those used in computer gaming are still represented in the MIDI (Yim et al., 1998). It can be seen that PDSPs are good candidates for the implementation of these audio signal processing algorithms.

2.2.2 Image/Video Processing

Existing image and video compression standards such as JPEG and MPEG are based on the DCT (discrete cosine transform) algorithm. The upcoming JPEG 2000 image coding standard will also include coding algorithms that are based on the discrete wavelet transform (DWT). These standards are often implemented in modern digital cameras and digital camcorders, in which PDSPs will play an important role. An example of using the TMS320C549 to implement a digital camera is reported in Illgner et al. (1999), where the PDSP can be upgraded later to incorporate the upcoming JPEG 2000 standard.

Low-bit-rate video coding standards include the ITU H.263+/H.263M and MPEG4 simple profile. In Budagavi et al. (1999), the potential applications of TMS320C54x family chips to implement low-power, low-bit-rate video coding algorithms are discussed. On the other hand, decoding of MPEG-II broadcasting grade video sequences using either the TMS320C80 (Bonomini et al., 1996) chip or the TMS320C6201 (Cheung et al., 1999) chip has been reported.

Medical imaging has become another fast-growing application area of PDSPs. Reported in Chou et al. (1997) is the use of TMS320C3x as a controller and on-line data processor for processing magnetic resonance imaging (MRI). It can perform real-time dynamic imaging such as cardiac imaging, angiography (examination of the blood vessels using x-rays following the injection of a radi-opaque substance), and abdominal imaging. Recently, an implementation of real-time data acquisition, processing, and display of ungated cardiac movies at 20 frames/sec using PDSPs was reported in Morgan et al. (1999).

2.2.3   Printing

The current printer consists of embedded processors to process various formats of page description languages (PDLs) such as PostScript. In Ganesh and Thakur (1999), a PDSP is used to interpret the PDL code, to create a list of elements to be displayed, and to estimate the time needed to render the image. Rendering is the process of creating the source pixel map. In this process, a common source map is 600×600 pixels per square inch, with four colors for each pixel, and eight bits for each color. Compression is necessary to store the output map when rendering and screening cannot be completed within the real-time requirement. This phase involves JPEG compression and matrix transformations and interpolations. Depending on the characteristics of the screened image and the storage memory available, the compressed image may be either lossless or lossy. Decompression of the bit-mapped image occurs in real time as the compressed image is fed to the print engine. The screening process converts the source pixel map into the appropriate output format. Because the process must be repeated for all pixels, the number of calculations is enormous for a high-resolution color image, especially in real time.

2.2.4   SAR Image Processing

Synthetic aperture radar (SAR) signal processing possesses a significant challenge due to its very large computation and data storage requirements. A sensor transmits pulses and receives the echoes in a direction approximately perpendicular to the direction of travel. The problem becomes 2D space-variant convolution using the range-Doppler algorithm, in which all the signals and coefficients are complex numbers with a precision of at least 16 bits. A heterogeneous architecture—vector/scalar architecture—is proposed and analyzed (Meisl, 1996). The vector processor (using Sharp LH9124 for FFTs) and the scalar processing unit (using eight SHARC 21060's connected in a mesh network) are chosen based on performance, scalability, flexibility, development cost, and repeat cost-evaluation criterion. The design is capable of processing SAR data at about one-tenth of the real-time rate.

2.2.5 Biometric Information Processing

Handwritten signature verification, one of the biometric authentication techniques, is inexpensive, reliable, and nonintrusive to the person being authorized. A DSP kernel for on-line verification using the TMS32010 with a 200-Hz sampling rate was developed (Dullink et al., 1995). The authentication kernel comprises a personalized table and some general-purpose procedures. This verification method can be part of a variety of entrance monitoring and security systems.

2.3 Control and Data Acquisition

As expected, PDSP has found numerous applications in modern control and data acquisition applications as well. Several control applications are implemented using Motorola DSP56000 PDSPs that function as both powerful microcontrollers and as fast digital signal processors. Its 56-bit accumulator (hence, the code name 56xxx) provides 8-bit extension registers in conjunction with saturation arithmetic to allow 256 successive consecutive additions without the need to check for overflow condition or limit cycles. The output noise power due to round-off noise of the 24-bit DSP56000/DSP56001 is 65,536 times less than that for 16-bit PDSPs and microcontrollers. Design examples include a PID (proportional-integral-derivative) controller (Stokes and Sohie, 1990) and an adaptive controller (Renard, 1992).

Another example of DSP system development is the Computer-Assisted Dynamic Data Monitoring and Analysis System (CADDMAS) project developed for the U.S. Air Force and NASA (Sztipanovits et al., 1998). It is applied to turbine engine stress testing and analysis. The project makes use of TMS320C40 for distributed-memory parallel PDSPs. An application-specific topology interconnects 30 different systems, with processor counts varying from 4 to 128 processors. More than 300 sensors are used to measure signals with a sampling rate in excess of 100 kHz. Based on measured signals, the system performs spectral analysis, autocorrelation and cross-correlation, tricoherence, and so forth.

2.4   DSP Applications of Hardware Programmable PDSP

There are a variety of FPGA implementation examples of specific DSP functions, such as the FIR (finite impulse response) digital filter DFT/FFT (discrete Fourier transform/fast Fourier transform) processor (Dick, 1996), image/video processing (Schoner et al., 1995), wireless CDMA (Code Division Multiple Access) rake receiver (Shankiti and Lesser, 2000), and Viterbi decoding (Goslin, 1995).

2.4.1   16-Tap FIR Digital Filter

A distributed arithmetic (DA) implementation of a 16-tap finite impulse response digital filter has been reported in Goslin (1995). The DA implementation of the multiplier uses look-up tables (LUTs). Because the product of two n-bit integers will have 2^{2n} different results; the size of the LUT increases exponentially with respect to the word length. For practical implementation, compromises must be made to trade additional computation time for smaller number of LUTs.

2.4.2   CORDIC-Based Radar Processor

The improvement of FPGA-based CORDIC arithmetic implementation is studied further in (Andraka, 1998). The iteration process of CORDIC can be unrolled so that each processing element always performs the same iteration. Unrolling the processor results in two significant simplifications. First, shift amounts become fixed and can be implemented in the wiring. Second, constant values for the angle accumulator are distributed to each adder in the angle accumulator chain and can be hardwired instead of requiring storage space. The entire processor is reduced to an array of interconnected adder–subtractors, which is strictly combinatorial. However, the delay through the resulting circuit can be substantial but can be shortened using pipelining without additional hardware cost. A 14-bit, 5-iteration pipelined CORDIC processor that fits in half of an Xilinx XC4013E-2 runs at 52 MHz. This design is used for high-throughput polar-to-Cartesian coordinate transformations in a radar target generator.

2.4.3   DFT/FFT

An FPGA-based systolic DFT array processor architecture is reported in Dick (1996). Each processing element (PE) contains a CORDIC arithmetic unit, which consists of a series of shift and adds to avoid the requirement for area-consuming multiplier. The timing analyzer *xdelay* determines the maximum clock frequency to be 15.3 MHz implemented on a Xilinx XC4010 PG191-4 FPGA chip.

2.4.4   Image/Video Signal Processor

In Schoner et al. (1995), the implementation of an FPGA-augmented low-complexity video signal processor was reported. This combination of ASIC and

Table 3 Achievable Frame Rate of Four Different Image
Processing Operations

Algorithms	Frames/sec	Latency (msec)
7 × 7 Mask 2D filter	13.3	75.2
8 × 8 Block DCT	55.0	18.2
4 × 4 Block vector, quantization at 0.5 bit/pixel	7.4	139.0
One-level wavelet transform	35.7	28.0

FPGA is flexible enough to implement four common algorithms in real time. Specifically, for 256 × 256 × 8-bit pictures, this device is able to achieve the frame rates presented in Table 3.

2.4.5 CDMA Rake Receiver

A CDMA rake receiver for a real-time underwater data communication system has been implemented using four Xilinx XC4010 FPGA chips (Shankiti and Lesser, 2000) with one multiplier on each chip. The final design of each multiplier occupies close to 1000 CLBs (configurable logic blocks) and is running at a clock frequency of 1 MHz.

2.4.6 Viterbi Decoder

Viterbi decoding is used to achieve maximum likelihood decoding of a binary stream of symbols. Because it involves bit-stream operations, it cannot be efficiently implemented using the word-parallel architecture of general-purpose microprocessors or PDSPs. It has been reported (Goslin, 1995) that a Xilinx XC4013E-based FPGA implementation of a Viterbi decoder achieves 2.5 times processing speed (135 nsec versus 360 nsec) compared to a dual-PDSP implementation of the same algorithm.

3 PERFORMANCE MEASURES

The comparison of the performance between PDSP and general-purpose microprocessors, between various PDSPs, and between PDSPs and dedicated hardware chip sets is a very difficult task. A number of factors contribute to this difficulty:

1. *A set of objective performance metrics is difficult to define for PDSPs.*
   It is well known that with modern superscalar instruction architecture,

the usual metrics such as MIPS (millions instructions per section) and FLOPS (floating-point operations per second) are no longer valid metrics to gauge the performance of these microprocessors. Some PDSPs also adopt such architecture. Hence, a set of appropriate metrics is difficult to define.

2. *PDSPs have fragmented architecture.* Unlike general-purpose microprocessors that have converged largely to a similar data format (32-bit or 64-bit architecture), PDSPs have a much more fragmented architecture in terms of internal or external data format and fixed-point versus floating-point operations. The external memory interface is varied on platform by platform basis. This is due to the fact that most PDSPs are designed for embedded applications and, hence, cross-platform compatibility is not of major concern for different manufacturers of PDSPs. Furthermore, PDSPs often have specialized hardware to accelerate a special type of operation. Such specialized hardware makes the comparison even more difficult.

3. *PDSP applications are often hand programmed with respect to a particular platform.* The performance of cross-platform compilers is still far from realistic. Hence, it is not meaningful to run the same high-level language benchmark program on different PDSP platforms.

Some physical parameters of PDSPs are summarized in Table 4.

Usually peak MIPS, MOPS, MFLOPS, MAC/sec, and MB/sec for a particular architecture are just the product of instructions, operations, floating-point operations, multiply-accumulate operations, and memory access in bytes executed in parallel multiplied by maximum clock frequency, respectively. They can be achieved instantaneously in real applications at certain clock cycle and some-

Table 4 Physical Performance Parameters

Parameters	Units
Maximum clock frequency	MHz
Power consumption	Absolute power, watts (W), power (W)/MIPS
Execution throughput, peak and sustained	MIPS, MOPS (million operations/sec), MACS (no. of MAC/sec), MFLOPS
Operation latency	Instruction cycles
Memory access	Clock cycles
Bandwidth	MB/sec (megabytes per second)
Latency	Clock cycle
Input/output	No. of ports

Table 5 Examples of DSP Benchmarks

Level	Algorithm/application names
Kernel	
General	FFT/IFFT, FIR, IIR, matrix/vector multiply, Viterbi decoding, LMS (least mean square) algorithm
Multimedia/graphic	DCT/IDCT, VLC (variable-length code) decoding, SAD
Application	
General	Radar (Bhargava et al., 1998)
Multimedia/graphic	MediaBench (Lee et al., 1997) G.722, JPEG, Image (Bhargava et al., 1998)

how misleading. From a user's perspective, the ultimate performance measure is the "execution time" (wall clock time) of individual benchmark.

Recently, efforts have been made to establish benchmark suites for PDSPs. The proposed benchmark suites (Bhargava et al., 1998; Lee et al., 1997) can be categorized into kernel and application levels. They can be classified into general DSP and multimedia/graphic. Because each kernel contributes to the run time of each application at some certain percentage of run time and each application may contain more than single DSP kernel, conducting benchmark tests at both levels gives more accurate results than just the raw number of some DSP kernels. A number of DSP benchmarks are summarized in Table 5.

4 MODERN PDSP ARCHITECTURES

In this section, several modern PDSP architectures will be surveyed. Based on different implementation methods, modern PDSPs can be characterized as PDSP chip, PDSP core, multimedia PDSPs, and NSP instruction set. The following aspects of these implementation approaches are summarized in terms of three general sets of characteristics:

1. Program (instruction) execution
2. Datapath
3. Physical implementation

Program execution of the PDSP is characterized by processing core (how the PDSP achieves parallelism), instruction width (bits), maximum number of instructions issued, and address space of program memory (bits). Its datapath is concerned with the number and bit width of datapath, pipelining depth, native data type (either fixed point or floating point), number of ALUs, shifters, multipliers, and bit manipulation units as well as their corresponding data precision/ accuracy and data/address registers. Finally, physical characteristics to be compared include maximum clock frequency, typical operating voltage, feature size and implementation technology, and power consumption.

4.1 PDSP Chips

Some of the recent single-chip PDSPs are summarized in Table 6.

4.1.1 DSP16xxx

The Lucent DSP16xxx (Bier, 1997) achieves ILP from parallel operations encoded in a complex instruction. These complex instructions are executed at a maximum rate of one instruction per clock cycle. Embedding up to 31 instructions following the Do instruction can eliminate overheads due to small loops. These embedded instructions can be repeated a specified number of times without additional overhead. Moreover, a high instruction/data I/O bandwidth can be achieved from a 60-kword (120-kbyte) dual-ported on-chip RAM, a dedicated data bus, and a multiplexed data/instruction bus.

4.1.2 TMS320C54xx

Characterized as a low-power PDSP, each TMS320C54xx (Texas Instruments, 1999a) chip is composed of two independent processor cores. Each core has a 40-bit ALU, including a 40-bit barrel shifter, two 40-bit accumulators, and a 17-bit $\times$ 17-bit parallel multiplier coupled with a 40-bit adder to facilitate single-cycle MAC operation. The C54 series is optimized for low-power communication applications. Therefore, it is equipped with a compare, select, and store unit (CSSU) for the add/compare selection of the Viterbi operator. Loop/branch overhead is eliminated using instructions such as `repeat`, `block-repeat`, and `conditional store`. Interprocessor communication is carried out via two internal eight-element first-in-first-out (FIFO) register.

4.1.3 TMS320C62x/C67x

TMS320C62x/C67x (Texas Instruments, 1999b, 1999c; Seshan, 1998) is a series of fixed-point/floating-point, VLIW-based PDSPs for high-performance applications. During each clock cycle, a *compact instruction* is fetched and decoded (decompressed) to yield a packet of eight 32-bit instructions that resemble those of conventional VLIW architecture. The compiler performs software pipelining, loop unrolling, and ''If'' conversion to a predicate execution. Furthermore, programmers from a high-level language such as C can access a number of special-

Table 6 Summary of Recent Single-Chip PDSPs

	DSP 16xxx	SHARC	TMS32054xx	TMS32062xx	TMS32067xx	TriCore
Family name:						
Model no.:	DSP16210	ADSP21160	TMS320VC5421	TMS320C 6203	TMS320C 6701	TC10GP
Company:	Lucent	Analog Device	Texas	Texas	Texas	Infineon
Processing core:	VLIW	Multiproc./SIMD	Multiprocessor	VLIW	VLIW	Superscalar
Instruction						
Width (bits)	16 & 32	32	16 & 32	256	256	16 & 32
Maximum issued	1	4	2	8	8	2
Address space (bits)	20	32	—	32	32	32
Datapath						
No. of datapaths	2	2	2	2	2	3
Width of datapath (bits)	16	32	16	32	32	32
Pipeline depth	3	3	—	11	17	4
Data type	Fixed-point	Floating-point	Fixed-point	Fixed-point	Floating-point	Fixed-point
Functional units						
ALUs	2 (40b)	2	2 (40b)	4	4	1
Shifters	2	2	2 (40b)	2	0	—
Multipliers	2 (16b × 16b)	2	2 (17b × 17b)	2	2	2 (16b × 16b)
Address generator	2	2	2 × 2	ALUs	ALUs	1
Bit manipulation unit	1 (40b)	Shifter	2 (40b)	Shifter	Shifter	1
Program control						
Hardware loop	Y	Y	Y	N	N	Y
Nesting levels	2	—	2	2	2	3
On-chip storage						
Data registers	8	2 × 16	2 × 2	2 × 16	2 × 16	16
Width (bits)	40	40	40	32	32	32
Address registers	21	2 × 8	2 × 8	—	—	16
Width (bits)	20	32	—	—	—	32
Performance						
Maximum clock (MHz)	150	100	—	300	167	66
Operating voltage (V)	3.0	—	1.8	1.5	1.8	2.5
Technology	CMOS	—	CMOS	CMOS (15C05)	CMOS (18C05)	CMOS
Feature size (µm)	—	—	—	0.15	0.18	0.35
Power consumption	294 mW at 100 MHz	—	162 mW at 100 MHz	—	—	—

purpose DSP instructions called intrinsic functions. This feature helps ease the programming task and improves the code performance.

4.1.4   ADSP 21160 SHARC

Analog Device's ADSP21160 SHARC (Super Harvard Architecture) (Analog Device, 1999) contains two PEs, both using a 40-bit extended precision floating-point format. Every functional unit in each PE is connected in parallel and performs single-cycle operations. Even though its name is abbreviated from Harvard architecture, its program memory can store both instruction and data. Furthermore, SHARC doubles its data memory bandwidth to allow the simultaneous fetch of both operands.

4.1.5   TriCore

TriCore TC10GP (TriCore, 1999) is a dual-issued superscalar load/store architecture targeted at control-oriented/DSP-oriented applications. Even though its instructions are mixed 16/32 bits wide for low-code density, its datapath is 32 bits wide to accommodate high-precision fixed-point and single-precision floating-point numbers.

4.2   PDSP CORES

In Table 7, we compare the features of four DSP cores reported in the literature: Carmel, R.E.A.L., StarCore, and V850.

4.2.1   Carmel

One of the distinguishing features of Carmel (Carmel, 1999; Eyre and Bier, 1998) is its configurable long instruction words (CLIW) that are user-defined VLIW-like instructions. Each CLIW instruction combines multiple predefined instructions into a 144-bit-long superinstruction:

```
CLIW name (ma1, ma2, ma3, ma4) {                      // CLIW reference line
  MAC1 || ALU1 || MAC2 || ALU2 || MOV1  || MOV2 // CLIW def
}
```

Programmers can indicate up to four execution units plus two data moves according to the position of individual instruction within the long CLIW instruction. However, up to four memory operands can be specified using ma1 through ma4 . The assembler stores 48-bit reference line in program memory and 96-bit definition in a separate CLIW memory (1024 $\times$ 96 bits). In addition to CLIW,

Table 7 Summary of PDSP Cores

	Carmel	R.E.A.L.	StarCore	V850
Family name:	Carmel	R.E.A.L.	StarCore	V850
Model no.:	DSP 10XX	—	SC140	NA853C
Company	Infineon	Philips	Lucent & Motorola	NEC
Processing core	VLIW	VLIW	VLIW	RISC
Instruction				
Width (bits)	24 & 48	16 & 32	16	16 & 32
Maximum issued	1/2	2	6	—
Address space (bits)	23	—	32	26
Datapath				
No. of datapaths	2	2	—	—
Width of datapath (bits)	16	16	16	16
Data type	Fixed-point	Fixed-point	Fixed-point	Fixed-point
Functional units				
ALUs	2 (40b)	4 (16b)	4	1 (32b)
Shifters	1 (40)	1 (40b)	ALU (40b)	1 (32b)
Multipliers	2 (17b × 17b)	2 (16 × 16b)	ALU (16b × 16b)	1 (32b × 32b)
Address generator	1	2	2	—
Bit manipulation unit	Shifter	—	ALU	Shifter
On-chip storage				
Data registers	16 + 6	8	16	32
Width (bits)	16/40	16	40	32
Address registers	10	16	24	—
Width (bits)	16	—	32	—
Performance				
Maximum clock (MHz)	120	85	300	33
Operating voltage (V)	2.5	2.5	1.5	3.3
Technology	CMOS	—	CMOS	Titanium silicide
Feature size (μm)	0.25	0.25	0.13	0.35
Power consumption	200 mW at 120 MHz	—	180 mW at 300 MHz	—

a specialized hardware is provided to support Viterbi decoding. Almost all instructions can use predicated execution by two conditional-execution registers.

4.2.2 R.E.A.L.

Similarly, the R.E.A.L. PDSP (Kievits et al., 1998) core allows users to specify a VLIW-like set of application-specific instructions (ASIs) to exploit full parallelism of the datapath. Up to 256 ASIs can be stored in a look-up table. A special class of 16-bit instructions with an 8-bit index field activates these instructions. Each ASI is 96 bits wide and has the following predicated form:

```
Cond (3) || XACU (11) || YACU (10) || MPY1 (3) || MPY0 (3) || ALUs
(62) || DSU (2) || BNU (2)
ASI [if(asi_cc)]alu3_op,alu2_op,alu1_op,alu0_op
  [mult1_op][,mult0_op][,dr_op][,xacu_op][,yacu_op];
ASI [if(asi_cc)]alu32_op, alu10_op [,mult1_op][,mult0_op][,dr_op]
  [,xacu_op][,yacu_op];
ASI [if(asi_cc)]lfsr [,mult1_op][,mult0_op][,xacu_op][,yacu_op];
```

Each ASI starts with a 3-bit condition code followed by an 11-bit X ALU opcode, 10-bit Y ALU, 3-bit multiplier 1 and 0's opcodes, 62-bit operands, and so on. In addition to the user-defined VLIW instruction, R.E.A.L. allows application-specific execution units (AXUs) to be defined by the customer, which can be placed anywhere in the datapath or address calculation units. Its application is a GSM baseband signal processor.

4.2.3 StarCore

StarCore (StarCore, 1999; Wolf and Bier, 1998) is a joint development between Lucent and Motorola for wireless software handset configurable terminals (radios) of third-generation wireless systems. It is expected to operate at a low voltage down to 0.9 V. A fetch set (8-word instruction set) is fetched from memory. A program sequencing (PSEQ) unit detects a portion of this set to be executed in parallel and dispatched to the appropriate execution unit. This feature is called a variable-length execution set (VLES). StarCore achieves maximum parallelism by allowing multiple address generation and data ALUs to execute multiple operations in a single cycle. StarCore is targeted at speech coding, synthesis, and voice recognition.

4.2.4 V850

The NEC NA853E (NEC, 1997) is a five-stage pipeline RISC (reduced instruction set computer) core suitable for real-time control applications. Not only is the instruction set a mixture of 16 and 32 bits wide, but it also includes intrinsic instructions for high-level language support to increase the efficiency of the object code generated by the compiler and to reduce the program size.

4.3 Multimedia PDSPs

Multimedia PDSPs are designed specifically for audio/video applications as well as 2D/3D graphics. Some of their common characteristics are as follows:

1. Multimedia input/output (I/O): This may include ports and codec (coder/decoder) for video, audio, as well as super-VGA graphics signals.
2. Multimedia-specific functional units such as a YUV to RGB converter for video display, variable-length decoder for digital video decoding, descrambler in TriMedia (Phillips, 1999), and motion estimation unit for digital video coding/compression in Mpact2 (Kala, 1998; Purcell, 1998).
3. High-speed host computer/memory interfaces such as PCI bus and RAMBUS DRAM interfaces.
4. Real-time kernel and operating system for MPACT and TriMedia, respectively.
5. Support of floating-point and 2D/3D graphic.

Examples of multimedia PDSPs include MPACT, TriMedia, TMS320C8x, and DDMP (Data-Driven Multimedia Processor) (Terada et al., 1999). Their architectural features are summarized in Table 8.

4.4 Native Signal Processing

Native signal processing (NSP) is the use of extended instruction sets in a general-purpose microprocessor to process signal processing algorithms. These are special-purpose instructions that often operate in a different manner than the regular instructions. Specifically, multimedia data formats usually are rather short (8 or 16 bits) compared to the 32-, 64-, and 128-bit native register length of modern general-purpose microprocessors. Therefore, up to eight samples may be packed into a single word and processed simultaneously to enhance parallelism at the subword level. Most NSP instructions operate on both integer (fixed-point) and floating-point numbers except the Visual Instruction Set (VIS) (Sun, 1997), which supports only fixed-point numbers. In general, NSP instructions can be classified as follows (Lee, 1996):

- Vector arithmetic and logical operations whose results may be vector or scalar
- Conditional execution using masking operations

Table 8 Summary of Multimedia PDSPs

Family name:	DDMP	MPACT	TMS320C8x	TriMedia
Model no.:	—	MPACT2/6000	TMS320C82	TM1300
Company:	Sharp	Chromatic	Texas	Philips
Processing core:	Dataflow	VLIW	Multiprocessor	VLIW
Instruction				
Width (bits)	72	81	32	16 to 224
Maximum issued	8	2	3	5
Address space (bits)	—	—	32	32
Datapath				
Width of datapath (bits)	12	72	32	32
Floating-point precision	NA[a]	Both	Single	Both
Functional units				
ALUs	2	2	3	7
Shifters	—	1	3	2
Multipliers	2 ALUs	1	3	2
Address generators	4	—	3	2
On-chip storage				
Data registers	4 Accumulators	512	48	128
Width (bits)	24	72	32	32
Performance				
Maximum clock (MHz)	120	125	60	166
Operating voltage (V)	2.5	—	—	2.5
Technology	CMOS (4 metal)	—	CMOS	CMOS
Feature size (μm)	0.25	0.35	—	0.25
Power Consumption	1.2 W at 120 MHz	—	—	3.5 W at 166 MHz

[a]NA = not available.

- Memory/cache access control such as cache prefetch to particular level of cache, nontemporal store, and so forth, as well as masked load/store
- Data alignment and subword rearrangement (i.e., permute, shuffle, etc.)

Most NSP instruction set architectures exhibit the following features:

1. Native signal processing instructions may share the existing functional units of regular instructions. As such, some overhead is involved when switching between NSP instructions and regular instructions. However, some NSP instruction sets have separate, exclusive execution units as well as register file.
2. Saturation and/or modulo arithmetic instructions are often implemented in hardware to reduce the overhead of dynamic range checking during execution, as illustrated in Section 1.2.3.
3. To exploit subword parallelism, manual or human optimization of NSP-based programs is often necessary for demanding applications such as image/video processing and 2D/3D graphics.

Common and distinguishing features of available NSPs are summarized alphabetically as presented in Table 9.

4.4.1 AltiVec

Motorola's AltiVec (Motorola, 1998; Tyler et al., 1999) features a 128-bit vector execution unit operating concurrently with the existing integer and floating-point units. There are totally 162 new instructions that can be divided into four major classes:

Intraelement arithmetic operations	Addition, subtraction, multiply–add, average, minimum, maximum, conversion between 32-bit integer and floating point
Intraelement nonarithmetic operations	Compare, select, logical, shift, and rotate
Interelement arithmetic operations	Sum of elements within a single vector register to a separate register
Interelement nonarithmetic operations	Wide field shift, pack, unpack, merge/interleave, and permute

AltiVec shows a significant amount of effort to exploit the maximum amount of parallelism. This results in a 32-entry, 128-bit-wide register file separating from the existing integer and floating-point register files. This is different from other NSP architectures that often share the NSP register file with the existing one. The purpose is to exploit additional parallelism through the superscalar dispatch of operations to multiple execution units; or through multithreaded

Table 9   Summary of Native Signal Processing Instruction Sets

Name:	AltiVec	MAX-2	MDMX	MMX/3D Now	MMX/SIMD	VIS
Company:	Motorola	HP	MIPS	AMD	Intel	Sun
Instruction set:	Power PC	PA RISC 2.0	MIPS-V	IA32	IA32	SPARC V.9
Processor:	MPC7400	PA RISC	R10000	K6-2	Pentium III	UltraSparc
Fixed point (integer)						
8-Bit	16	NA[a]	8	8	8	8
16-Bit	8	4	4	4	4	4
32-Bit	4	NA	NA	2	2	2
Floating point						
Single precision	4	2	2	2	4	Na
Fixed-point register file						
Size	$32 \times 128b$	$32 \times 64b$	$32 \times 64b$	$8 \times 64b$	$8 \times 64b$	$32 \times 64b$
Shared with	Dedicated	Integer reg.	FP reg.	Dedicated	FP reg.	FP reg.
Fixed-point accumulator						
Size	NA	NA	192	NA	NA	NA
Arithmetic						
Unsigned saturation	Y	Y	Y	Y	Y	Y
Modulo	Y	Y	Y	Y	Y	Y

Interelement arithmetic					
Multiply-Acc[a]	4	NA	4	2	2	NA
Fixed-point MAC precision[b]	32 += (16 × 16)	—	48 += (16 × 16)	32 += (16 × 16)	32 += (16 × 16)	—
Compare	Y	N	Y	Y	Y	Y
Min/max	Y	N	Y	N	Y	Y
Floating-point Multiply-Acc	4 single	2 single	2 single	2 single	4 single	N
Floating-point min/max	Y	N	N	Y	Y	N
Intraelement arithmetic					
Sum	Y	N	N	Y	Y	N
Floating-point Sum	Y	N	N	Y	N	N
Type conversion					
Pack	Y	Y	Y	Y	Y	Y
Unpack	Y	Y	Y	Y	Y	Y
Permute	Y	Y	Y	—	N	—
Merge	Y	Y	Y	Y	Y	Y
Special instructions	VREFP VRSQRTFP SPLAT VSEL	CACHE HINT DEPOSIT EXTRACT SHR PAIR	SELECT	FEMMS PFRCP PFRSQRT PREFETCH	EMMS DIVPS PREFETCH SFENCE	EDGE ARRAY PDIST BLOCK TRANSFER

[a] NA: not available.

[b] precision (bits): acc (bits) = acc (bits) + a (bits) × b (bits).

execution unit pipelines. Each instruction can specify up to three source operands and a single destination operand. Each operand refers to a vector register.
Target applications of AltiVec include multimedia applications as well as high-
bandwidth data communication, base station processing, IP telephony gateway,
multichannel modem, network infrastructure such as an Internet router, and a
virtual private network server.

4.4.2 MAX 2.0

Multimedia Acceleration eXtension (MAX) 2.0 (Lee, 1996) is an extension of
HP Precision Architecture RISC ISA on a PA8000 microprocessor with minimal
increased die area concern. Both 8-bit and 32-bit subwords are not supported due
to insufficient precision and insufficient parallelism compared to a 32-bit single-
precision floating point, respectively. Although pixels may be input and output
as 8 bits, using a 16-bit subword in intermediate calculations is preferred. The
additional hardware to support MAX2.0 is minimal because the integer pipeline already has two integer ALUs and shift merge units (SMUs), whereas the
floating-point pipeline has two FMACs and two FDIV, FSQRT units. MAX
special instructions are field manipulation instructions, as follows:

Cache hint	For spatial locality
Extract	Selects any field in the source register and places it right-aligned in the target
Deposit	Selects a right-aligned field from the source and places it anywhere in the target
Shift pair	Concatenates and shifts 64-bit or rightmost 32-bit contents of tow register into one result

4.4.3 MDMX

Based on MIPS' experience of designing Geometry Engine, Reality Engine, Maximum Impact, Infinite Reality, Nintendo64, O2, and Magic Carpet, the goal of
MIPS Digital Media Extension (MDMX) (MIPS Technology, 1997) is to improve
performance IEEE-compliant DCT accuracy. As a result, MDMX adds four- and
eight-element SIMD capabilities for an integer arithmetic through the definition
of these two data types:

Octal byte	Eight unsigned 8-bit integers with eight unsigned 24-bit accumulators
Quad half	Four unsigned 16-bit integers with four unsigned 48-bit accumulators

Note that both octal byte and quad half data types share a 192-bit accumulator,
which permits accumulation of $2^N N \times N$ multiples, where N is either 8 or 16

bits according to octal byte and quad half, respectively. MDMX's 32, 64-bit-wide registers and the 8-bit condition code coincide with the existing floating-point register file similar to the "paired-single"-precision floating-point data type. Data are moved between the shared floating-point register file and memory with a floating-point load/store double word and between floating-point and integer registers. In addition, MDMX has a unique feature with the vector arithmetic: It is able to operate on a specific element of a subword as an operand or as a constant immediate value. However, the reduction instruction (sum across) and sum of absolute difference (SAD) are judiciously omitted. In particular, SAD or L1 norm can be performed as an L2 norm without loss of precision using the 192-bit accumulator.

4.4.4 MMX 3DNow!

AMD 3DNow! (AMD, 1999; Oberman et al., 1999) is Intel's MMX-like multimedia extension, first implemented in the AMD K6-2 processor. Floating-point instructions are augmented to the integer-based MMX instruction set by introducing a new data type: single-precision floating-point to support 2D and 3D graphics. Similar to the MMX, applications must determine if the processor supports MMX or not. In addition, 3DNow! is implemented with a separate flat register file in contrast to the stack-based floating-point/MMX register file. Because no physical transfer of data between floating-point and multimedia unit register files is required, FEMMS (faster entry/exit of the MMX or floating-point state) is included to replace MMX EMMS instruction and to enhance the performance. Either the register X or Y execution pipeline can execute floating-point instructions for a maximum issue and execution rate of two operations per cycle (AMD, 1999). There are no instruction-decode or operation-issue pairing restrictions. All operations have an execution latency of two cycles and are fully pipelined. As long as two operations do not fall into the same category, both operations will start execution without delay. The 2 categories of the additional 21 instructions are as follows:

1. PFADD, PFSUB, PFSUBR, PFACC, PFCMPx, PFMIN, PFMAX, PI2FD, PFRCP, and PFRSQRT
2. PFMUL, PFRCPIT1, PFRSQIT1, and PFRCPIT2

Normally, all instructions should be properly scheduled so as to avoid delay due to execution resource contention or structural hazard by taking dependencies and execution latencies into account.

FEMMS	Similar to MMX's EMMS but faster because 3DNow! does not share MMX registers with those of floating point.
PFRCP	Scalar floating-point reciprocal approximation

| PFRSQRT | Scalar floating-point reciprocal square root approximation |
| PREFETCH | Loads 32 or greater number of bytes either nontemporal or temporal in the specified cache level |

4.4.5   MMX/SIMD

MMX (multimedia extension) is Intel's first native signal processing extension instruction set (Intel, 1999). Subsequently, additional instructions are augmented to the Streaming SIMD Extensions (SSE) (Intel, 1999) in Pentium III class processors. SIMD supports 4-way parallelism of 32-bit, single-precision floating-point for 2D and 3D graphics or 32-bit integer for audio processing. These new data types are held in a new separate set of eight 128-bit SIMD registers. Unlike MMX execution, traditional floating-point instructions can be mixed with SSE without the need to execute special instructions, such as EMMS. In addition, SIMD features explicit SAD instruction and introduces a new operating-system visible state:

EMMS (empty MMX state)	Must be used to empty the floating-point tag word at the end of an MMX routine before calling other routines executing floating-point instructions
DIVPS	Divides four pairs of packed, single-precision, floating-point operands
PREFETCH	Loads 32 or greater number of bytes either nontemporal or temporal in the specified cache level
SFENCE (store fence)	Ensures ordering between routines that produce weakly ordered results and routines that consume these data just like multiprocessor weak consistency; nontemporal stores implicitly weak ordered, no write-allocate, write combine/collapse so that cache pollution is minimized

4.4.6   VIS

Sun's VIS (Visual Instruction Set) (Sun, 1997; Tremblay et al., 1996) is the only NSP reviewed here that does not support parallelism of floating-point data type. However, the subword data share the floating-point register file with floating-point number, as indicated in Table 9. Some special instructions in VIS are Array, Pdist, and Block transfer:

| Array | Facilitates 3D texture mapping and volume rendering by computing a memory address for data look up |

	based on fixed-point x, y, and z; data laid out in a block fashion so that points which are near one another have their data stored in nearby memory locations
Edge	Computes a mask used for partial storage at an arbitrarily aligned start or stop address typically at boundary pixels
Pdist	Computes the sum of absolute value of difference of eight pixel pairs
Block transfer	Transfers 64 bytes of data between memory and registers

5 SOFTWARE PROGRAMMING TOOLS FOR PDSPs

5.1 Software Development Tools for Programming PDSPs

Since their introduction more than a decade ago, PDSPs have been incorporated in many high-performance embedded systems such as modems and graphic acceleration cards. A unique requirement of these applications is that they all demand high-quality (machine) code generation to achieve the highest performance while minimizing the size of the program to conserve premium on-chip memory space. Often, the difference of one or two extra instructions implies that either a real-time processing constraint may be violated, leaving the code generated useless, or an additional memory module may be needed, causing significant cost overrun.

High-level languages (HLLs) are attractive to PDSP programmers because they hide hardware-dependent details and simplify the task of programming. Unlike assembly codes, HLL programs are readable and maintainable and are more likely to be portable to other processors. In the case of an object-oriented HLL, such as C++, those programs are also more reliable and reusable. All these features contribute to reduce development time and cost.

Figure 1 depicts an example of typical software development for PDSPs— the TMS320C6x software development flowchart. There are three possible source programs: C source files, macro source files, and linear assembler source files. The latter sources are both at assembly program level. The assembly optimizer assigns registers and uses loop optimization to turn the linear assembly into a highly parallel assembly that takes advantage of software pipelining. The assembler translates assembly language source files into machine language object files. The machine language is based on the common object file format (COFF). Finally, the linker combines object files into a single executable object module. As it creates the executable module, it performs relocation and resolves external references. The linker also accepts relocatable COFF object files and object libraries as input.

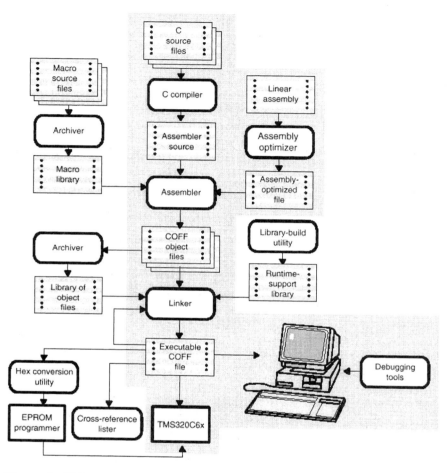

Figure 1 TMS320C6x software development flow. (From Texas Instruments, 1998a.)

To improve the quality of the code generated, C compilers are always equipped with extensive optimization options. Many of these compiler optimization strategies are based on GNU C Compiler (GCC) (see Table 10).

The debugger can usually be both simulator and profiler like the C source debugger (Texas Instruments, 1998b). The C source debugger is an advanced graphic user interface (GUI) to develop, test, and refine 'C6x C programs and assembly language programs. In addition to that, the 'C6x debugger accepts executable COFF files as input. It features the following capabilities that are common in other PDSP development environments:

- Multilevel debugging; user can debug both C and assembly language code.
- Dynamic profiling provides a method for collecting execution statistics and immediate feedback to identify performance bottlenecks within the code.
- Fully configurable graphical user interface.
- Comprehensive data displays.

5.2 On-Chip Emulation

The presence of the Joint Test Action Group (JTAG) test access module and enhanced on-chip emulation (EOnCE) module interface allows the user to insert the PDSP into a target system while retaining debug control. The EOnCE module, as shown in Figure 2, is used in PDSP devices to debug application software in real time. It is a separate on-chip block that allows nonintrusive interaction with the core. The user can examine the contents of registers, memory, or on-chip peripherals through the JTAG pins. Special circuits and dedicated pins on the core are defined, to avoid sacrificing user-accessible on-chip resources.

As applications grow in terms of both size and complexity, the EOnCE provides the user with many features, including the following:

- Breakpoints on data bus values
- Detection of events, which can cause a number of different activities configured by the user
- Nondestructive access to the core and its peripherals

Table 10 Compiler Optimization Options in DSP16000 Series

	Optimization performed	Targeted application
−O0	Default operation, no optimization	C level debug to verify functional correctness
−O1	Optimize for space	Optimize space for control code
−O2	Optimize for space and speed	Optimize space and speed for control code
−O	Equivalent to −O2	Equivalent to −O2
−O3	−O2 plus loop cache support, some loop unrolling	Optimize speed for control and loop code
−O4	Aggressive optimization with software pipeline	Optimize speed and space for control and loop code
−Os	Optimize for space	Optimize space for control and loop code

Source: Lucent, 1999.

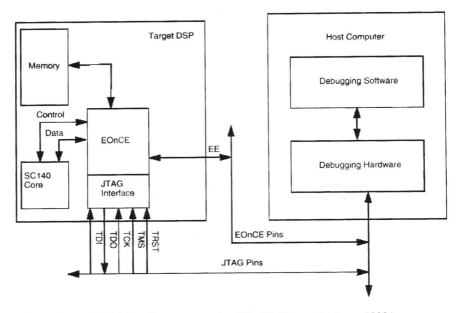

Figure 2   Typical debugging system using EOnCE. (From StarCore, 1999.)

- Various means of profiling
- Program tracing buffer

The EOnCE module provides system-level debugging for real-time systems, with the ability to keep a running log and trace of the execution of tasks and interrupts and to debug the operation of real-time operating systems (RTOS).

5.3   Optimizing Compiler and Code Generation for PDSP

The PDSP architecture evolves from an ad hoc heterogeneous resource toward a homogeneous resource like the general-purpose RISC microprocessor. One of the reasons is to make compiler optimization techniques less difficult. Classical PDSP architecture is characterized by the following:

- A small number and nonuniform register sets in which certain registers and memory blocks are specialized for specific usage
- Highly irregular datapaths to improve code performance and reduce code size
- Very specialized functional units
- Restricted connectivity and limited addressing to multipartitioned memory

Several techniques have been proposed based on a simplified architecture of TMS320C2X/5X [e.g., instruction selection and instruction scheduling (Yu and Hu, 1994) register allocation and instruction scheduling.

The success of RISC and its derivatives such as the superscalar architecture and the VLIW architecture have asserted significant impacts on the evolution of modern PDSP architecture. However, code density is a central concern in developing embedded DSP systems. This concern leads to the development of new strategies such as code compaction (Carmel, 1999) and user-defined long instruction word (Kievits et al., 1998). With a smaller code space in mind, code compaction using integer programming (Leupers and Marwedel, 1995) was proposed for applications to PDSPs that offer instruction-level parallelism such as VLIW. Later, an integer programming problem formulation for simultaneous instruction selection, compaction, and register allocation was investigated by Geboyts (Geboyts, 1997). It can be seen that earlier optimization techniques were focused on the optimization of basic code blocks. Thus, they can be considered a local optimization approach. Recently, the focus has shifted to global optimization issues such as loop unrolling and software pipelining. In Stotzer and Leiss (1999), results of implementing a software pipelining using modulo scheduling algorithm on the 'C6x VLIW DSP have been reported.

Artificial intelligence (AI) techniques such as planning were employed to optimize instruction selection and scheduling (Yu and Hu, 1994). With AI, concurrent instruction selection and scheduling yield code comparable to that of handwritten assembly codes by DSP experts. The instruction scheduler is a heuristic list-based scheduler. Both instruction scheduling and selection involve node coverage by pattern matching and node evaluation by heuristic search using means-end analysis and hierarchical planning. The efficiency is measured in terms of size and execution time of generated assembly code whose size is up to 3.8 times smaller than that of a commercial compiler.

Simultaneous instruction scheduling and register allocation for minimum cost based on a branch-and-bound algorithm are reported. The framework can be generalized to accumulator-based machines to optimize accumulator spilling, such as the TMS320C40. Their uses are likely intended to obtain more compact code.

Recently, in Leupers and Marwedel (1995) and Geboyts (1997), integer linear programming is shown to be effective in compiler optimization. The task of local code compaction in VLIW architecture is solved under a set of linear constraints such as a sequence of register transfers and maximum time budget. Because some DSP algorithms show more data flow and less control flow behavior (Leupers and Marwedel, 1995), code compaction exploits parallel register transfers to be scheduled into a single control step, resulting in a lower cycle count to satisfy the timing constraint. Of course, it is important to consider resource conflicts and dependencies, as well as the possible side effects of encoding restrictions and operations. Later effort has integrated instruction selection, compaction, and register allocation together (Geboyts, 1997). The targeted PDSP is defined using arc mappings and later into logical propositions to make it retargetable. These propositions are then translated into mathematical constraints to form the optimization model using integer linear programming. Code is generated and optimized for minimum code size and maximum performance in estimated energy dissipation.

In modern VLIW PDSPs, the architecture features homogeneous functional and storage resources, enabling global optimization by the compiler. Software pipelining is known as one of the most popular code scheduling techniques. It exploits the available instruction-level parallelism in various loop iterations. A software-pipelined loop consists of three components:

- A prolog: set up the loop initialization
- A kernel: execute pipelined loop body in steady state
- An epilog: drain the execution of the loop kernel

Modulo scheduling takes an innermost loop body and constructs a new schedule. The new schedule is equivalent to overlapping loop iterations. The algorithm utilizes a data precedence graph (DPG) and reservation table to construct a permissible schedule of the loop body under the available resource constraints. DPG is a directed graph (possibly cyclic) with nodes and edges representing operations and data flow dependencies of the original inner loop body. The resource requirements for an operation are modeled using a reservation table. Stotzer and Leiss (1999) report the result of software pipelining on a set of 40 loop kernels based on 'C6x architecture. However, the architectural features that impact performance gain of software pipelining are moderately sized register file, constraints on code size, and multiple assignment code.

6 DSP SYSTEM DESIGN METHODOLOGIES

Designing modern DSP systems requires more than just programming PDSP or processing cores. Instead, the system's performance must be the utmost performance criterion. The DSP system design methodologies are developed at different levels of abstraction. At the system level, the design scope includes task and data partitioning and software synthesis/simulation. At the architectural level, the focus is on architecture and compiler development. At the chip implementation level, hardware description languages such as VHDL and hardware/software codesign methodologies are quite important.

6.1 Application Development with Existing Hardware/Processing Core

A software engineering approach is incorporated to assist in the development of an application using a DSP array processor at Raytheon System Corporation

(Kelly and Oshana, 1998). Three performance measurements are used to gauge the quality of the design:

- Processor throughput rate
- Memory utilization
- I/O bandwidth utilization

A sensitivity analysis of these performance metrics is performed to examine trade-offs of various design approaches. Factors that affect the processor throughput rate include the quality of DSP algorithm formulation, the operation cost in processor cycles, the sustained throughput rate to peak throughput efficiency, and the expected speedup when it is upgraded to the next generation of PDSP. Regarding the memory utilization, it has been observed that the size of the data samples and the dynamic nature of memory usage patterns are the two most important factors. The I/O bandwidth utilization, on the other hand, depends on the algorithm as well as the hardware design. Several design tools used to develop the entire system and their factors that may degrade the performance during the design process are listed in Table 11.

Rate monotonic analysis (RMA) (Liu and Layland, 1973) is necessary to validate the schedulability of software architecture. In general, the following lessons have been learned through this design experience:

- Prototype early in the development cycle.
- Ignore processor marketing information (actual throughput is highly dependent on the application profile).
- Carefully analyze the most frequently executed function: task switching.
- Take inherent interface overheads such as interrupt handling, data packing, and data unpacking into account in estimating the throughput.

Another example of DSP system development is the Computer Assisted Dynamic Data Monitoring and Analysis System (CADDMAS) developed for the

Table 11 DSP System Development Tools and Factors That May Degrade Performance

Tools	Factors that may degrade performance
Code generation	Compiler efficiency
	Quality of generated assembly code
	Size of load image
Instruction level processor simulator	Cycle counts for elementary operation
Cycle-accurate device level VHDL model	External memory access time
	Instruction caching effects
	Resource contention between processor and DMA channels

U.S. Air Force and NASA (Sztipanovitz et al., 1998). Its details have been described in Section 2.3. An adaptive approach was necessary to allow the structure of the system to adapt to changing external requirements and sensor availabilities. This leads to the application of a reconfigurable controller for process *control* called *structurally adaptive signal processing*. Unlike parametric adaptation, where the topology of the graph is fixed and coefficients can change over time, a structurally adaptive signal processing system can change its computational structure on the fly. Therefore, its control functionality can be maintained even in the face of sensor failures; the performance will be gracefully degraded but correct control action is still present.

6.2   Application-Driven Design: Fine-Tuning the Processing Core

Two reasons contributing to the poor performance of HLL PDSP commercial compilers are, first, that compilers are developed after a target architecture has been established and, second, the inability to exploit DSP-specific architectural features in DSP compiler (Lee, 1994). The following application-driven design methodologies are adopted:

- A DSP architecture and its compiler are developed in parallel.
- Its dynamic statistics assesses the impact trade-offs on performance.
- An iterative analysis is undertaken to fine-tune the architecture and compiler.

The PDSP architecture is based on VLIW. As a result, an optimizing C compiler is necessary to exploit static instruction-level parallelism as well as DSP-specific hardware features. Those hardware features are modulo addressing, low overhead looping, and dual data-memory banks. Meanwhile, an instruction set simulator is developed to gather statistics on the run-time behavior of DSP programs. A suite of DSP benchmarks in terms of kernel and application are chosen to evaluate the system. The performance success of the compiler is due to the flexibility of the model VLIW architecture. The statistics indicate the areas of improvement to be fed back for fine-tuning the architecture. However, its drawback is the high instruction-memory bandwidth requirements that can be too expensive and impractical to implement.

As another means of DSP architecture development, machine description language (MDL) has been proposed to achieve rapid prototyping at architectural level. Recently, LISA (Peesl et al., 1999) was developed for the generation of bit and cycle accurate models of a PDSP. It includes instruction set architecture that enables automatic generation of simulators and assemblers. LISA is composed of resource and operation declarations. Resource declaration represents the storage objects of the hardware architecture (e.g., registers, memories, pipelines).

Declaration description collects the description of different properties of the system (i.e., the instruction set model, the behavioral model, the timing model, and necessary declarations). LISA supports cycle-accurate processor models, including constructs to specify pipelines and their mechanisms. It targets SIMD, VLIW, and superscalar architectures. Direct support for compiled simulation techniques and strong orientation on C programming language are contributed in LISA. The Texas Instruments' TMS320C6201 DSP, realized as a real-world example, was modeled on a cycle-by-cycle basis by only one designer and finished within 2 months.

6.3 Reconfigurable Computing: Hardware/Software Codesign for a Given Application

In the system design process, it has been traditional that the decision is made on a subtask-by-subtask basis to be implemented in either custom hardware or software running on PDSP(s). On one hand, custom hardware or ASIC can be customized to a particular subtask, resulting in relatively fast and efficient implementation. ASIC is physically programmed by patterning devices (transistors) and metal interconnection prior to fabrication process. Higher throughput and lower latency can be achieved with more space dedicated to particular functional units. On the other hand, PDSP is programmed later by software resulting in flexible but relatively slow and inefficient realization. Temporal or sequential operations can be accomplished by a set of instructions to program a processor after its fabrication.

Between these two extremes, reconfigurable computing (RC) architecture can be programmed to perform any specific function by a set of configuration bits. In other words, RC combines temporal programmability with spatial computation in hardware after it is fabricated at low overhead. The hardware/software boundaries can be altered by the RC paradigm (DeHon and Wawrzynek, 1999).

Reconfigurable computing is also known as the 90/10 rule of thumb, where 90% of run time is spent on 10% of the program, hardware/software partitioning is inspired by the higher percentage of run time of specialized computation: the greater improvement of cost/performance if it is implemented in hardware, and the more specialized computation dominate the application the more closely the specialized processor should be coupled with host processor. This rule of thumb has been successfully applied to the floating-point processing unit as well as RC.

In the heterogeneous system approach, RC is combined with the general-purpose processing capability of the traditional microprocessor. The interface between these two can be either closely or loosely coupled, depending on its applications. How frequently RC's functionality should be reconfigured dynamically must be determined based on the performance optimality of the particular hardware/software architecture.

Existing RC architectures are mostly of the form of a 1D or 2D array of configurable cells interconnected by programmable links. The array communicates with the outside world through peripheral I/O cells. The architecture can be characterized by the logic capability of each cell at two different granularities:

- The fine-grained configurable cell performs simple logic functions with the support of more complex functions such as the fast carry chain.
- The coarse-grained configurable cell performs word-parallel arithmetic functions such as addition, multiplication, and so forth with support of simple bit-level logic functions.

Either general-purpose computing or DSP application primarily drives the architecture of the configurable cell, either fine grained or coarse grained. Its configuration or *context* is stored locally like an SRAM-based FPGA. Its implementation is either FPGA-like or a custom-designed configurable cell based on current technology and scale of integration. Instead of being off-chip and loosely coupled to the host processor, RC is going toward an on-chip coprocessor as the technology advances to SoC.

Some existing RC architectures for DSP applications are presented in Table 12. CHAMP (Patriquin and Gurevich, 1995) is a system of 8 PEs interconnected in a 32-bit ring topology. Each PE consists of 2 Xilinx XC4013 FPGAs, dual-port memory, and 32-bit reconfigurable crossbar switch ports between PEs and memory.

DRLE (Dynamically Reconfigurable Logic Engine) (Fujii et al., 1999; Nishitani, 1999) is capable of real-time reconfiguration with several layers of configuration tables. An experimental chip is composed of an 4×12 configurable cells. Each cell can realize two different logic operations equivalent to 4-bit input to 1-bit output or 3-bit input to 2-bit output. Up to eight different configurations can be locally stored in each cell memory. With 0.25-μm CMOS technology, the chip contains 5.1 million transistors in a 10×10-mm^2 die area and consumes 500 mW at 70 MHz.

MATRIX (Mirsky and DeHon, 1996) is composed of a 2D array of identical 8-bit configurable cells overlaid with a configurable network. Each cell con-

Table 12 Some Reconfigurable Computing Architectures for DSP Applications

Fine grained	CHAMP (Patriquin and Gurevich, 1995), DRLE (Nishitani, 1999), Pleiades (Abnous and Rabaey, 1996)
Coarse grained	MATRIX (Mirsky and Delton, 1996), MorphoSys (Lu et al., 1999), REMARC (Miyamori and Olukotun, 1998)

sists of a 256 × 8-bit memory, 8-bit ALU and multiplier, and reduction control logic. The interconnect network supports three ranges of interconnection: nearest neighbor, bypass of length 4, and a global line with pipeline register. The configurable cell area is approximated to 29 million λ^2. Its cycle time is 10 nsec at 0.5-μm CMOS technology.

MorphoSys (Lu et al., 1999) is a 2D-mesh 8 × 8 reconfigurable cell array coprocessor. Each cell is similar to the datapath found in conventional processors consisting of ALU/multiplier, shifter, and four-entry register file. Moreover, bit-level application is also supported. Up to 32 contexts can be simultaneously resident in context memory. The whole array can be reconfigured in eight cycles (80 nsec at 100 MHz).

Pleiades (Abnous and Rabaey, 1996) is proposed as a heterogeneous system partitioned by control flow computing on microprocessor and data flow computing on RC for a future wireless embedded device. The RC array is composed of satellite (configurable) processors and a programmable interconnect to a main microprocessor. Data-flow-driven computing is implemented using global asynchronous and local synchronous clocking to reduce overhead. Therefore, operation starts only when all input data are ready.

7 CONCLUSION

In this chapter, we briefly surveyed the architecture, application, and programming methodologies of modern programmable digital signal processors. With a bit of stretch of the definition, we included in this survey the hardware programmable FPGA realization of DSP algorithms and the special multimedia extension instructions incorporated in general-purpose microprocessors to facilitate native signal processing. We also offered an overview of the existing applications of PDSPs. Finally, we summarized current software design methodologies and briefly mentioned future trends and open research issues.

REFERENCES

Abnous A, J Rabaey. Ultra-low-power domain-specific multimedia processors. VLSI Signal Process 9:461–470, 1996.
Alter JJ, JB Evins, JL Davis, DL Rooney. A programmable radar signal processor architecture. Proceedings of the 1991 IEEE National Radar Conference, 1991, pp 108–111.
AMD Inc. *3DNow!*™ *Technology Manual*, AMD, Inc., 1999, http://www.amd.com/K6/k6docs/pdf/21928.pdf
Analog Device Inc. ADSP-21160 SHARC DSP Hardware Reference, 1999.

Andraka R. A survey of CORDIC algorithms for FPGA based computers. Proceedings of the 1998 ACM/SIGDA Sixth International Symposium on Field Programmable Gate Arrays, 1998, pp 191–200.

Bhargava R, LK John, BL Evans, R Radhakrishnan. Evaluating MMX Technology Using DSP and Multimedia Applications, Proceedings of the IEEE Symposium on Microarchitecture (MICRO-31), Dallas, Texas, pp 37–46, Dec. 1998.

Bier JJ. DSP16xxx targets communication apps. Microprocessor Rep 11(12): 1997.

Bonomini F, F De Marco-Zompit, GA Mian, A Odorico, D Palumbo. Implementing an MPEG2 Video Decoder Based on the TMS320C80 MVP, September 1996.

Borkar S. Design challenges of technology scaling. IEEE Micro 23–29, July–Aug. 1999.

Budagavi M, W Rabiner, J Webb, R Talluri. Wireless MPEG-4 video on Texas Instruments DSP chips. Proc. 1999 IEEE International Conference on Acoustics, Speech, and Signal Processing, 1999, Vol. 4, pp 2223–2226.

CARMEL™ Development Chip, August 1999. http://www.carmeldsp.com/Pdf/CDEV2.2.pdf

Cheung N-M, T Kawashimea, Y Iwata, S Traunnumn, J Tsay, BI Pawate. Software Mpeg-2 video decoder on A DSP enhanced memory module. 1999 IEEE 3rd Workshop on Multimedia Signal Processing, 1999, pp 661–666.

Chishtie MA, U.S. digital cellular error-correction coding algorithm implementation on the TMS320C5x, Texas Instruments Report SPRA137, October 1994.

Chou P, H Jiang, Z-P Liang. Implementing real-time cardiac imaging using the TMS320C3x DSP, July 1997.

Daffara F, P Vinson. Improved search algorithm for fast acquisition in a DSP-based GPS receiver. 1998 URSI International Symposium on Signals, Systems, and Electronics, 1998, pp 310–314.

DeHon A, J Wawrzynek. Reconfigurable computing: What, why, and implications for design automation. Proceedings of the 36th ACM/IEEE Conference on Design Automation, 1999, pp 610–615.

Dick C. Computing the discrete Fourier transform on FPGA based systolic arrays. Proceedings of the 1996 ACM Fourth International Symposium on Field-Programmable Gate Arrays, 1996, pp 129–135.

Dullink H, B van Daalen, J Nijhuis, L Spaanenburg, H Zuidhof. Implementing a DSP Kernel for Online, 1995.

Eyre J, J Bier. Camel enables customizable DSP. Microprocessor Rep 12(17): 1998.

Fujii T, K-I Furuta, M Motomura, M Nomura, M Mizuno, K-I Anjo, K Wakabayashi, Y Hirota, Y-E Nakazawa, H Ito, M Yamashina. A dynamically reconfigurable logic engine with a multi-context/multi-mode unified-cell architecture. Solid-State Circuits Conference, 1999. Digest of Technical Papers. ISSCC. 1999 IEEE International, 1999, pp 364–365.

Ganesh S, V Thakur. Print screening with advanced DSPs. Texas Instruments Report SPIA004, 1999.

Geboyts CH. An efficient model for DSP code generation: Performance, code size, estimated energy. Proceedings of the Tenth International Symposium on System Synthesis, 1997, pp 41–47.

Golston J. Single-Chip H.324 Videoconferencing, IEEE Micro, 21–33, August 1996.

Goslin GR. A Guide to Using Field Programmable Gate Arrays (FPGAs) for Application-Specific Digital Signal Processing Performance. Xilinx, Inc., 1995.

Gottlieb AM. A DSP-based research prototype reverse channel transmitter/receiver for ADSL. 1994 IEEE International Conference on Acoustics, Speech, and Signal Processing, 1994, Vol 3, pp III/253–III/256.

Illgner K, H-G Gruber, P Gelabert, J Liang, Y Yoo, W Rabadi, R Talluri. Programmable DSP Platform for Digital Still Cameras. Texas Instruments, 1999.

Intel Corp. Intel® Architecture Optimization Reference Manual, http://developer.intel.com/design/pentiumii/manuals/245127.htm

Kalapathy P. Hardware–software interactions on Mpact. IEEE Micro 20, March/April 1997.

Kelly DP, S Oshana. Software performance engineering: A digital signal processing application. Proceedings of the First International Workshop on Software and Performance, 1998, pp 42–48.

Kibe SV, KA Shridhara, MM Jayalalitha. Software-based GIC/GNSS compatible GPS receiver architecture using TMS320C30 DSP processor. Fifth International Conference on Satellite Systems for Mobile Communications and Navigation, 1996, pp 36–39.

Kievits P, E Lambers, C Moerman, R Woudsma. R.E.A.L. DSP Technology for Telecom Baseband, Processing. Philips Semiconductors, 1999, http://www-us.semiconductors.philips.com/acrobat/literature/other/dsp/icspat98_pks.pdf

Knapp SK. Using Programmable Logic to Accelerate DSP Functions. Xilinx, Inc., 1995, http://www.xilinx.com/appnotes/dspintro.pdf

Ledger D, J Tomarakos. Using The Low Cost, High Performance ADSP-21065L Digital Signal Processor for Digital Audio Applications, Revision 1.0. Analog Device, Inc., 1998.

Lee C, M Potkonjak, WH Mangione-Smith. MediaBench: A tool for evaluating and synthesizing multimedia and communications systems. Proceedings of the Thirtieth Annual IEEE/ACM International Symposium on Microarchitecture, 1997, pp 330–335.

Lee RB. Subword parallelism with MAX-2. IEEE Micro, 51–59, August 1996.

Leupers R, P Marwedel. Time-constrained code compaction for DSPs. Proceedings of the Eighth International Symposium on System Synthesis, 1995, pp 54–59.

Liu CL, Layland JW. Scheduling algorithms for multiprogramming in a hard-real-time environment. J Assoc Computing Mach 20(1):46–61, 1973.

Lu G, H Singh, H-L Ming, N Bagherzadeh, FJ Kurdahi, EMC Filho, AV Castro. The MorphoSys dynamically reconfigurable system-on-chip. Proceedings of the First NASA/DoD Workshop on Evolvable Hardware, 1999, pp 152–160.

Lucent, Inc. DSP16000 C Compiler, December 1999, http://www.lucent.com/micro/dsp16000/pdf/AP99052.pdf

Meisl PG, MR Ito, IG Cumming. Parallel processors for synthetic aperture radar imaging. Proceedings of the 1996 International Conference on Parallel Processing, 1996, Vol 3; Software 2:124–131, 1996.

MIPS Technologies, Inc. MIPS extension for digital media with 3D, 1997, http://www.mips.com/Documentation/isa5_tech_brf.pdf

Mirsky E, A DeHon. MATRIX: A reconfigurable computing architecture with configurable instruction distribution and deployable resources, Proc. IEEE Symposium on FPGAs for Custom Computing Machines, 1996, pp 157–166.

Mitchell JL, et al., MPEG Video Compression standard. New York: Chapman & Hall, 1997.

Miyamori T, U Olukotun. A quantitative analysis of reconfigurable coprocessors for multimedia applications. Proc. IEEE Symposium on FPGAs for Custom Computing Machines, 1998, pp 2–11.

Morgan PN, RJ Iannuzzelli, FH Epstein, RS Balaban. Real-time cardiac MRI using DSPs. IEEE Trans Med Imaging MI-7:649–653, 1999.

Motorola, Inc. AltiVec Technology Programming Environments Manual, Rev. 0.1. Motorola, Inc., 1998.

NEC Inc. NA853C microcontroller overview, 1997, 12085e40.pdf

Nishitani T. An approach to a multimedia system on a chip, IEEE Workshop on Signal Processing Systems, 1999, pp 13–21.

Oberman S, G Favor, F Weber. AMD 3Dnow! technology: Architecture and implementations. IEEE Micro, 37–48, April 1999.

Owen RE, S Purcell. An enhanced DSP architecture for the seven multimedia functions: The Mpact 2 media processor. IEEE Workshop on Signal Processing Systems—Design and Implementation, 1997, pp 76–85.

Patriquin R, I Gurevich. An automated design process for the CHAMP module. Proceedings of the IEEE National Aerospace and Electronics Conference 1995, pp 417–424.

Pawate BI, PD Robinson. Implementation of an HMM-based, speaker-independent speech recognition system on the TMS320C2x and TMS320C5x, 1996.

Peesl S, A Hoffmann, V Zivojnovic, H Meyr. LISA—machine description language for cycle—accurate models of programmable DSP architectures. Proceedings of the 36th ACM/IEEE Conference on Design Automation Conference, 1999, pp 933–938.

Peleg A, S Wilkie, U Weiser. Intel MMX for Multimedia PCs, Communications of the ACM, Jan. 1997, Vol. 40, No. 1, pp. 25–38.

Pennebaker WB, JL Mitchell. JPEG Still Image Data Compression Standard. New York: Van Nostrand Reinhold, 1993.

Philips. TM1300 Preliminary Data Book, October 1999, http://www.us. semiconductors.philips.com/trimedia/products/#tm1300.

Purcell S. The impact of Mpact2. IEEE Signal Processing Mag 102–107, March 1998.

Renard P. Implementation of Adaptive Controllers on the Motorola DSP56000/DSP56001. Motorola, Inc., 1992, APR15.pdf

Resweber EJ. A DSP GMSK Modem for Mobitex and Other Wireless Infrastructures. Synetcom Digital Inc., 1996.

Robinson A, C Lueck, J Rowlands. Audio Decoding on the C54X, Texas Instruments, 1998.

Schoner B, J Villasenor, S Molloy, R Jain. Techniques for FPGA implementation of video compression systems. Proceedings of the Third International ACM Symposium on Field-Programmable Gate Arrays, 1995, pp 154–159.

Seshan N. High VelociTi processing. IEEE Signal Processing Mag. 86–101, March 1998.

Shankiti AM, M Leeser. Implementing a RAKE receiver for wireless communications on an FPGA-based computer system. Proceedings of the ACM/SIGDA International Symposium on Field Programmable Gate Arrays 2000, pp 145–151.

StarCore. SC140 DSP Core Reference Manual, Rev. 0, 1999, http://www.mot.com/pub/ SPS/DSP/LIBRARY/STARCORE/sc140dspcore_rm.pdf

Stokes J, GRL Sohie. Implementation of PID Controllers on the Motorola DSP56000/ DSP56001. Motorola, Inc., 1990, PR5.pdf

Stotzer E, E Leiss. Modulo scheduling for the TMS320C6x VLIW DSP. Proceedings of the ACM SIGPLAN 1999 Workshop on Languages, Compilers, and Tools for Embedded Systems, 1999, pp 28–34.

Sztipanovits J, G Karsai, T Bapty. Self-adaptive software for signal processing. Commun ACM 41(5):66–73, 1998.

Sun. VIS Instruction Set User's Manual, 1997, http://www.sun.com/microelectronics/ manuals/805-1394.pdf

Taipale D. Implementing Viterbi decoders using the VSL instruction on DSP families DSP56300 and DSP56600, APR40/D (Revision 0) May 1998.

Texas Instruments Europe Caller ID on TMS320C2xx. Texas Instruments Europe Report BPRA056, 1997.

Texas Instruments. TMS320C6x optimizing C Compiler User's Guide. Texas Instruments, 1998a.

Texas Instruments. TMS320C6x C Source Debugger User's Guide. Texas Instruments, 1998b.

Texas Instruments. TMS320C62x/C67x, Programmer's Guide. Texas Instruments, 1998c.

Texas Instruments. TMS320VC5421, Advance Information. Texas Instruments, 1999a.

Texas Instruments. TMSC320C6203 Product Preview. Texas Instruments, 1999b, sprs104.pdf

Texas Instruments. TMSC320C6701 Product Preview. Texas Instruments, 1999c, sprs067c.pdf

Terada H, S Miyata, M Iwata. DDMP's: Self-time super-pipelined data driven multimedia processors. Proc IEEE 87(2):282–296, 1999.

Tremblay M, JM O'Connor, V Narayan, L He. VIS speeds new media processing. IEEE Micro, 10–20, August 1996.

TriCore Architecture Manual, Infineon, Inc., 1999, http://www.infineon.com/us/micro/ tricore/arch/archman.pdf

Tyler J, J Lent, A Mather, N Huy, 1999 IEEE International Performance, Computing and Communications Conference, 1999, pp 437–44, AltiVec™: bringing vector technology to the PowerPC™ processor family.

Wolf O, J Bier. StarCore launches first architecture. Microprocessor Rep 12(14):1998.

Yao F, B Li, M Zhang. A fixed-point DSP implementation for a low bit rate vocoder. Proceedings 1998 5th International Conference on Solid-State and Integrated Circuit Technology, 1998, pp 365–368.

Yim S, Y Ding, EB George. Implementing Real-Time MIDI Music Synthesis Algorithms, ABS/OLA, and SMS for the TMS320C32 DSP, Texas Instruments, 1998.

Yiu H. Implementing V.32bis Viterbi Decoding on the TMS320C62xx DSP. Hong Kong: Texas Instruments Hong Kong Ltd, 1998.

Yu KH, YH Hu. Efficient scheduling and instruction selection for programmable digital signal processors. IEEE Trans Signal Process sp-42(12):3549–3552, 1994.

2

VLIW Processor Architectures and Algorithm Mappings for DSP Applications

Ravi A. Managuli and Yongmin Kim
University of Washington, Seattle, Washington

1 INTRODUCTION

In order to meet the real-time requirements of various applications, digital signal processors (DSPs) have traditionally been designed with special hardware features, such as fast multiply and accumulate units, multiple data memory banks and support for low-overhead looping that can efficiently execute DSP algorithms (Lee, 1988, 1989). These applications included modems, disk drives, speech synthesis/analysis, and cellular phones. However, as many forms of media [e.g., film, audio, three-dimensional (3D) graphics, and video] have become digital, new applications are emerging with the processing requirements different from what can be provided by traditional DSPs. Several examples of new applications include digital TV, set-top boxes, desktop video conferencing, multifunction printers, digital cameras, machine vision, and medical imaging. These applications have large computational and data flow requirements and need to be supported in real time. In addition, these applications are quite likely to face an environment with changing standards and requirements; thus the flexibility and upgradability of these new products, most likely via software, will play an increasingly important role.

Traditionally, if an application has a high computational requirement (e.g., military and medical), a dedicated system with multiple boards and/or multiple processors was developed and used. However, for multimedia applications requiring high computational power at a low cost, these expensive multiprocessor systems are not usable. Thus, to meet this growing computational demand at an affordable cost, new advanced processor architectures with a high level of on-chip parallelism have been emerging. The on-chip parallelism is being implemented mainly using both instruction-level and data-level parallelism. Instruction-level parallelism allows multiple operations to be initiated in a single clock cycle. Two basic approaches to achieving a high degree of instruction-level parallelism are VLIW (Very Long Instruction Word) and superscalar architectures (Patterson and Hennessy, 1996). Philips Trimedia TM1000, Fujitsu FR500, Texas Instruments TMS320C62 and TMS320C80, Hitachi/Equator Technologies MAP1000, IBM/Motorola PowerPC 604, Intel Pentium III, SGI (Silicon Graphics Inc.) R12000, and Sun Microsystems UltraSPARC III are few examples of recently developed VLIW/superscalar processors. With data-level parallelism, a single execution unit is partitioned into multiple smaller data units. The same operation is performed on multiple datasets simultaneously. Sun Microsystems' VIS (Visual Instruction Set), Intel's MMX, HP's MAX-2 (Multimedia Acceleration eXtensions-2), DEC's MAX (MultimediA eXtensions), and SGI's MIPS MDMX (MIPS Digital Media eXtension) are several examples of data-level parallelism.

Several factors make the VLIW architecture especially suitable for DSP applications. First, most DSP algorithms are dominated by data-parallel computation and consist of core tight loops (e.g., convolution and fast Fourier transform) that are executed repeatedly. Because the program flow is deterministic, it is possible to develop and map a new algorithm efficiently to utilize the on-chip parallelism to its maximum prior to the run time. Second, single-chip high-performance VLIW processors with multiple functional units (e.g., add, multiply and load/store) have become commercially available recently.

In this chapter, both architectural and programming features of VLIW processors are discussed. In Section 2, VLIW's architectural features are outlined, and several commercially-available VLIW processors are discussed in Section 3. Algorithm mapping methodologies on VLIW processors and the implementation details for several algorithms are presented in Sections 4 and 5, respectively.

2   VLIW ARCHITECTURE

A VLIW processor has a parallel internal architecture and is characterized by having multiple independent functional units (Fisher, 1984). It can achieve a high level of performance by utilizing instruction-level and data-level parallelisms. Figure 1 illustrates the block diagram for a typical VLIW processor with N functional units.

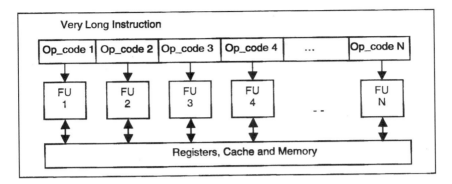

Figure 1   Block diagram for a typical VLIW processor with multiple functional units (FUs).

2.1   Instruction-Level Parallelism

The programs can be sped up by executing several RISC-like operations, such as load, stores, multiplications and additions, all in parallel on different functional units. Each very long instruction contains an operation code for each functional unit, and all the functional units receive their operation codes at the same time. Thus, VLIW processors typically follow the same control flow across all functional units. The register file and on-chip memory banks are shared by multiple functional units. A better illustration of instruction-level parallelism (ILP) is provided with an example. Consider the computation of

$$y = a_1x_1 + a_2x_2 + a_3x_3$$

on a sequential RISC processor

cycle 1: load a_1
cycle 2: load x_1
cycle 3: load a_2
cycle 4: load x_2
cycle 5: multiply z_1 a_1 x_1
cycle 6: multiply z_2 a_2 x_2
cycle 7: add y z_1 z_2
cycle 8: load a_3
cycle 9: load x_3
cycle 10: multiply z_1 a_3 x_3
cycle 11: add y y z_2

which requires 11 cycles. On the VLIW processor that has two load/store units, one multiply unit, and one add unit, the same code can be executed in only five cycles.

> cycle 1: load a_1
> load x_1
> cycle 2: load a_2
> load x_2
> Multiply $z_1\ a_1\ x_1$
> cycle 3: load a_3
> load x_3
> Multiply $z_2\ a_2\ x_2$
> cycle 4: multiply $z_3\ a_3\ x_3$
> add $y\ z_1\ z_2$
> cycle 5: add $y\ y\ z_3$

Thus, the performance is approximately two times faster than that of a sequential RISC processor. If this loop needs to be computed repeatedly (e.g., finite impulse response [FIR]), the free slots available in cycles 3, 4, and 5 can be utilized by overlapping the computation and loading for the next output value to improve the performance further.

2.2 Data-Level Parallelism

Also, the programs can be sped up by performing partitioned operations where a single arithmetic unit is divided to perform the same operation on multiple smaller precision data, [e.g., a 64-bit arithmetic and logic unit (ALU) is partitioned into eight 8-bit units to perform eight operations in parallel]. Figure 2 shows an example of `partitioned_add`, where eight pairs of 8-bit pixels are added in parallel by a single instruction. This feature is often called as multimedia extension (Lee, 1995). By dividing the ALU to perform the same operation on

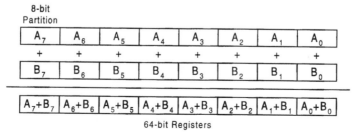

Figure 2   Example partition operation: `partitioned_add`.

multiple data, it is possible to improve the performance by two, four, or eight times depending on the partition size. The performance improvement using data-level parallelism is also best explained with an example of adding two arrays (*a* and *b*, each having 128 elements, with each array element being 8 bits), which is as follows:

```
/* Each element of array is 8 bits */
char a[128], b[128], c[128];
for (i = 0; i < 128; i++){
c[i] = a[i] + b[i];
}
```

The same code can be executed utilizing `partitioned_add`:

```
long a[16], b[16], c[16];
for (i = 0; i < 16; i++){
c[i] = partitioned_add(a[i], b[i]);
}
```

The performance with data-level parallelism is increased by a factor of 8 in this example. Because the number of loop iterations also decreases by a factor of 8, there will be an additional performance improvement due to the reduction of branch overhead.

2.3   Instruction Set Architecture

The data-level parallelism in multimedia applications can be utilized by a special subset of instructions, called Single Instruction Multiple Data (SIMD) instructions (Basoglu et al., 1998; Rathnam and Slavenburg, 1998). These instructions operate on multiple 8-, 16-, or 32-bit subwords of the operands. The current SIMD instructions can be categorized into the following groups:

- Partitioned arithmetic/logic instructions: `add`, `subtract`, `multiply`, `compare`, `shift`, and so forth.
- Sigma instructions: `inner-product`, sum of absolute difference (`SAD`), sum of absolute value (`SAM`), and so forth
- Partitioned select instructions: `min`/`max`, `conditional_selection`, and so forth
- Formatting instructions: `map`, `shuffle`, `compress`, `expand`, and so forth
- Processor-specific instructions optimized for multimedia, imaging and 3D graphics

The instructions in the first category perform multiple arithmetic operations in one instruction. The example of `partitioned_add` is shown in Figure 2, which performs the same operation on eight pairs of pixels simultaneously. These partitioned arithmetic/logic units can also saturate the result to the maximum positive or negative value, truncate the data, or round the data. The instructions in the second category are very powerful and useful in many DSP algorithms. Equation (1) is an `inner-product` example, whereas Eq. (2) describes the operations performed by the `SAD` instruction, where x and c are eight 8-bit data stored in each 64-bit source operand and the results are accumulated in y:

$$y = \sum_{i=0}^{i=7} c_i x_i \tag{1}$$

$$y = \sum_{i=0}^{i=7} |c_i - x_i| \tag{2}$$

The `inner-product` instruction is ideal in implementing convolution-type algorithms, and the `SAD` and `SAM` instructions are very useful in video processing (e.g., motion estimation). The third category of instructions can be used in minimizing the occurrence of `if`/`then`/`else` to improve the utilization of instruction-level parallelism. The formatting instructions in the fourth category are mainly used for rearranging the data in order to expose and exploit the data-level parallelism. An example of using `shuffle` and `combine` to transpose a 4×4 block is shown in Figure 3. Two types of `compress` are presented in Figure 4. `compress1` in Figure 4a packs two 32-bit values into two 16-bit values and stores them into a partitioned 32-bit register while performing the right-shift operation by a specified amount. `compress2` in Figure 4b packs four 16-bit values into four 8-bit values while performing the right-shift operation by a specified amount. `compress2` saturates the individual partitioned results after compressing to 0 or 255. In the fifth category of instructions, each processor has

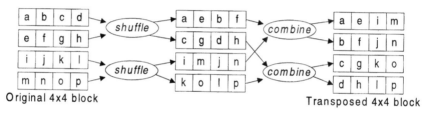

Figure 3   Transpose of a 4×4 block using `shuffle` and `combine`.

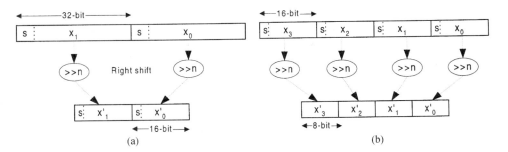

Figure 4   Partitioned 64-bit compress instructions.

its own instructions to further enhance the performance. For example, complex_multiply shown in Figure 5, which performs two partitioned complex multiplications in one instruction, is useful for implementing the FFT and autocorrelation algorithms.

Although these instructions are powerful, they take multiple cycles to complete, which is defined as latency. For example, partitioned arithmetic/logic instructions have a three-cycle latency, whereas sigma instructions have a latency of five to six cycles. To achieve the best performance, all the execution units need to be kept as busy as possible in every cycle, which is difficult due to these latencies. However, these instructions have a single-cycle throughput (i.e., another identical operation can be issued in the next cycle) due to hardware pipelining. In Section 4.2, loop unrolling and software pipelining are discussed, which tries to exploit this single-cycle throughput to overcome the latency problem.

Many improvements in the processor architectures and powerful instruction sets have been steadily reducing the processing time, which makes the task of bringing the data from off-chip to on-chip memory fast enough so as not to slow

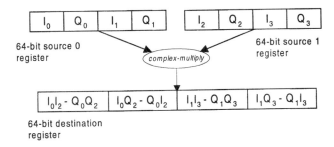

Figure 5   complex-multiply instruction.

down the functional units a real challenge. This problem gets exasperated with the growing speed disparity between the processor and the off-chip memory (e.g., the number of CPU cycles required to access the main memory doubles approximately every 6.2 years) (Boland and Dollas, 1994).

2.4 Memory I/O

There are several methods to move the data between slower off-chip memory and faster on-chip memory. The conventional method of handling data transfers in general-purpose processors has been via data caches (Basoglu et al., 1998), whereas the DSPs have been relying more on direct memory access (DMA) controllers (Berkeley Design Technology, 1996). Data caches have a nonpredictable access time. The data access time to handle a cache miss is at least an order of magnitude slower than that of a cache hit. On the other hand, the DMA controller has a predictable access time and can be programmed to hide the data transfer time behind the processing time by making it work independently of the core processor.

The real-time requirement of many DSP applications is one reason that the DSP architecture traditionally contains a DMA controller rather than data caches. The DMA can provide much higher performance with predictability. On the other hand, it requires some effort by the programmer; for example, the data transfer type, amount, location, and other information, including synchronization between DMA and processor, have to be thought through and specified by the programmer (Kim et al., 2000). In Section 4.6, DMA programming techniques to hide the data movement time behind the core processor's computing time are presented.

Many DSP programmers have developed their applications in assembly language. However, programming in assembly language is difficult to code, debug, maintain, and port, especially as applications become larger and more complex and processor architectures get very sophisticated. For example, the arrival of powerful VLIW processors with the complex instruction set and the need to perform loop unrolling and software pipelining has increased the complexity and difficulty of assembly language programming significantly. Thus, much effort is being made to develop intelligent compilers that can reduce or ultimately eliminate the burden and need of assembly language programming.

In Section 3, several commercially available VLIW processors are briefly reviewed. In Sections 4 and 5, how to program VLIW processors will be discussed in detail.

3   EXAMPLES OF VLIW PROCESSORS

Every VLIW processor tries to utilize both instruction-level and data-level parallelisms. They distinguish themselves in the number of banks and amount of on-chip memory and/or cache, the number and type of functional units, the way in which the global control flow is maintained, and the type of interconnections between the functional units. In this section, five VLIW processors and their basic architectural features are briefly discussed. Many of these processors have additional functional units to perform sequential processing, such as that required in MPEG's Huffman decoding.

3.1 Texas Instruments TMS320C62

The Texas Instruments TMS320C62 (Texas Instruments, 1999) shown in Figure 6 is a VLIW architecture with 256 bits per instruction. This DSP features two clusters, each with four functional units. Each cluster has its own 16, 32-bit registers with 2 read ports and 1 write port for each functional unit. There is one cross-cluster read port each way, so a functional unit in one cluster can access values stored in the register file of the other cluster. Most operations have a single-cycle throughput and a single-cycle latency except for a few operations. For example, a multiply operation has a single-cycle throughput and a two-cycle latency, whereas a load/store operation has a single-cycle throughput and a five-cycle latency. Two integer arithmetic units support partitioned operations, in that each 32-bit arithmetic and logic unit (ALU) can be split to perform two 16-bit additions or two 16-bit subtractions. The TMS320C62 also features a programmable DMA controller combined with two 32-kbyte on-chip data memory blocks to handle I/O data transfers.

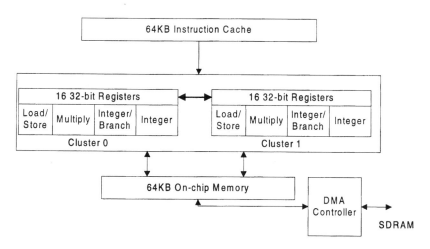

Figure 6   Block diagram of the Texas Instruments TMS320C62.

3.2 Fujitsu FR500

The block diagram of the Fujitsu FR500 (Fujitsu Limited, 1999) VLIW processor is shown in Figure 7. It can issue up to four instructions per cycle. It has two integer units, two floating-point units, a 16-kbyte four-way set-associative data cache, and a 16-kbyte four-way set-associative instruction cache. This processor has 64 32-bit general purpose registers and 64 32-bit floating-point registers. Integer units are responsible for double-word load/store, branch, integer multiply, and integer divide operations. They also support integer operations, such as rotate, shift, and AND/OR. All of these integer operations have a single-cycle latency except load/store, multiply, and divide. Multiply has a 2-cycle latency with a single-cycle throughput, divide has a 19-cycle latency with a 19-cycle throughput, and load/store has a 3-cycle latency with a single-cycle throughput. Floating-point units are responsible for single-precision floating-point operations, double-word load and SIMD-type operations. All of the floating-point operations have a three-cycle latency with a single-cycle throughput except load, divide, and square root. Floating-point divide and square root operations have a 10-cycle and 15-cycle latency, respectively, and they cannot be pipelined with another floating-point divide or square root operation because the throughput for both of these operations is equal to their latency. For load, latency is four-cycle, whereas throughput is single cycle. The floating-point unit also performs multiply and accumulate with 40-bit accumulation, partitioned arithmetic operations on 16-bit data, and various formatting operations. Partitioned arithmetic operations have either one- or two-cycle latency with a single-cycle throughput. All computing units support predicated execution for if/then/else-type statements. Be-

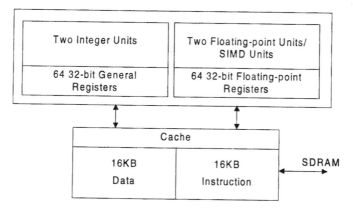

Figure 7 Block diagram of the Fujitsu FR500.

cause this processor does not have a DMA controller, it has to rely on a caching mechanism to move the data between on-chip and off-chip memory.

3.3 Texas Instruments TMS320C80

The Texas Instruments TMS320C80 (Guttag et al., 1992) incorporates not only instruction-level and data-level parallelisms but also multiple processors on a single chip. Figure 8 shows the TMS320C80's block diagram. It contains four Advanced Digital Signal Processors (ADSPs; each ADSP is a DSP with a VLIW architecture), a reduced instruction set computer (RISC) processor, and a programmable DMA controller called a transfer controller (TC). Each ADSP has its own 2-kbyte instruction cache and four 2-kbyte on-chip data memory modules that are serviced by the DMA controller. The RISC processor has a 4-kbyte instruction cache and a 4-kbyte data cache.

Each ADSP has a 16-bit multiplier, a three-input 32-bit ALU, a branch unit, and two load/store units. The RISC processor has a floating-point unit, which can issue floating-point multiply/accumulate instructions on every cycle. The programmable DMA controller supports various types of data transfers with complex address calculations. Each of the five processors is capable of executing multiple operations per cycle. Each ADSP can execute one 16-bit multiplication (which can be partitioned into two 8-bit multiply units), one 32-bit add/subtract (that can be partitioned into two 16-bit or four 8-bit units), one branch, and two load/

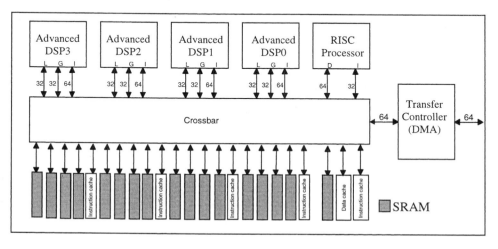

Figure 8   Block diagram of the Texas Instruments TMS320C80.

store operations in the same cycle. Each ADSP also has three zero-overhead loop controllers. However, this processor does not support some powerful operations, such as SAD or inner-product. All operations on the ADSP including load/store, multiplication and addition are performed in a single cycle.

3.4 Philips Trimedia TM1000

The block diagram of the Philips Trimedia TM1000 (Rathnam and Slavenburg, 1998) is shown in Figure 9. It has a 16-kbyte data cache, a 32-kbyte instruction cache, 27 functional units, and coprocessors to help the TM-1000 perform real-time MPEG-2 decoding. In addition, TM1000 has one peripheral component interface (PCI) port and various multimedia input/output ports. The TM1000 does not have a programmable DMA controller and relies on the caching mechanism to move the data between on-chip and off-chip memory. The TM1000 can issue 5 simultaneous operations to 5 out of the 27 functional units per cycle (i.e., 5 operation slots per cycle). The two DSP-arithmetic logic units (DSPALUs) can each perform either 32-bit or 8-bit/16-bit partitioned arithmetic operations. Each of the two DSP-multiplier (DSPMUL) units can issue two 16×16 or four 8×8 multiplications per cycle. Furthermore, each DSPMUL can perform an inner-product operation by summing the results of its two 16×16 or four 8×8 multiplications. In ALU, pack/merge (for data formatting) and select operations

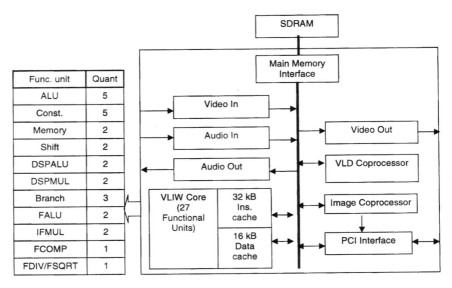

Figure 9 Block diagram of the Philips Trimedia TM1000.

are provided for 8-bit or 16-bit data in the 32-bit source data. All of the partitioned operations, including load/store and inner-product type operations, have a three-cycle latency and a single-cycle throughput.

3.5   Hitachi/Equator Technologies MAP1000

The block diagram of the Hitachi/Equator Technologies MAP1000 (Basoglu et al., 1999) is shown in Figure 10. The processing core consists of two clusters, a 16-kbyte four-way set-associative data cache, a 16-kbyte two-way set-associative instruction cache, and a video graphics coprocessor for MPEG-2 decoding. It has an on-chip programmable DMA controller called Data Streamer (DS). In addition, the MAP1000 has two PCI ports and various multimedia input/output ports, as shown in Figure 10. Each cluster has 64, 32-bit general registers, 16 predicate registers, a pair of 128-bit registers, an Integer Arithmetic and Logic Unit (IALU), and an Integer Floating-Point Graphics Arithmetic Logic Unit (IFGALU). Two clusters are capable of executing four different operations (e.g., two on IALUs and two on IFGALUs) per clock cycle. The IALU can perform either a 32-bit fixed-point arithmetic operation or a 64-bit load/store operation. The IFGALU can perform 64-bit partitioned arithmetic operations, sigma operations on 128-bit registers (on partitions of 8, 16, and 32), and various formatting operations on 64-bit data (e.g., map and shuffle). The IFGALU unit can also execute floating-point operations, including division and square root. Partitioned arithmetic operations have a 3-cycle latency with a single-cycle throughput, multiply and inner-

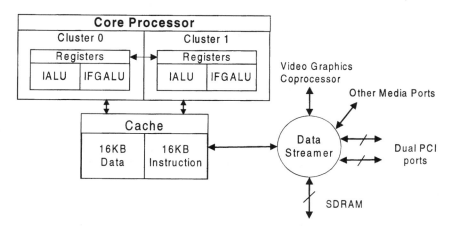

Figure 10   Block diagram of the Hitachi/Equator Technologies MAP1000.

product operations have a 6-cycle latency with a single-cycle throughput, and floating-point operations have a 17-cycle latency with a 16-cycle throughput. The MAP1000 has a unique architecture in that it supports both data cache and DMA mechanism. With the DMA approach, the 16-kbyte data cache itself can be used as on-chip memory.

The MAP-CA is a sister processor of MAP1000 with a similar architecture specifically targeting consumer appliances (Equator Technologies, 2000). The MAP-CA has a 32-kbyte data cache, a 32-kbyte instruction cache (instead of 16-kbytes each on the MAP1000), and one PCI unit (instead of two). It has no floating-point unit at all. Even though execution units of each cluster on MAP-CA are still called IALU and IFGALU, IFGALU unit does not perform any floating-point operations.

3.6   Transmeta's Crusoe Processor TM5400

None of the current general-purpose microprocessors are based on the VLIW architecture because in PC and workstation-based applications, the requirement that all of the instruction scheduling must be done during compilation could become a disadvantage because much of the processing are user-directed and cannot be generalized into a fixed pattern (e.g., word processing). The binary code compatibility (i.e., being able to run the binary object code developed for the earlier microprocessors on the more recent microprocessors) tends to become another constraint in the case of general-purpose microprocessors. However, one exception is the Transmeta's Crusoe processor. This is a VLIW processor, which when used in conjunction with Transmeta's ×86 code morphing software, it provides ×86-compatible software execution using dynamic binary code translation (Greppert and Perry, 2000). Systems based on this solution are capable of executing all standard ×86-compatible operating systems and applications, including Microsoft Windows and Linux.

The block diagram of the Transmeta's TM5400 VLIW processor is shown in Figure 11. It can issue up to four instructions per cycle. It has two integer units, a floating-point unit, a load/store unit, a branch unit, a 64-kbyte 16-way set-associative L1 data cache, a 64-kbyte 8-way set-associative instruction cache, a 256-kbyte L2 cache, and a PCI port. This processor has 64, 32-bit general-purpose registers. The VLIW instruction can be of size 64–128 bits and contain up to 4 RISC-like instructions. Within this VLIW architecture, the control logic of the processor is kept simple and software is used to control the scheduling of the instruction. This allows a simplified hardware implementation with a 7-stage integer pipeline and a 10-stage floating-point pipeline. The processor support partitioned operations as well.

In the next section, we discuss common algorithm mapping methods that can be utilized across several VLIW processors to obtain high performance. In Section 5, we discuss mapping of several algorithms onto VLIW processors.

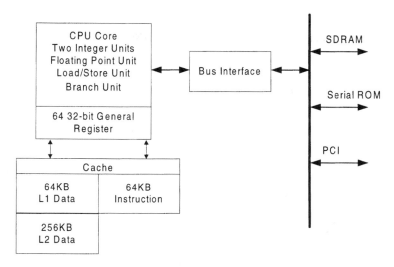

Figure 11 Block diagram of Transmeta's Crusoe TM5400.

4 ALGORITHM MAPPING METHODS

Implementation of an algorithm onto a VLIW processor for high-performance requires a good understanding of the algorithm, processor architecture, and instruction set. There are several programming techniques that can be used to improve the algorithm performance. These techniques include the following:

- Judicious use of instructions to utilize multiple execution units and data-level parallelism
- Loop unrolling and software pipelining
- Avoidance of conditional branching
- Overcoming memory alignment problems
- The utilization of fixed-point operations instead of floating-point operations
- The use of the DMA controller to minimize I/O overhead

In this section, these techniques are discussed in detail and a few example algorithms mapped to the VLIW processors utilizing these programming techniques are presented in Section 5.

4.1 Judicious Use of Instructions

Very long instruction word processors have optimum performance when all the functional units are utilized efficiently and maximally. Thus, the careful selection of instructions to utilize the underlying architecture to keep all the execution units busy is critical. For illustration, consider an example where a look-up table operation is performed {i.e., LUT (x[i])}:

```
char x[128], y[128];
for (i = 0; i < 128; I++)
y[i] = LUT(x[i]);
```

This algorithm mapped to the MAP1000 without considering the instruction set architecture requires 3 IALU operation (2 loads and 1 store) per data point, which corresponds to 384 instructions for 128 data points (assuming 1 cluster). By utilizing multimedia instructions so that both IALU and IFGALU are well used, the performance can be improved significantly, as shown in Figure 12. Here, four data points are loaded in a single `load` IALU instruction, and the IFGALU is utilized to separate each data point before the IALU performs LUT operations. After performing the LUT operations, the IFGALU is again utilized to pack these four data points so that a single `store` instruction can store all four results. This algorithm leads to six IALU operations and six IFGALU operations for every four data points. Because the IALU and IFGALU can run concurrently, this reduces the number of cycles per pixel to 1.5 compared to 3 earlier. This results in a performance improvement by a factor of 2. This is a simple example illustrating that it is possible

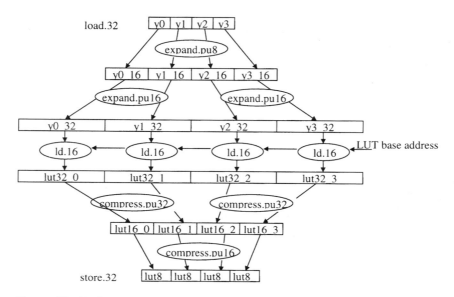

Figure 12   Performing LUT using IALU and IFGALU on the MAP1000.

to improve the performance significantly by carefully selecting the instructions in implementing the intended algorithm.

4.2 Loop Unrolling and Software Pipelining

Loop unrolling and software pipelining are very effective in overcoming the multiple-cycle latencies of the instructions. For illustration, consider an algorithm implemented on the MAP1000, where each element of an array is multiplied with a constant k. On the MAP1000, load/store has a five-cycle latency with a single-cycle throughput, and partitioned_multiply (which performs eight 8-bit partitioned multiplications) has a six-cycle latency with a single-cycle throughput. Figure 13a illustrates the multiplication of each array element (in_data) with a constant k, where k is replicated in each partition of the register for partitioned_multiply. The array elements are loaded by the IALU, and partitioned_multiply is performed by the IFGALU. Because load has a five-cycle latency, partitioned_multiply is issued after five

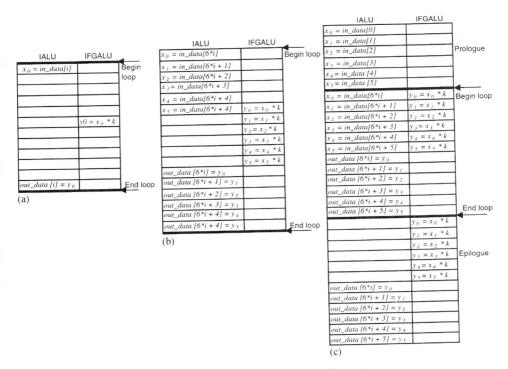

Figure 13 Example of loop unrolling and software pipelining.

cycles. The result is stored after another latency of six cycles because partitioned_multiply has a six-cycle latency. Only 3 instruction slots are utilized out of 24 possible IALU and IFGALU slots in the inner loop, which results in wasting 87.5% of the instruction issue slots and leads to a disappointing computing performance.

To address this latency problem and underutilization of instruction slots, loop unrolling and software pipelining can be utilized (Lam, 1988). In loop unrolling, multiple sets of data are processed inside the loop. For example, six sets of data are processed in Figure 13b. The latency problem is partially overcome by taking advantage of the single-cycle throughput and filling the delay slots with the unrolled instructions. However, many instruction slots are still empty because the IALU and IFGALU are not used simultaneously. Software pipelining can be used to fill these empty slots, where operations from different iterations of the loop are overlapped and executed simultaneously by the IALU and IFGALU, as shown in Figure 13c. By the IALU loading, the data to be used in the next iteration and the IFGALU executing the partitioned_multiply instructions using the data loaded in the previous iteration, the IALU and IFGALU can execute concurrently, thus increasing the instruction slot utilization. Few free slots available in the IFGALU unit can be utilized for controlling the loop counters. However, to utilize software pipelining, some preprocessing and postprocessing need to be performed, (e.g., loading in the prologue the data to be used in the first iteration and executing partitioned_multiply and store in the epilogue for the data loaded in the last iteration, as shown in Figure 13c). Thus, the judicious use of loop unrolling and software pipelining results in the increased data processing throughput when Figures 13a and 13c are compared [a factor of 5.7 when an array of 480 is processed (i.e., 720 cycles versus 126 cycles)].

4.3 Fixed Point Versus Floating Point

The VLIW processors predominantly have fixed-point functional units with some floating-point support. The floating-point operations are generally computationally expensive with longer latency and lower throughput than fixed-point operations. Thus, it is desirable to carry out computations in fixed-point arithmetic and avoid floating-point operations if we can.

While using fixed-point arithmetic, the programmer has to pay attention to several issues (e.g., accuracy and overflow). When multiplying two numbers, the number of bits required to represent the result without any loss in accuracy is equal to the sum of the number of bits in each operand (e.g., while multiplying two N-bit numbers $2N$ bits are necessary). Storing $2N$ bits is expensive and is usually not necessary. If only N bits are kept, it is up to the programmer to determine which N bits to keep. Several instructions on these VLIW processors provide a variety of options to the programmer in selecting which N bits to keep.

Overflow occurs when too many numbers are added to the register accumulating the results (e.g., when a 32-bit register tries to accumulate the results of 256 multiplications, each with 2 16-bit operands). One measure that can be taken against overflow is to utilize more bits for the accumulator (e.g., 40 bits to accumulate the above results). Many DSPs do, in fact, have extra headroom bits in the accumulators (TMS320C62 and Fujitsu FR500). The second measure that can be used is to clip the result to the largest magnitude positive or negative number that can be represented with the fixed number of bits. This is more acceptable than permitting the overflow to occur, which otherwise would yield a large magnitude and/or sign error. Many VLIW instructions can automatically perform a clip operation (MAP1000 and Trimedia). The third measure is to shift the product before adding it to the accumulator. A complete solution to the overflow problem requires that the programmer be aware of the scaling of all the variables to ensure that the overflow would not happen.

If a VLIW processor supports floating-point arithmetic, it is often convenient to utilize the capability. For example, in the case of computing the square root, it is advantageous to utilize a floating-point unit rather than using an integer unit with a large look-up table. However, to use floating-point operations with integer operands, some extra operations are required (e.g., converting floating-point numbers to fixed-point numbers and vice versa). Furthermore, it takes more cycles to compute in floating point compared with in fixed point.

4.4 Avoiding If/Then/Else Statements

There are two types of branch operations that occur in the DSP programming:

Loop branching: Most DSP algorithms spend a large amount of time in simple inner loops. These loops are usually iterated many times, the number of which is constant and predictable. Usually, the branch instructions that are utilized to loop back to the beginning of a loop have a minimum of a two-cycle latency and require decrement and compare instructions. Thus, if the inner loop is not deep enough, the overhead due to branch instructions can be rather high. To overcome this problem, several processors support the hardwired loop-handling capability, which does not have any delay slots and does not require any decrement and compare instructions. It automatically decrements the loop counter (set outside the inner loop) and jumps out of the loop as soon as the branch condition is satisfied. For other processors that do not have a hardwired loop controller, a loop can be unrolled several times until the effect of additional instructions (decrement and compare) becomes minimal.

If/then/else *branch*: Conditional branching inside the inner loop can severely degrade the performance of a VLIW processor. For example, the direct implementation of the following code segment on the MAP1000 (where X, Y, and S are 8-bit data points) would be Figure 14a, where the `branch-if-greater-than` (BGT) and `jump` (JMP) instructions have a three-cycle latency:

```
if (X > Y)
    S = S + X;
else
    S = S + Y;
```

Due to the idle instruction slots, it takes either 7 or 10 cycles per data point (depending on the path taken) because we cannot use instruction-level and data-level parallelisms effectively. Thus, to overcome this `if/then/else` barrier in VLIW processors, two methods can be used:

- *Use predicated instructions*: Most of the instructions can be predicated. A predicated instruction has an additional operand that determines whether or not the instruction should be executed. These conditions are stored either in a separate set of 1-bit registers called predicate registers or regular 32-bit registers. An example with a predicate register to handle the `if/then/else` statement is shown in Figure 14b. This method requires only 5 cycles (compared to 7 or 10) to execute the same code segment. A disadvantage of this approach is that only one data point is processed at a time; thus it cannot utilize data-level parallelism.

- *Use select instruction*: `select` along with `compare` can be utilized to handle the `if/then/else` statement efficiently, `compare` as illustrated in Figure 14c is utilized in comparing each pair of subwords in two partitioned source registers and storing the result of the test (i.e., TRUE or FALSE) in the respective subword in another partitioned destination register. This partitioned register can be used as a mask register

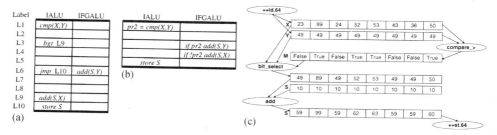

Figure 14 Avoiding branches while implementing `if/then/else` code.

(M) for the `select` instruction, which, depending on the content of each mask register partition, selects either the X or Y subword. As there are no branches to interfere with software pipelining, it only requires four cycles per loop. More importantly, because the data-level parallelism (i.e., partitioned operations) of the IFGALU is used, the performance increases further by a factor of 8 for 8-bit subwords (assuming the instructions are software pipelined).

4.5 Memory Alignment

To take advantage of the partitioned operations, the address of the data loaded from memory needs to be aligned. For example, if the partitioned register size is 64 bits (8 bytes), then the address of the data loaded from memory into the destination register should be a multiple of 8 (Peleg and Weiser, 1996). When the input data words are not aligned, extra overhead cycles are needed in loading two adjacent data words and then extracting the desired data word by performing shift and mask operations. An instruction called `align` is typically provided to perform this extraction. Figure 15 shows the use of `align`, where the desired nonaligned data, x_3 through x_{10}, are extracted from the two adjacent aligned data words (x_0 through x_7 and x_8 through x_{15}) by specifying a shift amount of 3.

4.6 DMA Programming

In order to overcome the I/O bottleneck, a DMA controller can be utilized, which can stream the data between on-chip and off-chip memories, independent of the core processor. In this subsection, two DMA modes frequently used are described: 2D block transfer and guided transfer. Two-dimensional block transfers are utilized for most applications, and the guided transfer mechanism is utilized for some special-purpose applications (e.g., look-up table) or when the required data are not consecutive.

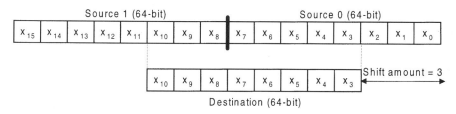

Figure 15 `align` instruction to extract the non-aligned eight bytes.

4.6.1   2D Block Transfer

In this mode, the input data are transferred and processed in small blocks, as shown in Figure 16. To prevent the processor from waiting for the data as much as possible, the DMA controller is programmed to manage the data movements concurrently with the processor's computation. This technique is illustrated in Figure 16 and is commonly known as *double buffering*. Four buffers, two for input blocks (`ping_in_buffer` and `pong_in_buffer`) and two for output blocks (`ping_out_buffer` and `pong_out_buffer`), are allocated in the on-chip memory. While the core processor computes on a current image block (e.g., block #2) from `pong_in_buffer` and stores the result in `pong_out_buffer`, the DMA controller moves the previously calculated output block (e.g., block #1) in `ping_out_buffer` to the external memory and brings the next input block (e.g., block #3) from the external memory into `ping_in_buffer`. When the computation and data movements are both completed, the core processor and DMA controller switch buffers, with the core processor starting to use the `ping` buffers and the DMA controller working on the `pong` buffers.

4.6.2   Guided Transfer

Whereas 2D block-based transfers are useful when the memory access pattern is regular, it is inefficient for accessing the nonsequential or randomly scattered data. The guided transfer mode of the DMA controller can be used in this case to efficiently access the external memory based on a list of memory address offsets from the base address, called *guide table*. One example of this is shown

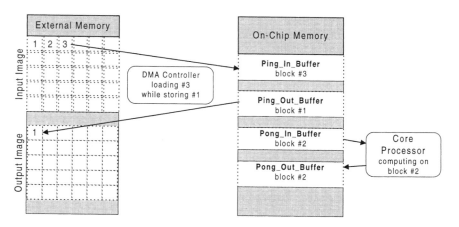

Figure 16   Double buffering with a programmable DMA controller.

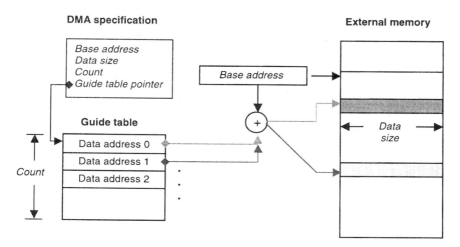

Figure 17 Guided transfer DMA controller.

in Figure 17. The guide table is either given before the program starts (off-line) or generated in the earlier stage of processing. The guided transfer is set up by specifying *base address*, *data size*, *count*, and *guide table pointer*. *data size* is the number of bytes that will be accessed from each guide table entry, and the guide table is pointed by *guide table pointer*.

5 MAPPING OF ALGORITHMS TO VLIW PROCESSORS: A FEW EXAMPLES

For VLIW processors, the scheduling of all instructions is the responsibility of the programmer and/or compiler. Thus, the assembly language programmers must understand the underlying architecture intimately to be able to obtain high performance in a certain algorithm and/or application. Smart compilers for VLIW processors to ease the programming burden are very important. Tightly coupled with the advancement of compiler technologies, there have been many useful programming techniques, as discussed in Section 4 and use of C intrinsics (Faraboschi et al., 1998). The C intrinsics can be used as a good compromise between the performance and programming productivity. A C intrinsic is a special C language extension, which looks like a function call, but directs the compiler to use a certain assembly language instruction. In programming TMS320C62, for example, the int_add2(int, int) C intrinsic would generate an ADD2 assembly instruction (two 16-bit partitioned additions) using two 32-bit integer arguments.

The compiler technology has advanced to the point that the compiled programs using C intrinsics in some cases have been reported to approach up to 60–70% of the hand-optimized assembly program performance (Seshan, 1998).

The use of C intrinsics improves the programming productivity because the compiler can relieve the programmer of register allocation, software pipelining, handling multicycle latencies, and other tasks. However, what instructions to use still depends on the programmer, and the choice of instructions decides the performance that can be obtained. Thus, careful analysis and design of an algorithm to make good use of powerful instructions and instruction-level parallelism is essential (Shieh and Papachristou, 1991). Algorithms developed without any consideration of the underlying architecture often do not produce the desired level of performance.

Some VLIW processors (e.g., MAP1000 and TM1000) have powerful and extensive partitioned instructions, whereas other processors (e.g., TMS320C80 and TMS320C62) have limited partitioned instructions. In order to cover the spectrum of instruction set architecture (extensive partitioned instructions to limited/no partitioned instructions), we will discuss algorithm mapping for TMS320C80, which has minimal partitioned operations and the MAP1000, which has extensive partitioned operations. The algorithm mapping techniques we are discussing are for 2D convolution, fast Fourier transform (FFT), inverse discrete cosine transform (IDCT), and affine warp. The detailed coverage on mapping these algorithms can be found elsewhere (Managuli et al., 1998; Managuli et al., 2000; Basoglu et al., 1997; Lee, 1997; Mizosoe et al., 2000; Evans and Kim, 1998; Chamberlain, 1997). The algorithm mapping techniques discussed for these two processors can be easily extended to other processors as well. In this section, we will describe how many cycles are needed to compute each output pixel using an assembly type of instructions. However, as mentioned earlier, if suitable instructions are selected, the C compiler can map the instructions to the underlying architecture efficiently, obtaining a performance close to that of assembly implementation.

5.1   2D Convolution

Convolution plays a central role in many image processing and digital signal processing applications. In convolution, each output pixel is computed to be a weighted average of several neighboring input pixels. In the simplest form, generalized 2D convolution of an $N \times N$ input image with an $M \times M$ convolution kernel is defined as

$$b(x, y) = \frac{1}{s} \sum_{i=x}^{x+M-1} \sum_{j=y}^{x+M-1} f(i, j)h(x - i, y - j) \qquad (3)$$

where f is the input image, h is the input kernel, s is the scaling factor, and b is the convolved image.

The generalized convolution has one division operation for normalizing the result as shown in Eq. (3). To avoid this time-consuming division operation, we multiply the reciprocal of the scaling factor with each kernel coefficient beforehand and then represent each coefficient in 16-bit sQ15 fixed-point format (1 sign bit followed by 15 fractional bits). With this fixed-point representation of coefficients, right-shift operations can be used instead of division. The right-shifted result is saturated to 0 or 255 for the 8-bit output; that is, if the right-shifted result is less than 0, it is set to zero, and if it is greater than 255, then it is clipped to 255; otherwise it is left unchanged.

5.1.1 Texas Instruments TMS320C80

Multiply and accumulate operations can be utilized to perform convolution. A software pipelined convolution algorithm on the TMS320C80 is shown in Table 1 for 3 × 3 convolution. In the first cycle (Cycle 1), a pixel (X_0) and a kernel coefficient (h_0) are loaded using the ADSP's two load/store units. In the next cycle (Cycle 2), a multiplication is performed with the previously loaded data ($M_0 = X_0 h_0$), whereas new data (X_1 and h_1) are loaded for the next iteration. In Cycle 3, the add/subtract unit can start accumulating the result of the previous multiplication ($A_0 = 0 + M_0$). Thus, from Cycle 3, all four execution units are kept busy, utilizing the instruction-level parallelism to the maximum extent. The load/store units in Cycles 10 and 11 and the multiply unit in Cycle 11 perform

Table 1   TMS320C80's Software Pipelined Execution of Convolution with a 3 × 3 Kernel

Cycle	Load/store unit 1	Load/store unit 2	Multiply unit	Add/subtract unit
1	Ld X_0	Ld h_0		
2	Ld X_1	Ld h_1	$M_0 = X_0 h_0$	
3	Ld X_2	Ld h_2	$M_1 = X_1 h_1$	$A_0 = 0 + M_0$
4	Ld X_3	Ld h_3	$M_2 = X_2 h_2$	$A_1 = A_0 + M_1$
5	Ld X_4	Ld h_4	$M_3 = X_3 h_3$	$A_2 = A_1 + M_2$
6	Ld X_5	Ld h_5	$M_4 = X_4 h_4$	$A_3 = A_2 + M_3$
7	Ld X_6	Ld h_6	$M_5 = X_5 h_5$	$A_4 = A_3 + M_4$
8	Ld X_7	Ld h_7	$M_6 = X_6 h_6$	$A_5 = A_4 + M_5$
9	Ld X_8	Ld h_8	$M_7 = X_7 h_7$	$A_6 = A_5 + M_6$
10	Ld X_0	Ld h_0	$M_8 = X_8 h_8$	$A_7 = A_6 + M_7$
11	Ld X_1	Ld h_1	$M_0 = X_0 h_0$	$A_8 = A_7 + M_8$

the necessary operations for the next output pixel in the following iteration. Four additional instructions are needed to saturate the result to 0 or 255, store the result, and perform other tasks. Thus, because there are four ADSPs on the TMS320C80, the ideal number of cycles required to perform 3×3 convolution is 3.75 per output pixel. The programmable DMA controller can be utilized to bring the data on-chip and store the data off-chip using the `double-buffering` mechanism described in Section 4.6.1.

5.1.2 Hitachi/Equator Technologies MAP1000

The generic code for the 2D convolution algorithm utilizing a typical VLIW processor instruction set is shown below. It generates eight output pixels that are horizontally consecutive. In this code, the assumptions are that the number of partitions is 8 (the data registers are 64 bits with eight 8-bit pixels), the kernel register size is 128 bits (eight 16-bit kernel coefficients) and the kernel width is less than or equal to eight.

```
for (i = 0; i < kernel_height; i++){
    /* Load 8 pixels of input data x₀ through x₇ and kernel coefficients c₀
    through c₇ */
    image_data_x₀_x₇ = *src_ptr;   kernel_data_c₀_c₇ = *kernel_ptr;
    /* Compute inner-product for pixel 0 */
    accumulate_0 += inner-product (image_data_x₀_x₇, kernel_data_c₀_c₇);
    /* Extract data x₁ through x₈ from x₀ through x₇ and x₈ through x₁₅ */
    image_data_x₈_x₁₅ = *( src_ptr + 1);
    image_data_x₁_x₈ = align(image_data_x₈_x₁₅: image_data_x₀_x₇, 1);
    /* Compute the inner-product for pixel 1 */
    accumulate_1 += inner-product (image_data_x₁_x₈:
    kernel_data_c₀_c₇);
    /* Extract data x₂ through x₉ from x₀ through x₇ and x₈ through x₁₅ */
    image_data_x₂_x₉ = align(image_data_x₈_x₁₅: image_data_x₀_x₇, 2);
    /* Compute the inner-product for pixel 2 */
    accumulate_2 += inner-product (image_data_x₂_x₉:
    kernel_data_c₀_c₇);
    .......
    accumulate_7 += inner-product (image_data_x₇_x₁₅:
    kernel_data_c₀_c₇);
    /* Update the source and kernel addresses */
    src_ptr = src_ptr + image_width;
    kernel_ptr = kernel_ptr + kernel_width;
}/* end for i */
/* Compress eight 32-bit values to eight 16-bit values with right-shift
operation */
result64_ps16_0 = compress1(accumulator_0: accumulator_1, scale);
result64_ps16_1 = compress1(accumulator_2: accumulator_3, scale);
result64_ps16_2 = compress1(accumulator_4: accumulator_5, scale);
result64_ps16_3 = compress1(accumulator_6: accumulator_7, scale);
```

```
/* Compress eight 16-bit values to eight 8-bit values. Saturate each in-
dividual value to 0 or 255 and store them in two consecutive 32-bit regis-
ters */
result32_pu8_0 = compress2(result64_ps16_0: result64_ps16_1, zero);\
result32_pu8_1 = compress2(result64_ps16_2: result64_ps16_3, zero);
/* Store 8 pixels present in two consecutive 32-bit registers and update
the destination address */
*dst_ptr++ = result32_pu8_0_and_1;
```

If the kernel width is greater than 8, then the kernel can be subdivided into several sections and the inner loop is iterated multiple times, while accumulating the multiplication results.

The MAP1000 has an advanced inner-product instruction, called srshin-prod.pu8.ps16, as shown in Figure 18. It can multiply eight 16-bit kernel coefficients (of partitioned local constant [PLC] register) by eight 8-bit input pixels (of partitioned local variable [PLV] register) and sum up the multiplication results. This instruction can also shift a new pixel into a 128-bit PLV register. x_0 through x_{23} represent sequential input pixels, and c_0 through c_7 represent kernel

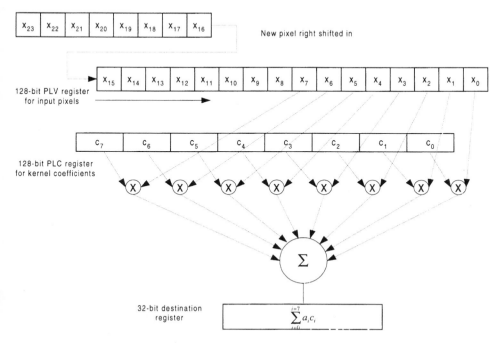

Figure 18   srshinprod.pu8.ps16 instruction using two 128-bit registers.

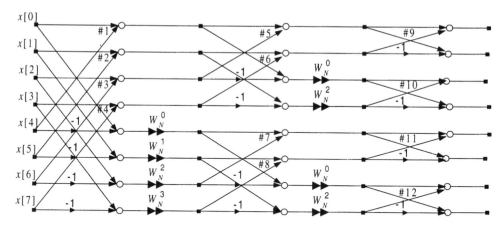

Figure 19 Flowgraph for a 1D eight-point FFT.

coefficients. After performing an inner-product operation shown in Figure 18, x_{16} is shifted into the leftmost position of the 128-bit register (PLV) and x_0 is shifted out. The next time this instruction is executed, the inner product will be performed between x_1-x_8 and c_0-c_7. This new pixel shifting-in capability eliminates the need of multiple align instructions used in the above code. An instruction called `setplc.128` sets the 128-bit PLC register with kernel coefficients. The MAP1000 also has `compress` instructions similar to the ones shown in Figure 4 that can be utilized for computing the convolution output. All of the partitioned operations can be executed only on the integer floating-point and arithmetic graphics unit (IFGALU), whereas ALU supports load/store and branch operations as discussed in Section 3.5. Thus, for 3×3 convolution, the ideal number of cycles required to process 8 output pixels are 33 [22 IALU (21 `load` and 1 `store`) instructions can be hidden behind 33 IFGALU (24 `srshinprod.pu8.ps16`, 3 `setplc.128`, 4 `compress1`, 2 `compress2`) instructions utilizing loop unrolling and software pipelining]. Because there are 2 clusters, the ideal number of cycles per output pixel is 2.1.

5.2 FFT

The fast Fourier transform (FFT) has made the computation of discrete Fourier transform (DFT) feasible, which is an essential function in a wide range of areas that employ spectral analysis, frequency-domain processing of signals, and image reconstruction. Figure 19 illustrates the flowgraph for a 1D 8-point FFT. Figure 20a shows the computation of butterfly and Figure 20b shows the detailed operations within a single butterfly. Every butterfly requires a total of 20 basic opera-

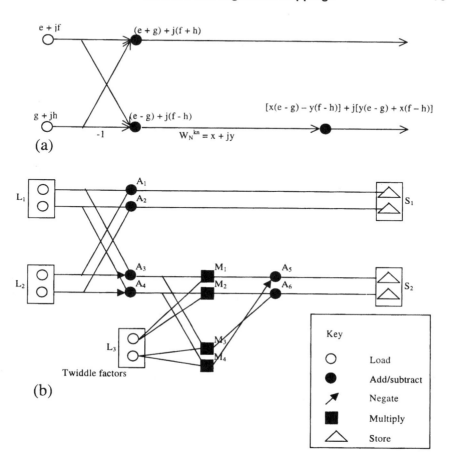

Figure 20 FFT butterfly.

tions: 4 real multiplications, 6 real additions/subtractions, 6 loads, and 4 stores. Thus, the Cooley–Tukey N-point 1D FFT algorithm with complex input data requires $2N \log_2 N$ real multiplications and $3N \log_2 N$ real additions/subtractions.

An $N \times N$ 2D FFT using the direct 2D algorithm with 2×2 butterflies requires $3N^2 \log_2 N$ real multiplications and $5.5N^2 \log_2 N$ real additions/subtractions (Dudgeon and Mersereau, 1984). Although computationally efficient, such a direct 2D FFT leads to data references that are highly scattered throughout the image. For example, the first 2×2 butterfly on a 512×512 image would require the following pixels: $x(0, 0)$, $x(0, 256)$, $x(256, 0)$, and $x(256, 256)$. The large distances between the data references make it difficult to keep the necessary data for a butterfly in the cache or on-chip memory. Alternatively, a 2D FFT can be decomposed by row–column 1D FFTs, which can be computed by performing the 1D FFT on all of the rows (rowwise FFT) followed by the 1D FFT on all of the columns (columnwise FFT) of the row FFT result as follows:

$$X[k, l] = \sum_{n=0}^{N-1} \left(\sum_{m=0}^{N-1} x\left(n, m \right) W_N^{lm} \right) W_N^{kn} \tag{4}$$

where x is the input image, W_N are the twiddle factors, and X is the FFT output. This method requires $4N^2 \log_2 N$ real multiplications and $6N^2 \log_2 N$ real additions/subtractions (Dudgeon and Mersereau, 1984), which is 33% more multiplications and 9.1% more additions/subtractions than the direct 2D approach. However, this separable 2D FFT algorithm has been popular because all of the data for the rowwise or columnwise 1D FFT being computed can be easily stored in the on-chip memory or cache. The intermediate image is transposed after the rowwise 1D FFTs so that another set of rowwise 1D FFTs can be performed. This is to reduce the number of SDRAM row misses, which otherwise (i.e., if 1D FFTs are performed on columns of the intermediate image) will occur many times. One more transposition is performed before storing the final result.

5.2.1   Texas Instruments TMS320C80

The dynamic range of the FFT output is $\pm 2^{M \log_2 N}$, where M is the number of bits in each input sample and N is the number of samples. Thus, if the input samples are 8 bits, any 1D FFT with N larger than 128 could result in output values exceeding the range provided by 16 bits. Because each ADSP has a single-cycle 16-bit multiplier, there is a need to scale the output of each butterfly stage so that the result can be always represented in 16 bits. In Figure 20, A_1 and A_2 have to be scaled explicitly before being stored, whereas the scaling of A_5 and A_6 can be incorporated into M_1–M_4. If all of the coefficients were prescaled by one-half, the resulting output would also be one-half of its original value. Because all of the trigonometric coefficients are precomputed, this prescaling does not require any extra multiplication operations.

A modified Cooley–Tukey 2-point FFT butterfly that incorporates scaling operations is shown in Figure 21, where 22 basic operations are required. Several observations on the flowgraph of Figure 21 lead us to efficient computation. First, multiplications are independent from each other. Second, two (A_1 and A_2) of the six additions/subtractions are independent of the multiplications M_1–M_4. Finally, if the real and imaginary parts of the complex input values and coefficients are kept adjacent and handled together during load and store operations, then the number of load and store operations can be reduced to three and two rather than

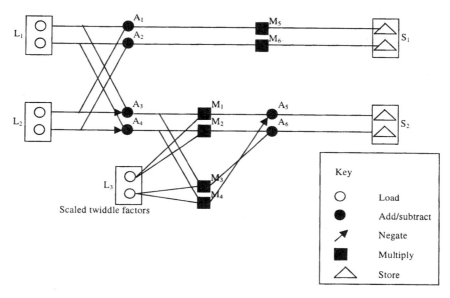

Figure 21 Modified FFT butterfly with scaling operations.

six and four, respectively, because both the real and imaginary parts (16 bits each) could be loaded or stored with one 32-bit load/store operation.

The software pipelined implementation of the butterfly on the TMS320C80 is shown in Table 2, where the 32-bit add/subtract unit is divided into 2 units to perform two 16-bit additions/subtractions in parallel. In Table 2, operations having the same background are part of the same butterfly, whereas operations within heavy black borders are part of the same tight-loop iteration. With this implementation, the number of cycles to compute each butterfly per ADSP is only six cycles (Cycles 5 through 10). All of the add/subtract and load/store operations are performed in parallel with these six multiply operations. Because there are 4 ADSPs working in parallel, the ideal number of cycles per 2-point butterfly is 1.5.

5.2.2 Hitachi/Equator Technologies MAP1000

Examining the flowgraph of the Cooley–Tukey FFT algorithm in Figure 19 reveals that within each stage, the butterflies are independent of each other. For example, the computation of butterfly #5 does not depend on the results of the butterflies #6–8. Thus, on architectures that support partitioned operations, multiple independent butterflies within each stage can be computed in parallel.

Table 2 Software Pipelined Execution of the Butterfly
on the TMS320C80

| Cycles | Multiply unit | 32-Bit add/ subtract unit | | Load/store unit #1 | Load/store unit #2 |
		16-bit unit #1	16-bit unit #2		
1				L_1	L_2
2		A_3	A_4	L_3	
3	M_1				
4	M_2				
5	M_3	A_1	A_2	L_1	L_2
6	M_4	A_3	A_4	L_3	
7	M_5	A_5	A_6		
8	M_6				
9	M_1			S_1	S_2
10	M_2				
11	M_3	A_1	A_2		
12	M_4				
13	M_5	A_5	A_6		
14	M_6				
15				S_1	S_2

The MAP1000 has the `complex_multiply` instruction shown in Figure 5, which can perform two 16-bit complex multiplications in a single instruction. Other instructions that are useful for FFT computations include `partitioned_add/subtract` to perform two 16-bit complex additions and subtractions in a single instruction and 64-bit `load` and `store`. Each butterfly (Fig. 20a) requires one complex addition, one complex subtraction, one complex multiplication, three loads, and two stores. Thus, three IFGALU and five IALU instructions are necessary to compute two butterflies in parallel (e.g., #1 and #3 together). Because both the IALU and IFGALU can execute concurrently, 40% of the IFGALU computational power is wasted because only three IFGALU instruction slots are utilized compared to five on the IALU. Thus, to balance the load between the IALU and IFGALU and efficiently utilize the available instruction slots, two successive stages of the butterfly computations can be merged together as the basic computational element of FFT algorithm (e.g., butterflies #1, #3, #5, and #7 are merged together as a single basic computational element). For the computation of this merged butterfly, the number of IALU instructions required is six and the number of IFGALU instructions required is also six, thus balancing the instruction slot utilization. If all the instructions are fully pipelined to overcome the latencies of these instructions (six for `complex_multiply`, three for `partitioned_add` and `partitioned_subtract`, and five for `load` and `store`), four butterflies can be computed in six cycles using a single cluster. Because `complex_multiply` is executed on 16-bit partitioned data, the intermediate results on the MAP1000 also require scaling operations similar to that of the TMS320C80. However, the MAP1000 partitioned operations have a built-in scaling feature, which eliminates the need for extra scaling operations. Because there are two clusters, ideally it takes 0.75 cycles for a two-point butterfly.

5.3 DCT and IDCT

The discrete cosine transform has been a key component in many image and video compression standards (e.g., JPEG, H.32X, MPEG-1, MPEG-2, and MPEG-4). There are several approaches in speeding up the DCT/IDCT computation. Several efficient algorithms [e.g., Chen's IDCT (CIDCT) algorithm] (Chen et al., 1977) have been widely used. However, on modern processors with a powerful instruction set, the matrix multiply algorithm might become faster due to their immense computing power. In this section, an 8×8 IDCT is utilized to illustrate how it can be efficiently mapped onto the VLIW processors.

The 2D 8×8 IDCT is given as

$$x_{ij} = \sum_{k=0}^{7} \frac{c(k)}{2} \left[\sum_{l=0}^{7} \frac{c(l)}{2} F_{kl} \cos\left(\frac{2j + 1)l\pi}{16}\right) \right] \cos\left(\frac{2i + 1)k\pi}{16}\right)$$

$$c(k) = \frac{1}{\sqrt{2}} \quad \text{for } k = 0; \qquad c(k) = 1 \quad \text{otherwise} \tag{5}$$

$$c(l) = \frac{1}{\sqrt{2}} \quad \text{for } l = 0; \qquad c(l) = 1 \quad \text{otherwise}$$

where F is the input data, $c(\cdot)$ are the scaling terms, and x is the IDCT result. It can be computed in a separable fashion by using 1D eight-point IDCTs. First, rowwise eight-point IDCTs are performed on all eight-rows, followed by columnwise eight-point IDCTs on all eight columns of the row IDCT result. Instead of performing columnwise IDCTs, the intermediate data after the computation of rowwise IDCTs are transposed so that another set of rowwise IDCTs can be performed. The final result is transposed once more before the results are stored. Because the implementation of DCT is similar to that of IDCT, only the IDCT implementation is discussed here.

5.3.1 Texas Instruments TMS320C80

Figure 22 illustrates the flowgraph for the 1D eight-point Chen's IDCT algorithm with the multiplication coefficients c_1 through c_7 given by $c_i = \cos(i\pi/16)$ for $i =$

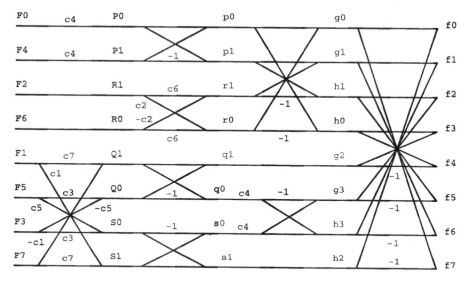

Figure 22 Chen's IDCT flowgraph.

1 through 7. When implemented with a basic instruction set, the CIDCT algorithm requires 16 multiplications and 26 additions. Thus, including 16 loads and 8 store operations, 66 operations are necessary. Table 3 shows the CIDCT algorithm implemented on the TMS320C80, where operations belonging to different 1D eight-point IDCTs (similar to FFT) are overlapped to utilize software pipelining. Variables with a single prime and double primes (such as F′ and F″) are the intermediate results. In Table 3, we are performing 32-bit additions/subtractions on the intermediate data because we are allocating 32 bits for the multiplications of two 16-bit operands to reduce quantization errors. Another description of implementing CIDCT can be found in (Lee, 1997), where 16 bits are utilized for representing multiplication results, thus performing 16-bit additions/subtractions on the intermediate data. The coefficients need to be reloaded because of lack of registers. Because there are 4 ADSPs, the ideal number of cycles per 8-point IDCT in our implementation is 6.5. Thus, it takes 104 cycles to compute one 8×8 2D IDCT.

5.3.2 Hitachi/Equator Technologies MAP1000

Table 4 illustrates the matrix-multiply algorithm to compute one eight-point IDCT, where each matrix element (i.e., A_{ux}) is equal to $C(u) \cos[\pi(2x + 1)u/16]$ with $C(u) = 1/\sqrt{2}$ for $u = 0$ and 1 otherwise. A 1D IDCT can be computed

Table 3 IDCT Implementation on the TMS320C80

Cycle	Multiply unit	Add/subtract unit	Load/store unit #1	Load/store unit #2
1	$F1'' = F1 * c1$	$p0 = P0 + P1$	store f4	store f5
2	$F7'' = F7 * c7$	$p1 = P0 - P1$	load c3	load c5
3	$F5' = F5 * c3$	$Q1 = F1' - F7'$		
4	$F3' = F3 * c5$	$S1 = F1'' - F7''$		
5	$F5'' = F5 * c5$	$Q0 = F5' - F3'$		
6	$F3'' = F3 * c3$	$q1 = Q1 + Q0$	load c2	load F2
7	$F2' = F2 * c2$	$S0 = F5'' + F3''$	load c6	load F6
8	$F6' = F6 * c6$	$q0 = Q1 - Q0$		
9	$F2'' = F2 * c6$	$r0 = F2' + F6'$		
10	$F6'' = F6 * c2$	$s0 = S1 - S0$	load c4	
11	$q0' = q0 * c4$	$s1 = S1 + S0$		
12	$s0' = s0 * c4$	$r1 = F2'' - F6''$		
13		$g0 = p0 + r0$		
14		$h0 = p0 - r0$		
15		$g1 = p1 + r1$		
16		$h1 = p1 - r1$		
17		$g3 = s0' - q0'$		
18		$h3 = s0' + q0'$		
19		$f0 = g0 + s1$		load c4
20		$f7 = g0 - s1$	store f0	load F0
21		$f1 = g1 + h3$	store f7	load F4
22		$f6 = g1 - h3$	store f1	load F1
23	$P0 = F0 * c4$	$f2 = h1 + g3$	store f6	load F7
24	$P1 = F4 * c4$	$f5 = h1 - g3$	load c7	store f3
25	$F1' = F1 * c7$	$f3 = h0 + q1$	load c1	load F5
26	$F7' = F7 * c1$	$f4 = h0 - q1$	store f2	load F3

Table 4 IDCT Using Matrix Multiply

$$
\begin{bmatrix} f0 \\ f1 \\ f2 \\ f3 \\ f4 \\ f5 \\ f6 \\ f7 \end{bmatrix}
=
\begin{bmatrix}
A_{00} & A_{10} & A_{20} & A_{30} & A_{40} & A_{50} & A_{60} & A_{70} \\
A_{01} & A_{11} & A_{21} & A_{31} & A_{41} & A_{51} & A_{61} & A_{71} \\
A_{02} & A_{12} & A_{22} & A_{32} & A_{42} & A_{52} & A_{62} & A_{72} \\
A_{03} & A_{13} & A_{23} & A_{33} & A_{43} & A_{53} & A_{63} & A_{73} \\
A_{04} & A_{14} & A_{24} & A_{34} & A_{44} & A_{54} & A_{64} & A_{74} \\
A_{05} & A_{15} & A_{25} & A_{35} & A_{45} & A_{55} & A_{65} & A_{75} \\
A_{06} & A_{16} & A_{26} & A_{36} & A_{46} & A_{56} & A_{66} & A_{76} \\
A_{07} & A_{17} & A_{27} & A_{37} & A_{47} & A_{57} & A_{67} & A_{77}
\end{bmatrix}
\times
\begin{bmatrix} F0 \\ F1 \\ F2 \\ F3 \\ F4 \\ F5 \\ F6 \\ F7 \end{bmatrix}
$$

as a product between the basis matrix A and the input vector F. Because the MAP1000 has an `srshinprod.ps16` instruction that can perform eight 16-bit multiplications and accumulate the results with a single-cycle throughput, only one instruction is necessary to compute one output element f_x:

$$f_x = \sum_{u=0}^{7} A_{ux} F_u$$

This instruction utilizes 128-bit PLC and PLV registers. The `setplc.128` instruction is necessary to set the 128-bit PLC register with IDCT coefficients. Because the accumulations of `srshinprod.ps16` are performed in 32 bits, to output 16-bit IDCT results, the MAP1000 has an instruction called `compress2_ps32_rs15` to compress four 32-bit operands to four 16-bit operands. This instruction also performs 15-bit right-shift operation on each 32-bit operand before compressing. Because there are a large number of registers on the MAP1000 (64,32-bit registers per cluster), once the data are loaded into the registers for computing the first 1D 8-point IDCT, the registers can be retained for computing the subsequent IDCTs, thus eliminating the need for multiple load operations. Ideally, an 8-point IDCT can be computed in 11 cycles (8 `inner-product`, 2 `compress2_ps32_rs15`, 1 `setplc.128`), and 2D 8 × 8 IDCT can be computed in $2 \times 11 \times 8 = 176$ cycles. Because there are two clusters, 88 cycles are necessary to compute a 2D 8 × 8 IDCT.

There are two sources of quantization error in IDCT when computed with a finite number of bits, as discussed in Section 4.3. The quantized multiplication coefficients are the first source, whereas the second one arises from the need to have the same number of fractional bits as the input after multiplication. Thus, to control the quantization error on different decoder implementations, the MPEG standard specifies that the IDCT implementation used in the MPEG decoder must comply with the accuracy requirement of the IEEE Standard 1180–1990. The simulations have shown that by utilizing 4 bits for representing the fractional part and 12 integer bits, the overflow can be avoided while meeting the accuracy requirements (Lee, 1997). MPEG standards also specify that the output x_{ij} in Eq. (5) must be clamped to 9 bits (-256 to 255). Thus, to meet these MPEG standard requirements, some preprocessing of the input data and postprocessing of the IDCT results are necessary.

5.4   Affine Transformation

The affine transformation is a very useful subset of image warping algorithms (Wolberg, 1990). The example affine warp transformations include rotation, scaling, shearing, flipping, and translation.

Mathematical equations for affine warp relating the output image to the input image (also called inverse mapping) are shown in Eq. (6), where x_o and y_o are the discrete output image locations, x_i and y_i are the inverse-mapped input locations, and a_{11}–a_{23} are the six affine warp coefficients:

$$x_i = a_{11}x_o + a_{12}y_o + a_{13}, \quad 0 \leq x_o < \texttt{image\_width}$$
$$y_i = a_{21}x_o + a_{22}y_o + a_{23}, \quad 0 \leq y_o < \texttt{image\_height}$$

(6)

For each discrete pixel in the output image, an inverse transformation with Eq. (6) results in a nondiscrete subpixel location within the input image, from which the output pixel value is computed. In order to determine the gray-level output value at this nondiscrete location, some form of interpolation (e.g., bilinear) has to be performed with the pixels around the mapped location.

The main steps in affine warp are (1) geometric transformation, (2) address calculation and coefficient generation, (3) source pixel transfer, and (4) 2 × 2 bilinear interpolation. While discussing each step, the details of how affine warp can be mapped to the TMS320C80 and MAP1000 are also discussed.

5.4.1 Geometric Transformation

Geometric transformation requires the computation of x_i and y_i for each output pixel (x_o and y_o). According to Eq. (6), four multiplications are necessary to compute the inverse mapped address. However, these multiplications can be easily avoided. After computing the first coordinate (by assigning a_{13} and a_{23} to x_i and y_i), subsequent coordinates can be computed by just incrementing the previously computed coordinate with a_{11} and a_{21} while processing horizontally and a_{12} and a_{22} while processing vertically. This eliminates the need for multiplications and requires only addition instructions to perform the geometric transformation. On the TMS320C80, the ADSP's add/subtract unit can be utilized to perform this operation, whereas on the MAP1000 `partitioned_add` instructions can be utilized. However, for each pixel, conditional statements are necessary to check whether the address of the inverse-mapped pixel lies outside the input image boundary, in which case the subsequent steps do not have to be executed. Thus, instructions in cache-based processors (e.g., TM1000) need to be predicated to avoid `if`/`then`/`else`-type coding. The execution of these predicated instructions depends on the results of the conditional statements (discussed in Sec. 4.4). However, on DMA-based processors, such as TMS320C80 and MAP1000, these conditional statements and predication of instructions are not necessary, as discussed in Section 5.4.3.

5.4.2 Address Calculation and Coefficient Generation

The inverse-mapped input address and the coefficients required for bilinear interpolation are generated as follows:

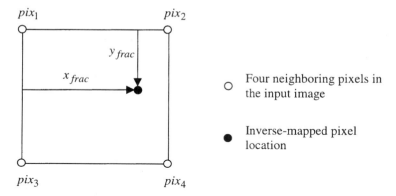

Figure 23 Bilinear interpolation to determine the gray-level value of the inverse-mapped pixel location.

$$\text{InputAddress} = \text{Source Address} + y_{\text{int}} * \text{pitch} + x_{\text{int}}$$
$$c_1 = (1 - x_{\text{frac}})(1 - y_{\text{frac}})$$
$$c_2 = x_{\text{frac}}(1 - y_{\text{frac}})$$
$$c_3 = (1 - x_{\text{frac}})y_{\text{frac}}$$
$$c_4 = x_{\text{frac}}y_{\text{frac}}$$

where pitch is the memory offset between two consecutive rows. The integer and fractional parts together (e.g., $x_i = x_{\text{int}}, x_{\text{frac}}$, and $y_i = y_{\text{int}}, y_{\text{frac}}$) indicate the nondiscrete subpixel location of the inverse-mapped pixel in the input image as, shown in Figure 23. The input address points to only the upper left image pixel pix_1 as in Fig. 23), and other three pixels required for interpolation are its neighbors. The flowgraph of computing c_1, c_2, c_3, and c_4 on the MAP1000 is shown in Figure 24 utilizing partitioned instructions, whereas on the TMS320C80, each coefficient needs to be computed individually.

5.4.3 Source Pixel Transfer

Affine warp requires irregular data accesses. There are two approaches in accessing the input pixel groups: cache based and DMA based. In cache-based processors, there are two penalties: (1) Because data accesses are irregular, there will be many cache misses and (2) the number of execution cycles is larger because conditional statements are necessary to check whether the address of every inverse-mapped location lies outside the input image boundary. These two disadvantages can be overcome by using a DMA controller.

In the DMA-based implementation, the output image can be segmented into multiple blocks, an example of which is illustrated in Figure 25. A given

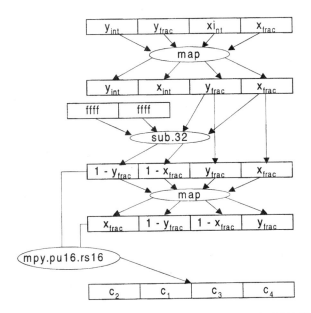

Figure 24   Computing bilinear coefficients on the MAP1000.

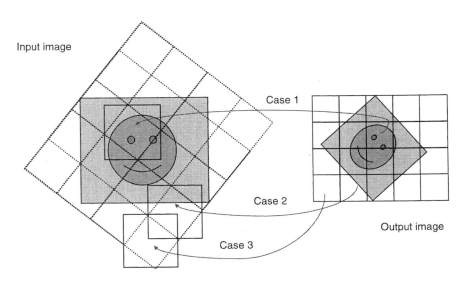

Figure 25   Inverse mapping of affine warp.

output block maps to a quadrilateral in the input space. The rectangular bounding block encompassing each quadrilateral in the input image is shown by the solid line. If the bounding block contains no source pixels (case 3 in Fig. 25), the DMA controller can be programmed to write zeros in the output image block directly without bringing any input pixels on-chip. If the bounding block is partially filled (case 2), then the DMA controller can be programmed to bring only valid input pixels on-chip and fill the rest of the output image block with zeros. If the bounding block is completely filled with valid input pixels (case 1), the whole block is brought on-chip using the DMA controller. In addition, these data movements are *double buffered* so that the data transfer time can be overlapped with the processing time. Thus, the DMA controller is very useful in improving the overall performance by overcoming the cache misses and conditional statements.

5.4.4   Bilinear Interpolation

The bilinear interpolation is accomplished by (1) multiplying the four pixels surrounding an inverse-mapped pixel location with four already-computed coefficients and (2) summing the results of these four multiplications. The first step in this stage is to load the four neighboring pixels of the input image from the location pointed to by the input address. The bilinear interpolation is similar to 2×2 convolution. Thus, on the TMS320C80, multiply and add/subtract units can be utilized to perform multiplications and accumulation similar to that shown for convolution in Table 1. On the MAP1000, `inner-product` is used to perform multiplications and accumulation. However, before `inner-product` is issued, all four 8-bit pixels loaded in different 32-bit registers need to be packed together into a single 64-bit register using the `compress` instructions of Figure 4.

The total number of cycles required for computing affine warp for one output pixel on the TMS320C80 per ADSP is 10, whereas the number of cycles required on the MAP1000 is 8. Because the TMS320C80 has 4 ADSPs and the MAP1000 has 2 clusters, the effective number of cycles per output pixel for affine warp on the TMS320C80 is 2.5, whereas it is 4 on the MAP1000.

5.5   Extending Algorithms to Other Processors

The algorithm mapping discussed for TMS320C80 and MAP1000 can be easily extended to other processors as well. In this subsection, we present how convolution algorithm mapped to TMS320C80 and MAP1000 can be extended to TMS320C6X and TM1000. The reader can follow similar leads for mapping other algorithm to other processors.

5.5.1 Texas Instruments TMS320C6x

The TMS320C6x has two load/store units, two add/subtract units, two multiply units, and two logical units. In Table 1, we utilized two load/store units, one add/subtract unit, and one multiply unit while implementing convolution for the TMS320C80. Thus, the same algorithm can be mapped to TMS320C6x utilizing two load/store units, one add/subtract unit, and one multiply unit. Because TMS320C6x does not have hardware loop controllers, other add/subtract unit can be used for branch operation (for decrementing the loop counters). Ideally, the number of cycles required for a 3×3 convolution on the TMS320C6x is 15 (4 additional cycles for clipping the result between 0 and 255, as discussed in Section 1.1).

5.5.2 Philips Trimedia TM1000

The pseudocode described in Section 1.2 for convolution on the MAP1000 can be easily extended to TM1000. On the Philips Trimedia TM1000, `inner-product` is available under the name `ifir16`, which performs two 16-bit multiplications and accumulation, and `align` is available under the name of `funshift` (Rathnam and Slavenburg, 1998). Two `ifir16` can be issued each cycle on the TM1000, executing four multiplications and accumulations in a single cycle. Instead of specifying a shift amount for `align`, the TM1000 supports several variations of `funshift` to extract the desired aligned data (e.g., `funshift1` is equivalent to `align` with a one-pixel shift and `funshift2` is equivalent to `align` with a two-pixel shift). The TM1000 does not have instructions to saturate the results between 0 and 255. Thus, `if/then/else` types of instruction are necessary to clip the results. Ideally, with these instructions, we can execute 7×7 convolution in 18 cycles on the TM1000.

6 SUMMARY

To meet the growing computational demand arising from the digital media at an affordable cost, new advanced digital signal processors architectures with VLIW have been emerging. These processors achieve high performance by utilizing both instruction-level and data-level parallelisms. Even with such a flexible and powerful architecture, to achieve good performance necessitates the careful design of algorithms that can make good use of the newly available parallelism. In this chapter, various algorithm mapping techniques with real examples on modern VLIW processors have been presented, which can be utilized to implement a variety of algorithms and applications on current and future DSP processors for optimal performance.

REFERENCES

Basoglu C, W Lee, Y Kim. An efficient FFT algorithm for superscalar and VLIW processor architectures. Real-Time Imaging 3:441–453, 1997.

Basoglu C, RJ Gove, K Kojima, J O'Donnell. A single-chip processor for media applications: The MAP1000. Int J Imaging Syst Technol 10:96–106, 1999.

Basoglu C, D Kim, RJ Gove, Y Kim. High-performance image computing with modern microprocessors. Int J Imaging Syst Technol 9:407–415, 1998.

Berkeley Design Technology (BDT). DSP processor fundamentals, http://www.bdti.com/products/dsppf.htm, 1996.

Boland K, A Dollas. Predicting and precluding problems with memory latency. IEEE Micro 14(4):59–67, 1994.

Chamberlain A. Efficient software implementation of affine and perspective image warping on a VLIW processor. MSEE thesis, University of Washington, Seattle, 1997.

Chen WH, CH Smith, SC Fralick. ''A fast computational algorithm for the discrete cosine transform,'' IEEE Trans. on Communications, vol. 25, pp. 1004–1009, 1977.

Dudgeon DE, RM Mersereau. Multidimensional Digital Signal Processing. Englewood Cliffs, NJ: Prentice-Hall, 1984.

Equator Technologies, ''MAP-CA Processor,'' http://www.equator.com, 2000.

Evans O, Y Kim. Efficient implementation of image warping on a multimedia processor. Real-Time Imaging 4:417–428, 1998.

Faraboschi P, G Desoli, JA Fisher. The latest world in digital media processing. IEEE Signal Processing Mag 15:59–85, March 1998.

Fisher JA. The VLIW machine: A multiprocessor from compiling scientific code. Computer 17:45–53, July 1984.

Fujitsu Limited. FR500, http://www.fujitsu.co.jp, 1999.

Greppert L, TS Perry. Transmeta's magic show. IEEE Spectrum 37(5):26–33, 2000.

Guttag K, RJ Gove, JR VanAken. A single chip multiprocessor for multimedia: The MVP. IEEE Computer Graphics Applic 12(6):53–64, 1992.

Kim D, RA Managuli, Y Kim. Data cache vs. direct memory access (DMA) in programming mediaprocessors. IEEE Micro, in press.

Lam M. Software pipelining: An effective scheduling technique for VLIW machines. SIGPLAN 23:318–328, 1988.

Lee EA. Programmable DSP architecture: Part I. IEEE ASSP Mag 5:4–19, 1988.

Lee EA. Programmable DSP architecture: Part II. IEEE ASSP Mag 6:4–14, 1989.

Lee R. Accelerating multimedia with enhanced microprocessors. IEEE Micro 15(2):22–32, 1995.

Lee W. Architecture and algorithm for MPEG coding. PhD dissertation, University of Washington Seattle, 1997.

Managuli R, G York, D Kim, Y Kim. Mapping of 2D convolution on VLIW mediaprocessors for real-time performance. J Electron Imaging 9:327–335, 2001.

Managuli RA, C Basoglu, SD Pathak, Y Kim. Fast convolution on a programmable mediaprocessor and application in unsharp masking. SPIE Med Imaging 3335:675–683, 1998.

Mizosoe H, Y Jung, D Kim, W Lee, Y Kim. Software implementation of MPEG-2 decoder on VLIW mediaprocessors. SPIE Proc 3970:16–26, 2000.

Patterson DA, JL Hennessy. Computer Architecture: A Quantitative Approach. San Francisco: Morgan Kaufman, 1996.

Peleg A, U Weiser. MMX technology extension to the Intel architecture. IEEE Micro 16(4):42–50, 1996.

Rathnam S, G Slavenburg. Processing the new world of interactive media. The Trimedia VLIW CPU architecture. IEEE Signal Process Mag 15(2):108–117, 1998.

Seshan N. High VelociTI processing. IEEE Signal Processing Mag 15(2):86–101, 1998.

Shieh JJ, CA Papachristou. Fine grain mapping strategy for multiprocessor systems. IEE Proc Computer Digital Technol 138:109–120, 1991.

Texas Instruments. TMS320C6211, fixed-point digital signal processor, http://www.ti.com/sc/docs/products/dsp/tms320c6211.html, 1999.

Wolberg G. Digital Image Warping. Los Alamitos, CA: IEEE Computer Society Press, 1990.

3

Multimedia Instructions in Microprocessors for Native Signal Processing

Ruby B. Lee and A. Murat Fiskiran
Princeton University, Princeton, New Jersey

1 INTRODUCTION

Digital signal processing (DSP) applications on computers have typically used separate DSP chips for each task. For example, one DSP chip is used for processing each audio channel (two chips for stereo); a separate DSP chip is used for modem processing, and another for telephony. In systems already using a general-purpose processor, the DSP chips represent additional hardware resources. Native signal processing is DSP performed in the microprocessor itself, with the addition of general-purpose multimedia instructions. Multimedia instructions extend native signal processing to video, graphics, and image processing, as well as the more common audio processing needed in speech, music, modem, and telephony applications. In this study, we describe the multimedia instructions that have been added to current microprocessor instruction set architectures (ISAs) for native signal processing or, more generally, for multimedia processing.

Multimedia information processing is becoming increasingly prevalent in the general-purpose processor's workload [1]. Workload characterization studies on multimedia applications have revealed interesting results. More often than not, media applications do not work on very high-precision data types. A pixel-oriented application, for example, rarely needs to process data that are wider than 16 bits. A low-end digital audio processing program may also use only 16-bit fixed-point numbers. Even high-end audio applications rarely require any precision beyond a 32-bit single-precision (SP) floating point (FP). Common usage

R_a: | 11111111 | 00001111 | 11110000 | 00000000 |

Figure 1 Example of a 32-bit integer register holding four 8-bit subwords. The subword values are 0xFF, 0x0F, 0xF0, and 0x00, from the first * to the fourth subword respectively.

of low-precision data in such applications translates into low computational efficiency on general-purpose processors, where the register sizes are typically 64 bits. Therefore, efficient processing of low-precision data types on general-purpose processors becomes a basic requirement for improved multimedia performance.

Media applications exhibit another interesting property. The same instructions are often used on many low-precision data elements in rapid succession. Although the large register sizes of the general-purpose processors are more than enough to accommodate a single low-precision data, the large registers can actually be used to process many low-precision data elements in parallel.

Efficient parallel processing of low-precision data elements is therefore a key for high-performance multimedia applications. To that effect, the registers of general-purpose processors can be partitioned into smaller units called *subwords*. A low-precision data element can be accommodated in a single subword. Because the registers of general-purpose processors will have multiple subwords, these can be processed in parallel using a single instruction. A *packed data type* will be defined as data that consist of multiple subwords packed together.

Figure 1 shows a 32-bit integer register that is made up of four 8-bit subwords. The subwords in the register can be pixel values from a gray-scale image. In this case, the register will be holding four pixels with values 0xFF, 0x0F, 0xF0, and 0x00. Similarly, the same 32-bit register can also be partitioned into two 16-bit subwords, in which case, these subwords would be 0xFF0F and 0xF000. One important point is that the subword boundaries do not correspond to a physical boundary in the register file. Whether data are packed or not does not make any difference regarding its representation in a register.

If we have 64-bit registers, the useful subword sizes will be bytes, 16-bit half-words, or 32-bit words. A single register can then accommodate eight, four, or two of these subwords respectively. The processor can carry out parallel

---

* Through this chapter, the subwords in a register will be indexed from 1 to n, where n will be the number of subwords in that register. The first subword (index = 1) will be in the most significant position in a register, whereas the last subword (index = n) will be in the least significant position. In the figures, the subword on the left end of a register will have index = 1 and therefore will be in the most significant position. The subword on the right end of a register will have index = n and therefore be in the least significant position.

operations on these subwords with a single instruction, as in single instructions–multiple data (SIMD) parallelism. SIMD parallelism is said to exist when a single instruction operates on multiple data elements in parallel. In the case of subword parallelism, the *multiple data* elements will correspond to the subwords in the packed register.

Traditionally, however, the term SIMD was used to define a situation in which a single instruction operated on multiple registers, rather than on the subwords of a single register. To address this difference, the parallelism exploited by the use of subword parallel instructions is defined as microSIMD parallelism [2]. Thus, an *add* instruction operating on packed data, can be viewed as a microSIMD instruction, where the *single instruction* is the *add* and the *multiple data* elements are the *subwords* in the packed source registers.

For a given processor, the ISA needs to be enhanced to exploit microSIMD parallelism (see Fig. 2). New instructions are added to allow parallel processing of packed data types. Minor modifications to the underlying functional units will also be necessary. Fortunately, the register file and the pipeline structure need not be changed to support packed data types.

We define *packed instructions* as the instructions that are specifically designed to operate on packed-data types. A *packed add*, for example, is an *add* instruction with the regular definition of addition, but it operates on packed data types. *Packed subtract* and *packed multiply* are other obvious instructions needed to efficiently manipulate packed data types.

All of the architectures in this chapter include varieties of packed instructions. More often than not, they also include other instructions that cannot be classified as packed arithmetic operations. As we shall see shortly, the introduction of subword parallelism to an ISA actually requires that new instructions

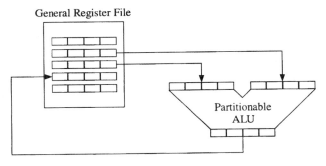

Figure 2 MicroSIMD parallelism uses packed data types and a partitionable arithmetic and logic unit structure.

other than the packed arithmetic instructions are also added. In this study, the multimedia instructions will be classified and discussed in the following order:

- Packed add/subtract operations
- Packed special arithmetic operations
- Packed multiply operations
- Packed compare/minimum/maximum operations
- Packed shift/rotate operations
- Data/subword packing and rearrangement operations
- Approximation operations

The above operations will be discussed in Sections 2–8, respectively. To provide examples, we will be referring to the following multimedia extensions/architectures:

- IA-64 [3], MMX [4], and SSE-2 [5] from Intel
- MAX-2 [6,7] from Hewlett-Packard
- 3DNow! [8,9] from AMD*
- AltiVec [10] from Motorola

Of these architectures, MAX-2 and MMX include only integer microSIMD extensions. SSE-2 and 3DNow! include only FP microSIMD extensions. IA-64 and AltiVec have both integer and FP microSIMD extensions.

1.1 Historical Overview

Prior to the ones we discuss in this chapter, there have been other notable multimedia extensions introduced to the general-purpose processors [11]. All of these earlier attempts had the same underlying idea as today's more recent extensions. They were based on subword parallelism: operating in parallel on lower-precision data packed into higher-precision words.

The first multimedia extensions came from Hewlett-Packard with their introduction of the PA-7100LC processor in January 1994 [12]. This processor featured a small set of multimedia instructions called MAX-1, which was the first version of the "Multimedia Acceleration Extensions" for the 32-bit PA-RISC instruction set architecture [13]. MAX-2, although designed simultaneously with MAX-1, was introduced later with the 64-bit PA-RISC 2.0 architecture. The application that best illustrated the performance of MAX-1 was the MPEG-1 video and audio decoding at real-time rates of up to 30 frames/sec [14]. For the first time, this performance was made possible on a general-purpose processor

---

* 3DNow! may be considered as having two versions. In June 2000, 25 new instructions were added to the original 3DNow! specification. In this text, we will actually be considering this extended 3DNow! architecture.

in a low-end desktop computer [15]. Until then, such video performance was not achievable without using specialized hardware or high-end workstations.

Next, Sun introduced VIS [16], which was an extension for the UltraSparc processors. Unlike MAX-1, VIS did not have a minimalist approach; thus, it was a much larger set of multimedia instructions. In addition to packed arithmetic operations, VIS provided specialized instructions that were designed for algorithms that manipulated visual data.

MAX-2 [7], which we discuss in this chapter, was Hewlett-Packard's multimedia extension for its 64-bit PA-RISC 2.0 processors [6]. MAX-2 included a few new instructions; especially subword permutation instructions over MAX-1, to better exploit the increased subword parallelism in 64-bit registers. Like MAX-1, MAX-2 was also a minimalist set of general-purpose media acceleration primitives. Neither included very specialized instructions found in other multimedia extensions.

All multimedia extensions referred in this chapter have the same basic goal: to allow high-performance media processing, or native signal processing on a general-purpose processor. The key idea shared by all of these extensions to achieve this goal is the use of subword parallelism. The instructions included in the extensions are commonly based on operating in parallel on packed data types. As we will address in the following sections, significant differences exist among different ISAs and extensions, in the types and the sizes of the subwords, as well as for the support provided for these subwords.

1.2 A Note on Instruction Formatting

Throughout this chapter, we assume that all the instructions (with the possible exclusion of loads and stores) use registers for operand and target fields. The first register in an instruction is the target register and all the remaining registers are the source registers. We index the registers so that the highest index always corresponds to the target register, whereas the source registers appear in increasing indices, starting from a. For example, we may represent an add operation as follows:

ADD R_c, R_a, R_b

R_c is the target register, whereas R_a and R_b are the first and the second source registers, respectively. For AltiVec and IA-64, where some instructions may have one target and three source fields, R_d is used to represent the target register. VSUMMBM, which will be explained in Section 4, is such an AltiVec instruction and it is represented as follows:

VSUMMBM R_d, R_a, R_b, R_c

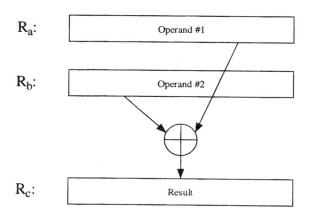

Figure 3   ADD R_c,R_a,R_b: In a typical *add* operation, two source registers are added and the sum is written to the target register.

R_d is the target register, whereas R_a, R_b and R_c are the first, second, and third source registers, respectively.

Our initial assumption that all the instructions use registers for source and target fields is not always true. MMX and SSE-2 are two important exclusions. Multimedia instructions in these extensions may use a memory location as a source operand. Thus, using our default representation for such instances will not be conforming to the rules of that particular architecture. However, to keep the notation simple and consistent, this distinction will not be observed, except for being noted here. For the exact instruction formatting and source–target register ordering, the reader is referred to the architecture manuals listed in the references.

2   PACKED ADD/SUBTRACT OPERATIONS

Packed add/subtract operations are nothing but regular *add/subtract* operations operating in parallel on the subwords of two source registers. Regular (i.e., nonpacked) and *packed add* operations are shown in Figures 3 and 4, respectively.* The *packed add* operation in Figure 4 uses source registers with four subwords each. The corresponding subwords from the two source registers

---

* For details on instruction formatting used in this discussion, please refer to the last paragraph of Section 1.2.

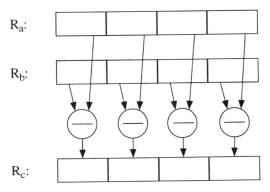

Figure 4   PADD R_c,R_a,R_b: *Packed add* operation. Each register has four subwords.

are summed up, and the four sums are written to the target register. Figure 5 shows a *packed subtract* operation that operates on registers holding four subwords each.

2.1   Implementing Packed Instructions

Very minor modifications to the underlying functional units are needed to implement *packed add* and *subtract* operations. Assume that we have an arithmetic and logic unit (ALU) with 32-bit integer registers, and we want to extend this ALU to perform a *packed add* operation that will operate on four 8-bit subwords in parallel. Because subwords are independent, the carry bits generated by the

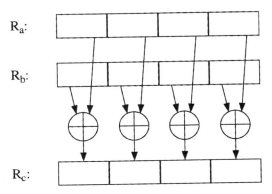

Figure 5   PSUB R_c,R_a,R_b: *Packed subtract* operation. Each register has four subwords.

addition of a particular subword pair should not be allowed to affect the sums of other subword pairs. Therefore, to implement this *packed add* operation, it is necessary and sufficient to block the carry propagation across the subword boundaries.

In Figure 6, the packed integer register R_a = [0xFF|0x0F|0xF0|0x00] is being added to another packed register R_b = [0x00|0xFF|0xFF|0x0F]. The result is written to the target register R_c. If the regular *add* instruction is used to add these packed registers, the overflows generated by the addition of the second and third subwords will propagate into the first two sums. The correct sums, however, can be achieved easily by blocking the carry bit propagation across the subword boundaries, which are spaced 8 bits apart from one another.

As shown in Figure 7, a 2-to-1 multiplexer placed at the subword boundaries of the adder can be used to control the propagation or the blocking of the carry bits. If the instruction is a *packed add*, the multiplexer control is set such that a 0 is propagated into the next subword. If the instruction is a regular *add*, the multiplexer control is set such that the carry from the previous stage is propagated. By placing such a multiplexer at each subword boundary and adding the control logic, the support for *packed add* operations will be added to this ALU. If multiple subword sizes must be supported, more multiplexers may be required. In this case, the multiplexer control gets more complicated; nevertheless, the area cost is still very insignificant for the performance provided by such microSIMD instructions.

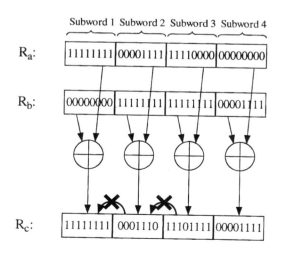

Figure 6   To get the correct results in this *packed add* operation, the carry bits are not propagated into the first and second sums.

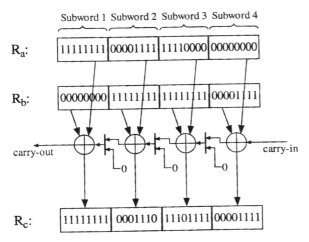

Figure 7 In a *packed add* instruction, the multiplexers propagate 0. In a regular *add* instruction, the multiplexers pass on the carry-out of the previous stage into the carry-in input of the next stage.

2.2 Packed Subtract Operations

By using 3-to-1 multiplexers instead of 2-to-1 multiplexers, we can also implement *packed subtract* instructions. In this case, the multiplexer control is set as follows:

- For *packed add* instructions, 0 is propagated into the next stage.
- For *packed subtract* instructions, 1 is propagated into the next stage.
- For regular *add/subtract* instructions, the carry bit from the previous stage is propagated into the next stage.

Propagating a 0 through a subword boundary in a *packed add* operation is equivalent to ignoring any overflow that might have been generated. In Figure 6, the two overflows generated in the second and the third subword boundaries were ignored. Similarly, propagating a 1 through a subword boundary in a *packed subtract* operation is equivalent to ignoring any borrow that might have been generated.

Ignoring overflows translates into the use of modular arithmetic in *add* operations. Although this can be desirable, there are times when the carry bits should not be ignored and have to be handled differently. The next section addresses these needs and proposes an interesting solution, known as *saturation arithmetic*.

2.3 Handling Parallel Overflows

How the overflows are handled in *packed add/subtract* operations is an important issue. Whenever an overflow is generated, any one of the following actions can be taken:

- The overflow may be ignored (modular arithmetic).
- A flag* bit may be set if at least one overflow is generated.
- Multiple flag bits (i.e., one flag bit for each addition operation on the subwords) may be set.
- A software trap can be taken.
- The results may be limited to within a certain range. If the outcome of the operation falls outside this range, the corresponding limiting value will be the result. This is the basis of saturation arithmetic, which will be explained in detail in Section 2.4.

Most nonpacked integer *add/subtract* instructions choose to ignore overflows and perform modular arithmetic. In modular arithmetic, the numbers wrap around from the largest representable number to the smallest representable number. For example, in 8-bit modular arithmetic, the operation 254 + 2 will give out 0 as a result. The expected result, 256, is larger than the largest representable number, which is 255, and therefore is wrapped around to the smallest representable number, which is 0.

Even though modular arithmetic may be an option in *packed add/subtract* operations as well, there can be specific applications where it cannot be used, and the overflows have to be handled differently. If the numbers in the previous example were pixel values in a gray-scale image, by wrapping the values from 255 down to 0, we would have essentially converted white pixels into black ones. This would be an example where modular arithmetic could not be used. One solution to this problem is to use overflow traps, which are implemented in software.

An overflow trap can be used to *saturate* the results so that:

---

* A flag bit is an indicator bit that is set or cleared depending on the outcome of a particular operation. In the context of this discussion, an overflow flag bit is an indicator that is set when an *add* operation generates an overflow. There are occasions where the use of the flag bits are desirable. Consider a loop that iterates many times and, in each iteration, performs many *add* operations. In this case, it is not desirable to handle overflows (by taking overflow trap routines) as soon as they occur, as this would negatively impact the performance by interrupting the execution of the loop body. Rather, the overflow flag can be set when the overflow occurs, and the program flow may be resumed as if the overflow did not occur. At the end of each iteration however, this overflow flag can be checked and the overflow trap can be executed if the flag turns out to be set. In this way, the program flow would not be interrupted while the loop body executes.

- Any result that is greater than the largest representable value is replaced by that largest value.
- Any result that is less than the smallest representable value is replaced by that smallest value.

The downside of the overflow trap approach is that, because it is handled in software, it may take many clock cycles to complete its execution. This can only be acceptable if the overflows are infrequent.

For nonpacked *add/subtract* operations, overflows can be rare enough to justify the use of software traps. The generation of an overflow on a 64-bit register by adding up 8-bit quantities will be rare. In such a case, a software overflow trap will work well. On the other hand, with modular arithmetic implemented in hardware, packed arithmetic operations are much more likely to generate multiple overflows frequently. Generating an overflow in an 8-bit subword is much more likely than in a 64-bit register. Moreover, because a 64-bit register may hold eight 8-bit subwords, there is actually a chance of multiple overflows in a single execution cycle. In such cases, handling the overflows by software traps may severely degrade performance. The time required to process software traps could easily exceed the time saved by executing packed operations. The use of saturation arithmetic comes up as a remedy to this problem.

2.4 Saturation Arithmetic

Saturation arithmetic implements in hardware the work done by the overflow trap described in Section 2.3. The results falling outside the allowed numeric ranges are saturated to the upper and lower limits by hardware. This can handle multiple parallel overflows efficiently without any operating system intervention, which can degrade performance.

There can be two types of overflows in arithmetic operations:

- A *positive overflow* occurs when the result is larger than the largest value that is in the defined range for that result.
- A *negative overflow* occurs when the result is smaller than the smallest value that is in the defined range for that result.

If saturation arithmetic is used in an operation, the result is clipped to the maximum value in its defined range if a positive overflow occurs, and to the minimum value in its defined range if a negative overflow occurs.

For a given instruction, multiple saturation options may exist, depending on whether the operands and the result are treated as signed or unsigned integers. For an instruction that uses three registers (two for source operands and one for the result), there can be eight* different saturation options. However, not all of

---

* Each one of the three registers can be treated as containing either a signed or an unsigned integer, which gives 2^3 possible combinations.

these eight options are necessary or even useful; and in practice, multimedia extensions typically prefer to use the following three options:

1. *sss* (signed result–signed first operand–signed second operand): In this saturation option, the result and the two operands are all treated as signed integers (or signed fixed-point numbers). If the operands and the result are 16-bit integer subwords, their most significant bits are considered as the sign bits. The result and the operands are defined in the range $[-2^{15}, 2^{15} - 1]$. If a negative or positive overflow occurs, the result is saturated to either -2^{15} or to $2^{15} - 1$ respectively. Consider an addition operation that uses the *sss* saturation option. Because the operands are signed numbers, a positive overflow is possible only when both operands are positive. Similarly, a negative overflow is possible only when both operands are negative.

2. *uuu* (unsigned result–unsigned first operand–unsigned second operand): In this saturation option, the result and the two operands are all treated as unsigned integers (or unsigned fixed-point numbers). Considering 16-bit integer subwords, the result and the operands are defined in the range $[0, 2^{16} - 1]$. If a negative or positive overflow occurs, the result is saturated to either 0 or to 2^{16}, respectively. Consider an addition operation that uses the *uuu* saturation option. Because the operands are unsigned numbers, negative overflow is not a possibility. However, for a subtraction operation using the *uuu* saturation, negative overflow is possible, and any negative result will be clamped to 0.

3. *uus* (unsigned result–unsigned first operand–signed second operand): In this saturation option, the result and the first operand are treated as unsigned integers (or unsigned fixed-point numbers), and the second operand is treated as a signed integer (or as a signed fixed-point number). This option is useful because it allows the addition of a signed increment to an unsigned pixel. As we will see in examples, it also allows negative numbers to be clipped to 0.

2.5 Uses of Saturation Arithmetic

In addition to efficient handling of parallel overflows, saturation arithmetic also facilitates several other useful computations:

• Saturation arithmetic can be used to clip results to arbitrary maximum or minimum values. Without saturation arithmetic, these operations could normally take up to five instructions. That would include instructions to check for bounds and then to perform the clipping. Using saturation arithmetic, however, this effect can be achieved in as few as two instructions (Table 1).

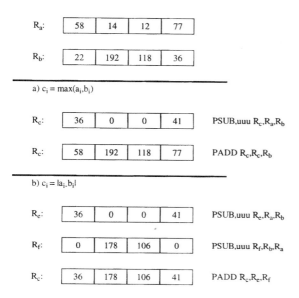

Figure 8 (a) *Packed maximum* operation using saturation arithmetic. (b) *Packed absolute difference* operation using saturation arithmetic. If no saturation option is given, modular arithmetic is assumed.

- By using saturation arithmetic, conditional statements can be evaluated in two or three instructions, without the need for conditional branch instructions. Some examples are *packed maximum* operation shown in Figure 8a and *packed absolute difference* operation shown in Figure 8b.

Table 1 contains examples of operations that can be performed using saturation arithmetic. All of the instructions in the table use three registers. The first register is the target register. The second and the third registers hold the first and the second operands, respectively. PADD/PSUB operations are *packed add/subtract* operations. The three-letter field after the instruction mnemonic specifies which saturation option is to be used. If this field is empty, modular arithmetic is assumed. All the examples in the table operate on 16-bit integer subwords.

2.6 Comparison of the Architectures

As far as the *packed add/subtract* instructions are concerned, differences among architectures are in the register and subword sizes and supported saturation options:

Table 1 Examples of Operations That Are Facilitated by Saturation Arithmetic

Operation	Instruction sequence	Notes
Clip a to an arbitrary maximum value v_{max} $[v_{max} < (2^{15} - 1)]$	PADD.sss R_a,R_a,R_b	R_b contains the value $(2^{15} - 1 - v_{max})$. If $a > v_{max}$, this operation clips a to $2^{15} - 1$ on the high end. a is at most v_{max}.
Clip a to an arbitrary minimum value v_{min} $[(v_{min} > -2^{15})]$	PSUB.sss R_a,R_a,R_b PSUB.sss R_a,R_a,R_b	R_b contains the value $(-2^{15} + v_{min})$. If $a < v_{min}$, this operation clips a to -2^{15} at the low end. a is at least v_{min}.
Clip a to within the arbitrary range $[v_{min}, v_{max}]$ $[-2^{15} < v_{min} < v_{max} < (2^{15} - 1)]$	PADD.sss R_a,R_a,R_b PADD.sss R_a,R_a,R_b PSUB.sss R_a,R_a,R_d PADD.sss R_a,R_a,R_c	R_b contains the value $(2^{15} - 1 - v_{max})$. This operation clips a to $2^{15} - 1$ on the high end. R_d contains the value $(2^{15} - 1 - v_{max} + 2^{15} - v_{min})$. This operation clips a to -2^{15} at the low end. R_c contains the value $(-2^{15} - v_{min})$. This operation clips a to v_{max} at the high end and to v_{min} at the low end.
Clip the signed integer a to an unsigned integer within the range $[0, v_{max}]$ $[0 < v_{max} < (2^{15} - 1)]$	PADD.sss R_a,R_a,R_b PSUB.uus R_a,R_a,R_b	R_b contains the value $(2^{15} - 1 - v_{max})$. This operation clips a to $2^{15} - 1$ at the high end. This operation clips a to v_{max} at the high end and to 0 at the low end.
Clip the signed integer a to an unsigned integer within the range $[0, v_{max}]$ $[(2^{15} - 1) < v_{max} < 2^{16} - 1]$	PADD.uus $R_a,R_a,0$	If $a < 0$, then $a = 0$, else $a = a$. If a was negative, it gets clipped to 0, else remains same.
$c = max(a, b)$ Maximum operation: It writes the greater subword to the target register.	PSUB.uuu R_c,R_a,R_b PADD R_c,R_b,R_c	If $a > b$, then $c = (a - b)$, else $c = 0$. If $a > b$, then $c = a$, else $c = b$.
$c = |a - b|$ Absolute difference operation: It writes the absolute value of the difference of the two subwords to the target register.	PSUB.uuu R_c,R_a,R_b PSUB.uuu R_f,R_b,R_a PADD R_c,R_e,R_f	If $a > b$, then $e = (a - b)$, else $e = 0$. If $a \leq b$, then $f = (b - a)$, else $f = 0$. If $a > b$, then $c = (a - b)$, else $c = (b - a)$.

Note: This table describes the contents of the registers (e.g. a or b) as if they contained a single value for simplicity. The same description applies to all the subwords in the registers. Initial contents of R_a and R_b are a and b unless otherwise noted.

Table 2 Summary of the Integer Register and Subword Sizes
for the Different Architectures

Architecture	IA-64	MAX-2	MMX	SSE-2	AltiVec
Size of integer registers (bits)	64	64	64	128	128
Number of registers	128	32	8	8	32
Supported subword sizes (bytes)	1, 2, 4	2	1, 2, 4	1, 2, 4, 8	1, 2, 4
Modular arithmetic	Y	Y	Y	Y	Y
Supported saturation options	*sss, uuu, uus*	*sss, uus*	*sss, uuu*	*sss, uuu*	*sss, uuu*

Note: Not every saturation option indicated is applicable to every subword size. 3DNow! does not have an entry in this table because it does not have integer microSIMD extensions.

- IA-64 architecture* has 64-bit integer registers. *Packed add/subtract* operations are supported for subword sizes of 1, 2, and 4 bytes. Modular arithmetic is defined for all subword sizes, whereas the saturation options (*sss, uuu,* and *uus*) exist for only 1- and 2-byte subwords.
- PA-RISC MAX-2 architecture has 64-bit integer registers. *Packed add/subtract* instructions operate on only 2-byte subwords. MAX-2 instructions support modular arithmetic and *sss* and *uus* saturation options.
- IA-32 MMX architecture defines eight 64-bit registers for use by the multimedia instructions. Although these registers are given special names, these are indeed aliases to eight registers in the FP data register stack. Supported subword sizes are 1, 2, and 4 bytes. Modular arithmetic is defined for all subword sizes, whereas the saturation options (*sss* and *uuu*) exist for only 1- and 2-byte subwords.
- IA-32 SSE-2 technology introduces a new set of eight 128-bit FP registers to the IA-32 architecture. Each of the 128-bit registers can accommodate four SP or two double-precision (DP) FP numbers. Moreover, these registers can also be used to accommodate packed integer data types. Integer subword sizes can be 1, 2, 4, or 8 bytes. Modular arithmetic is defined for all subword sizes, whereas the saturation options (*sss* and *uuu*) exist for only 1- and 2-byte subwords.
- PowerPC Altivec architecture has 32 128-bit registers. *Packed add/subtract* operations are defined for 1, 2, and 4-byte subwords. *Packed adds* can use either modular arithmetic or *uuu* or *sss* saturation. *Packed subtracts* can use only modular arithmetic or *uuu* saturation.

Table 2 contains a summary of the register and subword sizes and the saturation options for the architectures we discuss.

---

* All of the discussions in this chapter consider IA-64 as the base architecture. Evaluations of the other architectures are generally carried out by comparisons to IA-64.

2.7   Other Instructions

AltiVec architecture includes two operations for picking up the carry bits from a packed add or subtract operation. The VADDCUW instruction performs a *packed add* operation and writes the carry-out bits to a right-aligned field in the target register. Figure 9 shows this instruction. The VSUBCUW instruction performs a similar operation using *packed subtract* instead of *packed add.*

Some instructions of SSE-2 are classified as *scalar* instructions. These instructions operate on packed data types, but only the least significant subwords of the two source operands are involved in the operation. The other subwords of the target register are directly copied from the first source operand. An example of a *scalar FP add* operation is shown in Figure 10.

2.8   Packed Floating-Point Add/Subtract Operations

IA-64, SSE-2, 3DNow!, and AltiVec include *packed FP add/subtract* instructions. Due to the format of FP numbers, there are no issues like modular arithmetic or saturation options for these instructions. The only difference that exists for *packed FP add/subtract* operations across various architectures is in the precision levels (Table 3).

IA-64 does not have dedicated instructions for *packed FP add* and *subtract.* Instead, these operations are realized by using FPMA (Floating-Point Parallel

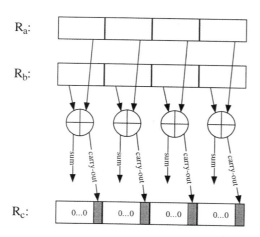

Figure 9   VADDCUW R_c,R_a,R_b: VADDCUW instruction of AltiVec writes the carry-out bits of the parallel adds to the least significant bits of the corresponding subwords of the target register. The sums are ignored.

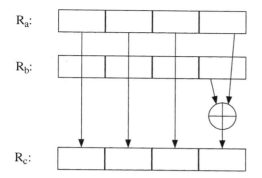

Figure 10 Scalar ADD R_c,R_a,R_b: *Scalar add* instruction (ADDSS) as defined by SSE and SSE-2 architectures. This instruction uses registers with four subwords each.

Multiply Add) and FPMS (Floating-Point Parallel Multiply Subtract) instructions as explained below.

IA-64 architecture specifies 128 FP registers, which are numbered FR0 through FR127. Of these registers, FR0 and FR1 are special. FR0 always returns the value +0.0 when sourced as an operand, and FR1 always reads +1.0. When FR0 or FR1 are used as source operands, the FPMA and FPMS instructions can be used to realize *packed FP add/subtract* and *packed FP multiply* operations.

Example

The format of the FPMA instruction is FPMA R_d,R_a,R_b,R_c and the operation it performs is $R_d = R_a * R_b + R_c$. If FR1 is used as the first or the second source operand (FPMA $R_d,FR1\ R_b,R_c$), a *packed FP add* operation is realized ($R_d = FR1 * R_b + R_c = 1.0 * R_b + R_c = R_b + R_c$). Similarly, a FPMS instruction can be used to realize a *packed FP subtract* operation. Using FR0 as the third source operand in FPMA or FPMS (FPMA R_d,R_a,R_b, FR0) results in a *packed FP multiply* operation ($R_d = R_a * R_b + FR0 = R_a * R_b + 0.0 = R_a * R_b$).

Table 3 Supported Precision Levels for the Packed FP Add/Subtract Operations

Architecture	IA-64	SSE-2	3DNow!	AltiVec
FP register size	82 bits	128 bits	128 bits	128 bits
Allowed packed FP data types	2 SP	4 SP or 2 DP	4 SP	4 SP

Note: SP and DP FP numbers are 32 and 64 bits long, respectively, as defined by the IEEE-754 FP number standard.

Table 4 *Packed Add/Subtract* Operations

	IA-64	MAX-2	MMX	SSE-2	3DNow!	AltiVec
Integer operations						
$c_i = a_i + b_i$	$\checkmark$	$\checkmark$	$\checkmark$	$\checkmark$		$\checkmark$
$c_i = a_i + b_i$ (with saturation)	$\checkmark$	$\checkmark$	$\checkmark$			$\checkmark$
$c_i = a_i - b_i$	$\checkmark$	$\checkmark$	$\checkmark$	$\checkmark$		$\checkmark$
$c_i = a_i - b_i$ (with saturation)	$\checkmark$	$\checkmark$	$\checkmark$			$\checkmark$
$lsbit(c_i) = carryout(a_i + b_i)$						$\checkmark$
$lsbit(c_i) = carryout(a_i - b_i)$						$\checkmark$
FP Operations						
$c_i = a_i + b_i$	$\checkmark$[a]			$\checkmark$[b]	$\checkmark$	$\checkmark$
$c_i = a_i - b_i$	$\checkmark$[c]			$\checkmark$[b]	$\checkmark$	$\checkmark$

[a] This operation is realized by using the FPMA instruction.
[b] A scalar version of this instruction also exists.
[c] This operation is realized by using the FPMS instruction.

3DNow! has two *packed subtract* instructions for FP numbers. The only difference between these two instructions is in the order of the operands. The PFSUB instruction subtracts the second packed source operand from the first, whereas the PFSUBR instruction subtracts the first packed source operand from the second.

Table 4 gives a summary of the *packed add/subtract* operations discussed in this section. In Table 3, the first column contains the description of the operations. The symbols a_i and b_i represent the subwords from the two source registers. The symbol c_i represents the corresponding subword in the target register. A shaded background indicates a packed FP operation.

3   PACKED SPECIAL ARITHMETIC OPERATIONS

This section describes variants of the *packed add* instructions that are generally designed to further increase performance in multimedia applications.

3.1   Packed Averaging

All of the architectures we refer to include instructions to support a *packed average* operation. *Packed average* operations are very common in media applications such as pixel averaging in MPEG-2 encoding, motion compensation, and video scaling.

In a *packed average* operation, the pairs of corresponding subwords in the two source registers are added to generate intermediate sums. From this point, two different paths may be taken, depending on which rounding option is used. These are addressed next.

3.1.1 Round Away from Zero

A 1 is added to the intermediate sums, and then the sums are shifted to the right by one bit position. If carry bits were generated during the addition operation, they are inserted into the most significant bit position during the shift-right operation. Figure 11 shows a *packed average* operation that uses this rounding option.

3.1.2 Round to Odd

Instead of adding 1 to the intermediate sums, a much simpler *or* operation is used. The intermediate sums are directly shifted right by one bit position, and the last two bits of each of the subwords of the intermediate sums are ORed to give the least significant bit of the final result. This makes sure that the least significant bit of the final results is set to 1 (odd) if at least one of the two least significant bits of the intermediate sums are 1.

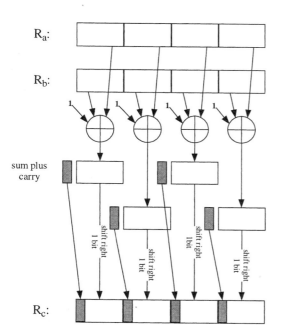

Figure 11   PAVG R_c, R_a, R_b: *Packed average* operation using the *round away from zero* option.

If the intermediate result is uniformly distributed over the range of possible values, then for half of the time, the bit shifted out is 0, and the result remains unchanged with rounding. The other half of the time, the bit shifted out is 1. If the next least significant bit is 1, then the result loses −0.5, but if the next least significant bit is a 0, then the result gains +0.5. Because these cases are equally likely with uniform distribution of the result, this round to odd option tends to cancel out the cumulative averaging errors that may be generated with repeated use of the averaging instruction. Hence, it is said to provide *unbiased rounding*. Figure 12 shows a *packed average* operation that uses this rounding option.

3.2 Accumulate

AltiVec provides an instruction to facilitate the accumulation of streaming data. This instruction performs an addition of the subwords in the same register and places the sum in the upper half of the target register, while repeating the same process for the second source register and using the lower half of the target register (Figure 13).

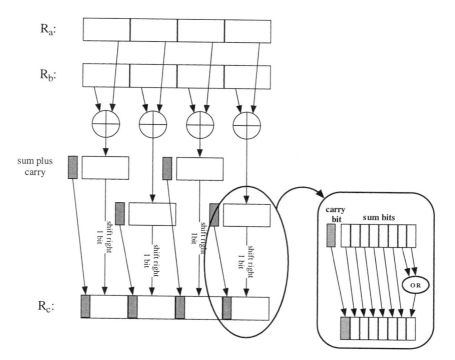

Figure 12 PAVG R_c,R_a,R_b: *Packed average* operation using the *round to odd* option.

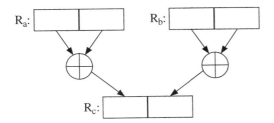

Figure 13 ACC R_c, R_a, R_b: *Accumulate* operation working on registers with two sub-words.

3.3 Sum of Absolute Differences

Sum of absolute differences (SAD) is another operation that proves useful in media applications. The motion estimation kernel in the MPEG-2 video encoding application is one example that can benefit from a SAD operation. In a SAD operation, the two packed operands are subtracted from one another. Absolute values of the resulting differences are then summed up. This sum is placed into the target register. Figure 14 shows how the SAD operation works. Most architectures feature a special instruction for this operation.

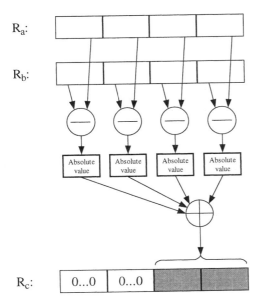

Figure 14 SAD R_c, R_a, R_b: *Sum of absolute differences* operation.

Table 5 Packed Special Arithmetic Operations

	IA-64	MAX-2	MMX	SSE-2	3DNow!	AltiVec
Integer operations						
$c_i = avg\ (a_i, b_i)$	√	√		√	√	√
$c_i = neg\ \_avg(a_i, b_i)$	√					
$c = \sum \|a_i - b_i\|$	√			√	√	
Accumulate Integer						√
FP Operations						
$c_i = -a_i$	√					
$c_i = \|a_i\|$	√					
$c_i = -\|a_i\|$	√					

Although useful, the SAD instruction is a multicycle instruction, with a typical latency of three cycles. This can complicate the pipeline control of otherwise single-cycle integer pipelines. Hence, minimalist multimedia instruction sets like MAX-2 do not have SAD instructions. Instead, MAX-2 uses generic PADDs and PSUBs with saturation arithmetic to perform the SAD operation (see Fig. 8b and Table 1).

Table 5 gives a summary of the packed special arithmetic operations discussed in this section. In Table 5, the first column contains the description of the operations. The symbols a_i and b_i represent the subwords from the two source registers. The symbol c_i represents the corresponding subword in the target register. A shaded background indicates a packed FP operation.

4 PACKED MULTIPLY OPERATIONS

We begin this section by explaining the multiplication of a packed integer register by a constant number. Next, we consider the more general case of multiplying two packed registers. Finally, we will give some examples of some more specialized instructions that involve packed multiplication operations.

4.1 Multiplication of a Packed Integer Register by an Integer Constant

Consider the multiplication of an unpacked integer register by an integer constant. This can be accomplished by a sequence of *shift left* instructions and *add* instructions. Shifting a register left by n bits is equivalent to multiplying it by 2^n. Because a constant number can be represented as a binary sequence of 1's and 0's, using this number as a multiplier is equivalent to a left shift of the multiplicand of n bits for each nth position where there is a 1 in the multiplier and an add of each shifted value to the result register.

As an example, consider multiplying the unpacked integer register R_a with the constant $C = 11$. The following instruction sequence performs this multiplication. Assume $R_a = 6$.

Initial values are $C = 11 = 1011_2$, $R_a = 6 = 0110_2$

Instruction	Operation	Result
Shift left by 1 R_b,R_a	$R_b = R_a \ll 1$	$R_b = 1100_2 = 12$
Add R_b,R_b,R_a	$R_b = R_b + R_a$	$R_b = 1100_2 + 0110_2 = 010010_2 = 18$
Shift left by 3 R_c,R_a	$R_c = R_a \ll 3$	$R_c = 0110_2 * 8 = 110000_2 = 48$
Add R_b,R_b,R_c	$R_b = R_b + R_c$	$R_b = 010010_2 + 110000_2 = 1000010_2 = 66$

This sequence can be shortened by combining the *shift left* instruction and *add* instruction into one new *shift left and add* instruction. The following new sequence performs the same multiplication in half as many instructions and uses one less register.

Initial values are $C = 11 = 1011_2$, $R_a = 6 = 0110_2$

Instruction	Operation	Result
Shift left by 1 and add R_b,R_a,R_a	$R_b = R_a \ll 1 + R_a$	$R_b = 18$
Shift left by 3 and add R_b,R_a,R_b	$R_b = R_a \ll 3 + R_b$	$R_b = 66$

Multiplication of packed integer registers by integer constants uses the same idea explained above. The *shift left and add* instruction becomes a *packed shift left and add* instruction to support the packed data types. As an example consider multiplying the subwords of the packed integer register $R_a = [1|2|3|4]$ by the constant $C = 11$. The instructions to perform this operation are

Initial values are $C = 11 = 1011_2$, $R_a = [1|2|3|4] = [0001|0010|0011|0100]_2$

Instruction	Operation	Result
Shift left by 1 and add R_b,R_a,R_a	$R_b = R_a \ll 1 + R_a$	$R_b = [0011|0110|1001|1100]_2$ $= [3|6|9|12]$
Shift left by 3 and add R_b,R_a,R_b	$R_b = R_a \ll 3 + R_b$	$R_b = [1011|10110|100001|1010100]_2$ $= 11|22|33|44]$

The same reasoning used for multiplication by constants applies to multiplication by fractional constants as well. Arithmetic right shift of a register by n bits is equivalent to dividing it by 2^n. Using a fractional constant as a multiplier is equivalent to an arithmetic right shift of the multiplicand by n bits for each nth position where there is a 1 in the multiplier and an add of each shifted value to the result register. By using a *packed arithmetic shift right and add* instruction, the *shift* and the *add* operations can be combined into one to further speed such computations. For instance, multiplication of a packed register by the fractional constant 0.011_2 ($= 0.375$) can be performed by using only two *packed arithmetic shift right and add* instructions.

Initial values are $C = 0.375 = 0.011_2$, $R_a = [1|2|3|4] = [0001|0010|0011|0100]_2$

Instruction	Operation	Result
Arithmetic Shift right by 3 and add $R_b, R_a, 0$	$R_b = R_a \gg 3 + 0$	$R_b = [0.001|0.01|0.011|0.1]_2$ $= [0.125|0.25|0.375|0.5]$
Arithmetic Shift right by 2 and add R_b, R_a, R_b	$R_b = R_a \gg 2 + R_b$	$R_b = [0.011|0.11|1.001|1.1]$ $= [0.375|0.75|1.125|1.5]$

These three examples demonstrate the pathlength reduction that can be achieved by the use of multimedia extensions in a general-purpose processor. Only two instructions were necessary to multiply four integer subwords by a constant number. Without subword parallelism, the same operations would take at least four instructions. Considering that a 128-bit register can contain 16, 8-bit subwords, the extent of pathlength reduction becomes more pronounced.

In the previous example, it was assumed that the *packed shift right and add* instruction shifted each of the subwords of the source operand by the same amount specified in the immediate operand. It may be that the shift amount is specified in a register or memory operand instead of being specified in an immediate field. This, however, would require the use of three source registers by the *packed shift right and add* instruction. A different *packed shift left* instruction may shift each of the subwords in the source operand by a different amount, which may be given in a register or an immediate operand. *Packed shift left* instructions may cause overflows, which should be detected and handled using one of the ways explained in Section 2. *Packed shift right* instructions do not have this problem. For a *shift right* operation, the shift can be arithmetic or logical. All of these options create many possibilities for how a particular *packed shift* operation will function. Section 6 will address these details.

The next subsection discusses different ways of multiplying two packed integer registers. As we shall see shortly, when both of the operands are registers, packed multiplication operations become tricky to handle, because each product is now twice the length of each multiplicand.

4.2 Multiplication of Two Packed Integer Registers

The most straightforward way to define a *packed multiply* is to multiply each subword in the first source register by the corresponding subword in the second source register. This, however, is impractical because the target register now has to be twice the size of each of the source registers. Two same-size multiplicands will produce a product with a size that is double the size of each of the multiplicands.

Consider the case in Figure 15, in which the register size is 64 bits and the subwords are 16 bits. The result of the *packed multiply* will be four 32-bit products, which cannot be accommodated in a target register, which is only 64 bits wide.

The architectures approach this problem in different ways. MMX has the PMULHW instruction that only places the most significant upper halves of the products into the target register. Similar to this instruction, SSE-2 has the PMULHUW, AltiVec has the VMUL, and the 3DNow! has the PMULHRW and PMULHUW instructions. All these instructions pick the higher-order bits from the products and place them in the target register. These instructions are depicted in Figure 16 for the case of four subwords.

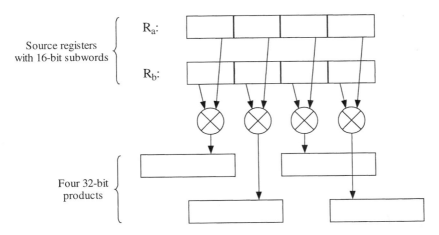

Figure 15 *Packed multiply* operation using four 16-bit subwords per register. Not all of the full-sized products can be accommodated in a single target register.

Lee and Fiskiran

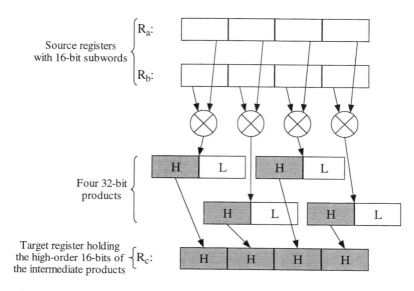

Figure 16 PMUL.high R_c,R_a,R_b: *Packed multiply* operation using four subwords per register. Only the high-order bits of the intermediate products are written to the target register.

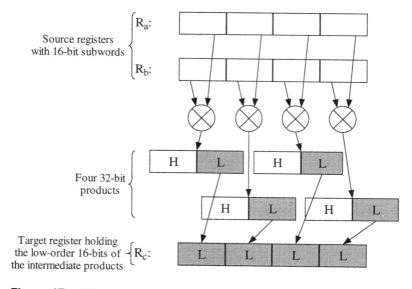

Figure 17 PMUL.low R_c,R_a,R_b: *Packed multiply* operation using four subwords per register. Only the low-order bits of the intermediate products are written to the target register.

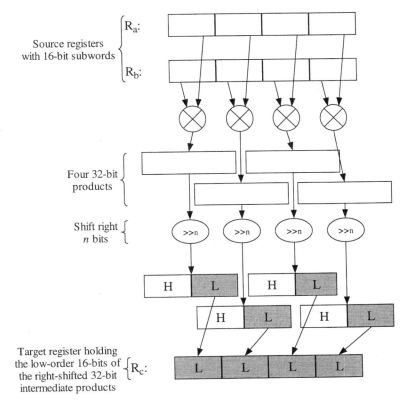

Figure 18   PMPYSHR R_c, R_a, R_b: The PMPYSHR instruction of the IA-64 architecture. The shift amount, which is given in the immediate register, is limited to 0, 7, 15, and 16 bits.

The other approach can be to keep the least significant halves of the products in the target register. Some examples to this are the MMX's PMULLW and AltiVec's VMUL. These instructions work as shown in Figure 17.

IA-64 architecture comes up with a generalization of this, with its PMPYSHR instruction. This instruction lifts the limit that one has to choose either the upper or the lower half of the products to be put into the target register. PMPYSHR instruction does a *packed multiply* followed by a *shift right* operation. This allows four* possible 16-bit fields (IA-64 has a 64-bit register size) from each of the 32-bit products to be chosen and be placed in the target register. The PMPYSHR instruction is shown in Figure 18.

---

* The right-shift amounts are limited to 0, 7, 15, or 16 bits. This limitation allows a reduction in the number of bits necessary to encode the instruction.

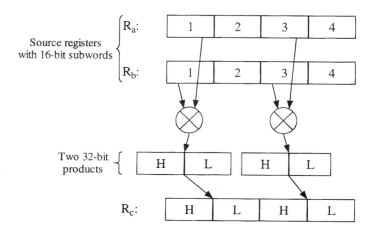

Figure 19 PMUL.odd R_c, R_a, R_b: *Packed multiply* operation where only the odd indexed subwords of the two source registers are multiplied.

All of the instructions we have seen thus far performed a full multiplication on all of the subword pairs of the source operands and then decided how to handle the large products. However, instead of truncating the products, the source subwords that will participate in the multiplication may be selected so that the final product is never larger than can be accommodated in a single target register.

IA-64 has the PMPY instruction, which has two variants. PMPY allows only half of the pairs of the source subwords to go into multiplication. Either the odd or the even indexed subwords are multiplied. This makes sure that only as many full products as can be accommodated in one target register are generated. The two variants of the PMPY instruction are depicted in Figures 19 and 20.

4.3   Other Variants of Packed Integer Multiplication Operations

Intel's MMX technology has an instruction that performs a packed multiplication followed by an addition. This PMADDWD instruction generates four 32-bit intermediate product terms in the packed multiply stage. Later, the first product is added to the second one, and the 32-bit sum is placed into the first half of the target register. The third and the fourth products are also summed and this sum is placed into the second half of the target register (Fig. 21).

Instructions in the AltiVec architecture may have up to three source registers. This allows AltiVec to realize operations that require three source operands,

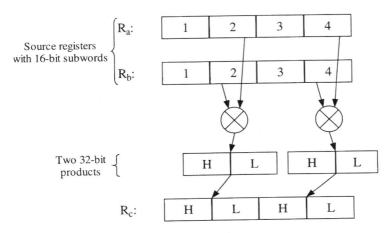

Figure 20   PMUL.even R_c,R_a,R_b: *Packed multiply* operation where only the even in-dexed subwords of the two source registers are multiplied.

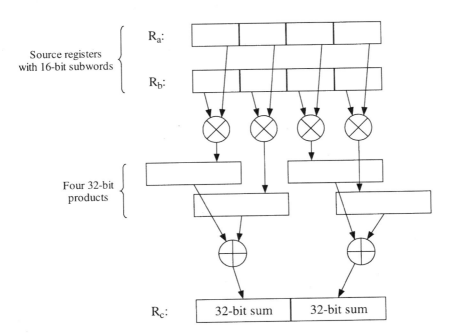

Figure 21   Multiply and accumulate R_c,R_a,R_b: The *multiply and accumulate* operation of the MMX architecture.

such as the *packed multiply and add*. The instructions (VMHxADD and VMLxADD) start just like *packed multiply* instructions, select either the higher or lower order bits of the full-sized products, and then perform a *packed add* between the values from a third register and the result of the multiplication operation. These two instructions are shown in Figures 22 and 23.

The very specialized VSUMMBM instruction of AltiVec performs a vector multiplication and a scalar addition using three packed source operands. First, all of the bytes within the corresponding words are multiplied in parallel and 16-bit products are generated. Later, all of these products are added to each other

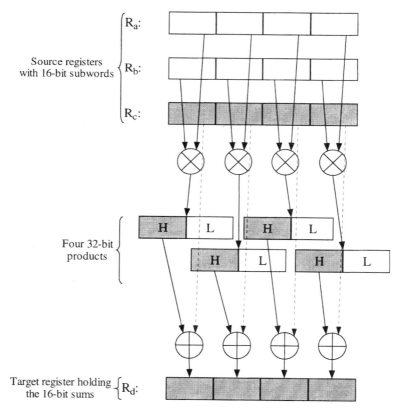

Figure 22 Multiply high and add R_d, R_a, R_b, R_c: The *multiply and add* instruction of the AltiVec architecture. In this variant of the instruction, only the high-order bits of the intermediate products are used in the addition.

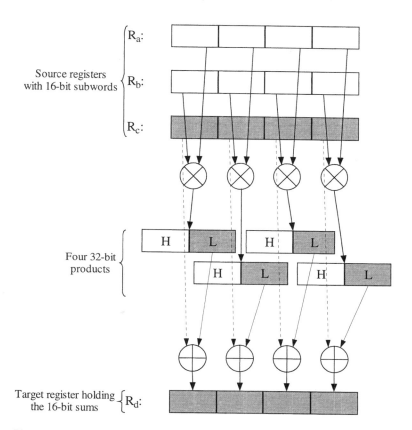

Figure 23 Multiply low and add R_d, R_a, R_b, R_c: The *multiply and add* instruction of the AltiVec architecture. In this variant of the instruction, only the low-order bits of the intermediate products are used in the addition.

to generate the *sum of products*. A third word from the third source operand is added to this *sum of products*. The final sum is placed in the corresponding word field of the target register. This process is repeated for each of the four words (Fig. 24).

4.4 A Note About Multiplication of Packed FP Registers

Packed FP multiplication does not cause the problems encountered in packed integer multiplication. In integer multiplication, the size of the product term is

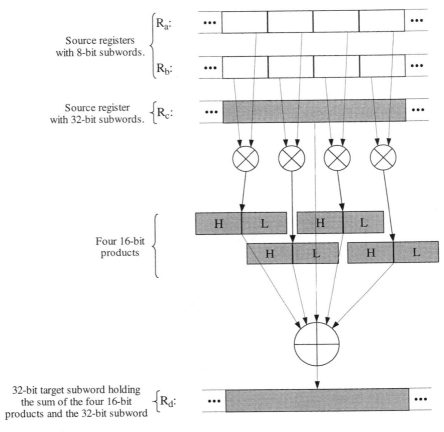

Figure 24 VSUMMBM R_d,R_a,R_b,R_c: AltiVec's VSUMMBM instruction proves useful in matrix multiplication operations. In this figure, only one-fourth of the instruction is shown. Each box represents a byte. This process is carried out for each 32-bit word in the 128-bit source registers.

always greater than either of the multiplicands. This does not allow all of the product terms to be written into the target register. The special format of FP numbers does not cause such a size problem. The same number of bits* is used to represent a FP number regardless of how large the number is. In this respect, multiplication of packed FP registers is similar to the addition of packed FP

---

* In general, 32 and 64 bits are used to represent SP and DP FP numbers, respectively.

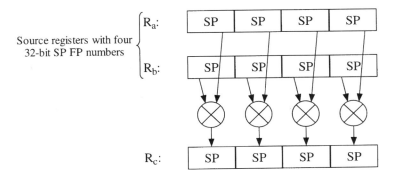

Source registers with four 32-bit SP FP numbers

R_a:

R_b:

R_c:

Figure 25 FP PMUL R_c,R_a,R_b: Packed FP multiplication does not introduce the size problems like the packed integer multiplication. This figure shows two packed FP integers being multiplied. The packed registers have two FP numbers each.

registers. Figure 25 shows the multiplication of two packed FP registers, each containing four SP FP numbers.

Table 6 gives a summary of the *packed multiply* operations discussed in this section. In Table 6, the first column contains the description of the operations. For instructions with three registers, the symbols a_i and b_i represent the subwords from the two source registers. The symbol c_i represents the corresponding subword in the target register. For instructions with four registers, the symbols a_i, b_i, and c_i represent the subwords from the three source registers. The symbol d_i represents the corresponding subword in the target register. A shaded background indicates a packed FP operation.

5 PACKED COMPARE/MAXIMUM/MINIMUM OPERATIONS

In a *packed compare* operation, pairs of subwords are compared according to the relation specified by the instruction. If the condition is true for a subword pair, the corresponding field in the target register is written with a 1-mask. If the condition is false, the corresponding field in the target register is written with a 0-mask. Some of the architectures have *compare* instructions that allows comparison of two numbers for all of the 12 possible relations,* whereas some architec-

---

* Two numbers a and b can be compared for one of the following 12 possible relations: equal, less-than, less-than-or-equal, greater-than, greater-than-or-equal, unordered, not-equal, not-less-than, not-less-than-or-equal, not-greater-than, not-greater-than-or-equal, ordered. Typical notation for these relations are as follows: $==$, $<$, $<=$, $>$, $>=$, ?, $!=$, $!<$, $!<=$, $!>$, $!>=$, $!?$, respectively.

Table 6 *Packed Multiply* Operations

	IA-64	MAX-2	MMX	SSE-2	3DNow!	AltiVec
Integer operations						
Packed shift left and add[a]						
$c_i = (a_i \ll n) + b_i$	✓	✓				
Packed shift right and add[b]						
$c_i = (a_i \gg n) + b_i$	✓	✓				
$c_i = lower\_half[(a_i * b_i) \gg n]$	✓[c]					
$c_i = lower\_half(a_i * b_i)$	✓		✓			✓
$c_i = upper\_half(a_i * b_i)$	✓		✓			✓
Multiply even						
$[c_{2i}, c_{2i+1}] = a_{2i} * b_{2i}$	✓					✓
Multiply odd						
$[c_{2i}, c_{2i+1}] = a_{2i} * b_{2i+1}$			✓			
Multiply and accumulate						
$[c_{2i}, c_{2i+1}] = a_{2i} * b_{2i} + a_{2i+1} * b_{2i+1}$				✓		✓
$d_i = upper\_half(a_i * b_i) + c_i$				✓		✓
$d_i = lower\_half(a_i * b_i) + c_i$						✓
VMSUMxxx instructions of AltiVec (general form)						✓
$[d_{2i}, d_{2i+1}] = a_{2i} * b_{2i} + a_{2i+1} * b_{2i+1} + [c_{2i}, c_{2i+1}]$						
VSUMMBM instruction of AltiVec						
$[d_{4i}, d_{4i+1}, d_{4i+2}, d_{4i+3}] = [c_{4i}, c_{4i+1}, c_{4i+2}, c_{4i+3}] + \sum_{j=1}^{4} a_{4i+j} * b_{4i+j}$						
FP operations						
$c_i = a_i * b_i$	✓			✓[d]	✓	✓
$d_i = -a_i * b_i$	✓					
$d_i = a_i * b_i + c_i$	✓					✓
$d_i = a_i * b_i - c_i$	✓					
$d_i = -a_i * b_i + c_i$	✓					

[a] For use in multiplication of a packed register by an integer constant.
[b] For use in multiplication of a packed register by a fractional constant.
[c] Shift amounts are limited to 0, 7, 15, or 16 bits.
[d] Scalar versions of these instructions also exist.

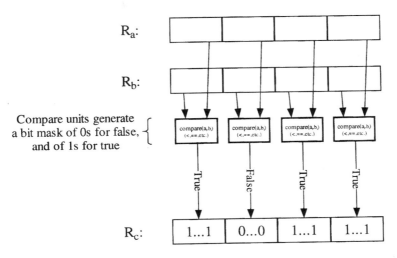

Figure 26 PCOMP R_c,R_a,R_b: *Packed compare* instruction. Bit masks are generated as a result of the comparisons made.

tures allow for a more limited subset of relations. A typical *packed compare* instruction is shown in Figure 26 for the case of four subwords.

In the *packed maximum/minimum* operations, the greater/smaller of the subwords in the compared pair gets written to the corresponding field in the target register. Figures 27 and 28 illustrate the *packed maximum/minimum* operations.

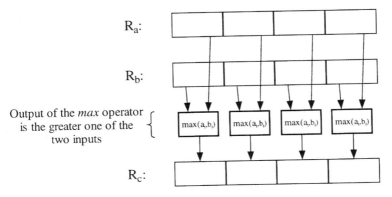

Figure 27 PMAX R_c,R_a,R_b: *Packed maximum* operation. The greater source subword is written to the corresponding location in the target register.

R_a:

R_b:

Output of the *min* operator
is the smaller one of the
two inputs

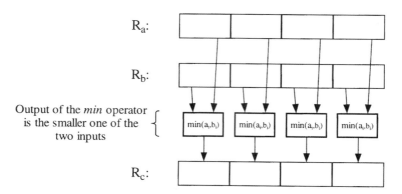

R_c:

Figure 28   PMIN R_c,R_a,R_b: *Packed minimum* operation. The smaller source subword
is written to the corresponding location in the target register.

In practice, architectures either include dedicated instructions or make use of
other existing instructions to perform a *packed maximum/minimum* operation.
MAX-2, for instance, is in the second category and performs *packed max/min*
operations by using *packed add/subtract* instructions with saturation arithmetic.
How the saturation arithmetic can be used to realize *packed max/min* operations
is detailed in Section 2. See Figure 8 for an example of the *packed maximum*
operation realized by using saturation arithmetic.

An interesting *packed compare* instruction is the VCMPBFP (Vector Compare Bounds Floating Point) instruction of AltiVec. This instruction compares
corresponding FP number pairs from the two packed source registers, and depending on the relation between the compared numbers, it generates a 2-bit result,
which is written to the target register. The resulting 2-bit field indicates the relation between the two compared FP numbers. For the instruction VCMPBFP
R_c,R_a,R_b, the FP number pairs (a_i, b_i) are compared, and a 2-bit field is written
into c_i, such that the following hold:

- Bit 0 of the 2-bit field is cleared if $a_i \leq b_i$ and is set otherwise.
- Bit 1 of the 2-bit field is cleared if $a_i \leq (-b_i)$ and is set otherwise.
- Both bits are set if any of the compared FP numbers is *not a number*
  (NaN).

The two-bit result field is written to the high-order two bits of c_i; the remaining bits of c_i are cleared to 0. Table 7 gives examples of input pairs that
result in each of the four different possible outputs for this instruction.

Table 8 gives a summary of the *packed compare/maximum/minimum* operations discussed in this section. In Table 8, the first column contains the descrip-

Table 7 Result of the VCMPBFP
Instruction for Different Input Pairs

Input		Output	
a_i	b_i	Bit 0	Bit 1
3.0	5.0	0	0
−8.0	5.0	0	1
8.0	5.0	1	0
3.0	−5.0	1	1

tion of the operations. The symbols a_i and b_i represent the subwords from the two source registers. The symbol c_i represents the corresponding subword in the target register. A shaded background indicates a packed FP operation.

6 PACKED SHIFT/ROTATE OPERATIONS

Most of the architectures have instructions that support *packed shift/rotate* operations on packed data types. These instructions prove very useful in multimedia, arithmetic, and encryption applications. There are usually great differences be-

Table 8 *Packed Compare/Maximum/Minimum* Operations

	IA-64	MAX-2	MMX	SSE-2	3DNow!	AltiVec
Integer operations						
$c_i = compare(a_i, b_i)$	✓		✓			✓
$c_i = max(a_i, b_i)$	✓	✓[a]		✓	✓	✓
$c_i = min(a_i, b_i)$	✓	✓[a]		✓	✓	✓
FP operations						
$c_i = compare(a_i, b_i)$	✓			✓[b]	✓	✓
$c_i = max(a_i, b_i)$	✓			✓[b]	✓	✓
$c_i = min(a_i, b_i)$	✓			✓[b]	✓	✓
$c_i = max(|a_i|, |b_i|)$	✓					
$c_i = min(|a_i|, |b_i|)$	✓					
$c_i = VCMPBFP(a_i, b_i)$[c]						✓

[a] This operation is realized by using saturation arithmetic.
[b] Scalar versions of these instructions also exist.
[c] This instruction is explained in the text.

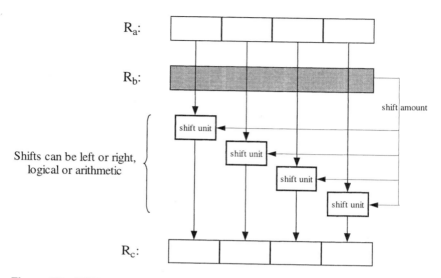

Figure 29   PSHIFT R_c,R_a,R_b: *Packed shift* operation. Shift amount is given in the second operand. Each subword is shifted by the same amount.

tween the *packed shift/rotate* instructions of different architectures, as there are several options to be considered:

- A *packed shift/rotate* instruction shifts/rotates the subwords in a packed register.
- For the *packed shift* command, one has to decide if the shift will be logical (0's substituted for vacated bits) or algebraic (sign bit is replicated for vacated bits on the left).
- Saturation arithmetic may or may not be used during the *packed shift* operations.
- The shift/rotate amount can be given by an immediate operand or a register operand.
- Each subword may have to be shifted/rotated by the same amount or the instruction may be sophisticated so that each subword in a packed register can be shifted/rotated by a different amount.

Given so many options, almost all architectures come up with their own solutions to the problem. The *packed shift/rotate* instructions are shown in Figures 29–32.

Table 9 gives a summary of the *packed shift/rotate* operations discussed in this section. In Table 9, the first column contains the description of the opera-

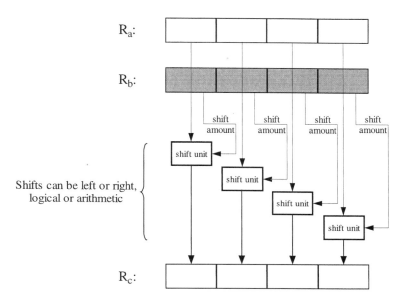

Figure 30 PSHIFT R_c, R_a, R_b: *Packed shift* operation. Shift amount is given in the second operand. Each subword can be shifted by a different amount.

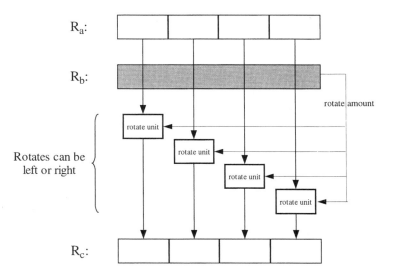

Figure 31 PROT R_c, R_a, R_b: *Packed rotate* operation. Rotate amount is given in the second operand. Each subword is rotated by the same amount.

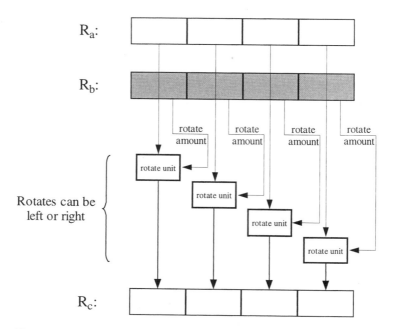

Figure 32 PROT R_c, R_a, R_b: *Packed rotate* operation. Rotate amount is given in the second operand. Each subword can be rotated by a different amount.

Table 9 *Packed Shift/Rotate* Operations

Integer operations	IA-64	MAX-2	MMX	SSE-2	3DNow!	AltiVec
$c_i = a_i \ll n$	$\checkmark$	$\checkmark$	$\checkmark$			
$c_i = a_i \ll b$	$\checkmark$		$\checkmark$			
$c_i = a_i \ll b_i$						$\checkmark$
$c_i = a_i \gg n$	$\checkmark$	$\checkmark$	$\checkmark$			
$c_i = a_i \gg b$	$\checkmark$		$\checkmark$			
$c_i = a_i \gg b_i$						$\checkmark$
$c_i = (a_i \ll n) + b_i$	$\checkmark$	$\checkmark$				
$c_i = (a_i \gg n) + b_i$	$\checkmark$	$\checkmark$				
$c_i = a_i \lll n$						
$c_i = a_i \lll b$						
$c_i = a_i \lll b_i$						$\checkmark$

tions. The symbols a_i and b_i represent the subwords from the two source registers. The symbol c_i represents the corresponding subword in the target register. Instruction formatting used in the table is as follows:

- n is used to represent a shift or rotate amount that is specified in the immediate field of an instruction. Hence, in the operation denoted as $c_i = a_i \ll n$, each subword of a is shifted to the left by n bits. The results are placed in c.
- Similarly, in the operation $c_i = a_i \ll b$, each subword of a is shifted to the left by the amount specified in the source register b. The results are placed in c.
- In $c_i = a_i \ll b_i$, each subword of a is shifted to the left by the amount specified in the corresponding subword of the source register b. The results are placed in c.
- $c_i = (a_i \ll n) + b_i$ represents a *shift left and add* operation. Each subword of a is shifted to the left by n bits. Corresponding subwords from the source register b are added to the shifted values. The sums are placed in their respective locations in c.
- In $c_i = a_i \lll n$, each subword of a is rotated left by n bits. The results are placed in c. None of the architectures have this operation.
- In $c_i = a_i \lll b$, each subword of a is rotated to the left by the amount specified in the source register b. The results are placed in c. None of the architectures have this operation.
- In $c_i = a_i \lll b_i$, each subword of a is rotated left by the amount specified in the corresponding subword of the source register b. The results are placed in c.

7 DATA/SUBWORD PACKING AND REARRANGEMENT OPERATIONS

The instructions for data/subword packing and rearrangement are most interesting and have the widest variety among different architectures.

7.1 Pack Instructions

All architectures include instructions for conversion between different packed data types. In general, the *pack* instructions are used to create packed data types from unpacked data types. A *pack* instruction can also be used to further pack an already-packed data type. Figure 33 shows how a packed data type can be created from two unpacked operands. Figure 34 shows how two packed data types can be packed further using a *pack* instruction. Differences between any

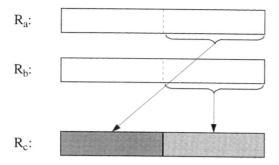

Figure 33 PACK R_c,R_a,R_b: *Pack* operation is used to create packed data types from unpacked data types.

two *pack* instructions are generally in the size of the supported subwords and in the saturation options that can be used.

7.2 Unpack Instructions

Unpack instructions are used to unpack the packed data types. The subwords in the two source operands are split and written to the target register in alternating order. Because only one-half of each of the source registers can be used, the *unpack* instructions always come with two variants: *high* or *low unpack*. These options allow the user to select which subwords in the source operand will be

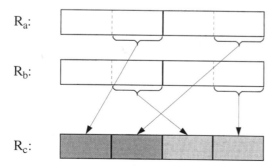

Figure 34 PACK R_c,R_a,R_b: *Pack* operation can be used to further pack already-packed data types.

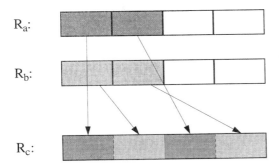

Figure 35 UNPACK.high R_c, R_a, R_b: *Unpack high* operation.

unpacked to the target register. The *high/low unpack* instructions select and unpack the high/low-order subwords of the source operands. (See Figs. 35 and 36.)

7.3 Permutation Instructions

Ideally, it is desirable to be able to perform all of the possible permutations on a packed data type. This is only possible when the subwords in the packed data type are not very many. (See Fig. 37.) When the number of subwords increases beyond a certain value, the number of control bits required to specify *arbitrary permutations* becomes too many to be encoded in the opcodes. For the case of n subwords, the number of control bits used to specify a particular permutation of these n subwords is calculated as $n \log_2(n)$. Table 10 shows how many control bits are required to specify *arbitrary permutations* for different number of subwords.

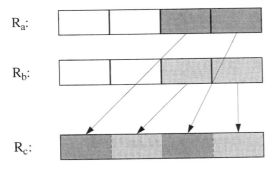

Figure 36 UNPACK.low R_c, R_a, R_b: *Unpack low* operation.

R_a:

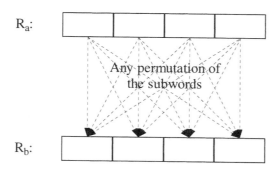

R_b:

Figure 37 PERMUTE R_b,R_a: *Arbitrary permutation* on a register with four subwords.

As Table 10 indicates, when the number of subwords is 16 or more, the number of required control bits exceeds the number of the bits available in the opcodes, which is typically 32. Therefore, it becomes necessary to use a second register* to contain the control bits used to specify the permutation. By using this second register, it is possible to get any *arbitrary permutation* of up to 16 subwords in 1 instruction.

AltiVec architecture takes an additional step to use the three source regis-

---

* This second register needs to be at least 64 bits wide to fully accommodate the 64 control bits needed for 16 subwords.

Table 10 Number of Control Bits Required to Specify an *Arbitrary Permutation* for a Given Number of Subwords

No. of subwords	No. of control bits required
2	2
4	8
8	24
16	64
32	160
64	384
128	896

ters it can have in one instruction. The VPERM instruction uses two registers to hold data, and the third register to hold the control bits. Thus, it allows any *arbitrary permutation* of 16 of the 32 bytes in the 2 source registers in a single instruction. The number of control bits required to specify this permutation (16 subwords out of 32) is calculated as $16 \log_2(32) = 80$.

Due to the problem explained above, only a small subset of all the possible permutations is realizable in practice. There is a great flexibility in the selection of this subset from the set of all of the possible permutations. It is sensible to select permutations that can be used as primitives to realize other permutations. One other distinction needs to be made between types of permutation. An instruction can use either one or two source operands for a permutation. In the latter case, only half of the subwords in the two source operands may actually appear in the target register. Examples to these two cases are the MUX and MIX instructions in IA-64, respectively, which correspond to PERMUTE and MIX instructions in MAX-2.

MIX is one useful operation that performs a permutation on two source registers. A MIX instruction picks alternating subwords from two source registers and places them into the target register. Because MIX uses two source registers, it appears in two variants. The first variant (Fig. 38) is called the *mix left* and uses the odd indexed subwords of the source registers in the permutation, starting from the leftmost subword. The other variant, *mix right* (Fig. 39), uses the even indexed subwords of the source registers, ending with the rightmost subword.

IA-64 architecture has the MUX instruction that can be used to perform permutations on 8- or 16-bit subwords. For 16-bit subwords, any arbitrary permutation is allowed. The immediate field is used to select 1 of the 256 possible

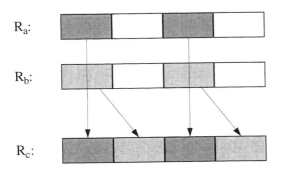

Figure 38 MIX left R_c, R_a, R_b: *Mix left* operation.

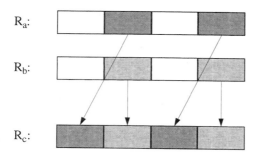

Figure 39 MIX right R_c,R_a,R_b: *Mix right* operation.

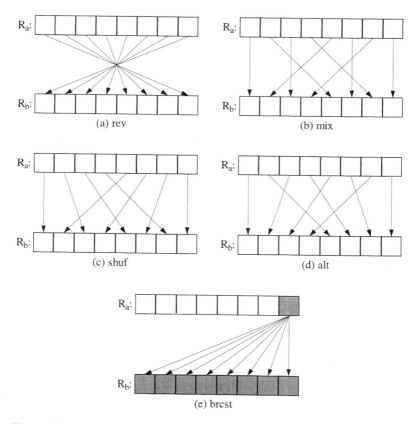

Figure 40 MUX.option R_b,R_a: MUX instruction of IA-64 has five permutation options for 8-bit subwords: rev, mix, shuf, alt, and brcst. These options are shown in (a) to (e), respectively.

permutations of the four 16-bit subwords, with or without repetitions of any subword. For the 8-bit subwords, only the following five permutations are allowed (Fig. 40):

- MUX.rev: Reverses the order of bytes.
- MUX.mix: Mixes the upper and lower 32-bit fields of the 64-bit register with byte granularity.
- MUX.shuf: Performs a perfect shuffle on the bytes.
- MUX.alt: Selects every other byte placing the even* indexed bytes on the left half of the result register followed by the odd indexed bytes.
- MUX.brcst: Replicates the least significant byte into all the byte locations.

7.4 Extract/Deposit Instructions

An *extract* instruction picks an arbitrary contiguous bit field from the source operand and places it right aligned into the target register. *Extract* instructions may be limited to work on subwords instead of arbitrarily long bit fields. In general, *extract* instructions clear the upper bits of the target register. Figures 41 and 42 show some possible *extract* instructions.

A *deposit* instruction picks an arbitrarily long right-aligned contiguous bit field from the source register and patches it into an arbitrary location in the target

---

* The bytes are indexed from 0 to 7. Index 0 corresponds to the most significant byte, which is on the left end of the registers.

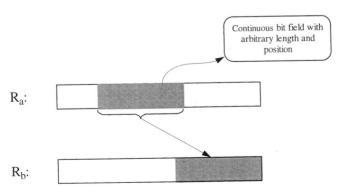

Figure 41 EXTRACT R_b,R_a: *Extract* instructions can be used to extract contiguous bit fields with arbitrary lengths and locations.

R_a:

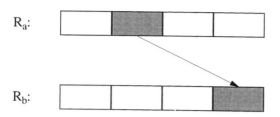

R_b:

Figure 42 EXTRACT R_b,R_a: A more limited *extract* instruction that can only extract subwords.

register. The remaining bits of the target register are either zeroed or unchanged. *Deposit* instructions may be limited to work on subwords instead of arbitrarily long bit fields and arbitrary patch locations. Figures 43 and 44 show some possible *deposit* instructions.

7.5   Moving a Mask of Most Significant Bits (Move Mask)

3DNow! includes the PMOVMSKB instruction for this operation. During the *move mask*, the most significant bit from each subword is picked, and placed in order into the target register, to a right aligned field (Fig. 45). Remaining bits in the target register are cleared.

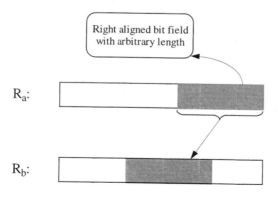

Figure 43 DEPOSIT R_b,R_a: *Deposit* instructions can be used to patch arbitrarily long, right-aligned contiguous bit fields from the source register into any location in the target register.

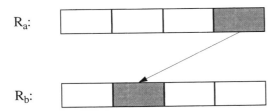

Figure 44 DEPOSIT R_b,R_a: A more limited *deposit* instruction that can only deposit subwords.

Table 11 gives a summary of the data/subword packing and rearrangement operations discussed in this section. In Table 11, the first column contains the description of the operations. PACK, UNPACK, MIX, MUX.option and Arbitrary Permutation operations are explained in the text.

In the FP PACK instruction of the IA-64 architecture, the two source operands are FP numbers in the 82-bit register format (this is a nonstandard format IA-64 uses for internal calculations). In the operation, the two numbers are first converted to standard 32-bit SP representation. These two SP FP numbers are then concatenated and the result is stored in the significand field (which is 64 bits) of the 82-bit target FP register. The exponent field of the target register (for its 82-bit register format) is set to the biased exponent for 2.0^{63}, and the sign bit is set to 0, indicating a positive number.

The FP UNPACK instructions of the SSE-2 architecture are identical to their integer counterparts, except that SP and DP FP numbers are used instead of integer subwords.

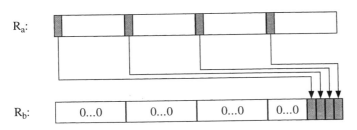

Figure 45 Move mask R_b,R_a: *Move mask* operation on a register with four subwords. The most significant bit of each subword is written in order to the least significant byte of the target register.

Table 11 Data/Subword Packing and Rearrangement Operations

	IA-64	MAX-2	MMX	SSE-2	3DNow!	AltiVec
Integer operations						
Mix Left	√	√				√
Mix Right	√	√				√
MUX.rev	√					
MUX.mix	√					
MUX.shuf	√					
MUX.alt	√					
MUX.brcst	√					
Arbitrary Permutation of n subwords	√ ($n = 4$)	√ ($n = 4$)		√ ($n = 4$)	√ ($n = 4$)	√ ($n = 16, 32$)[a]
PACK	√		√	√		√
UNPACK (high/low)	√		√			√
Move Mask					√	
FP operations						
Mix Left	√					
Mix Right	√					
PACK	√					
UNPACK(high/low)				√		
Arbitrary Permutation of n FP numbers				√ ($n = 2, 4$)		

[a]This is the VPERM instruction and it has some limitations for $n = 32$. See text for more details on this instruction.

8 APPROXIMATION OPERATIONS

Some multimedia applications (e.g., graphics) may be very intensive in computations like $1/x$ or $\sqrt{x}$. Typically, such computations are handled by the FP-ALU and take more execution cycles to complete compared to an operation handled in the Integer-ALU.

Computation of $1/x$ in the FP-ALU involves calculation of the result to infinite precision. This intermediate result is later rounded to either a SP or DP FP number. This process may be too slow for some applications that have stringent time constraints. A real-time graphics application involving intensive $1/x$ computations could be one example.

On the other hand, a different application may not even require SP accuracy. For such applications, waiting for many execution cycles for SP or DP results to complete would degrade performance. To address these problems, multimedia extensions include what is called *approximation instructions*. Approximation instructions return less precise results (than a SP FP number); however, they execute faster than a full computation, which returns a SP or DP accuracy.

Even in instances that require full SP or DP accuracy, it is undesirable to have a reciprocal instruction that takes many more cycles than a typical FP multiplication or addition. The goal, then, is to break these long operations down into a sequence of simpler operations, each of which takes about the same time (e.g., three to four cycles) as a *FP multiply*, *add*, or *multiply and accumulate* operation. The first operation in such a sequence is typically an approximation, which returns a low-precision estimate of the desired result. If low-precision results are acceptable, no further operations are necessary and the result can be used at that point. If a higher precision is required, the next operation in the sequence is used. This second operation is typically a *refine* operation. It uses the low-precision estimate generated in the first step and refines this value to a higher accuracy. Similarly, the next (third) operation in the sequence can be used to refine the output of the second operation, and this process can be repeated until the desired accuracy is reached.

To summarize, approximation instructions (and further instructions to refine the results) can be used whenever the following occur:

1. Using a full computation that gives SP or DP results may be too slow for acceptable performance.
2. A less accurate result (than a SP FP number) may be acceptable.
3. A SP or DP result is desired with a sequence of shorter operations rather than a single long operation.

For use in such instances, IA-64, SSE-2, 3DNow!, and AltiVec architectures have instructions that approximate the results of the $1/x$ and $1/\sqrt{x}$ operations. AltiVec also includes approximation instructions for $\log_2(x)$ and 2^x operations.

All these instructions operate on packed FP data types and take SP FP numbers as operands. The results of these instructions are less accurate than a SP FP number. 3DNow! also includes instructions to refine the results of its reciprocal and square root approximation instructions to SP accuracy. The IA-64 architecture does not include any separate instructions to refine the results of its approximation instructions. If standard accuracy is desired (either SP or DP), the approximation instruction is signaled through control bits to continue the computation until the IEEE-754-compliant result is reached.

Table 12 gives a summary of the approximation operations discussed in this section. In Table 12, the first column contains the description of the operations. The symbols a_i and b_i represent the subwords from the two source registers. The symbol c_i represents the corresponding subword in the target register. Other functions used in the table are as follows:

- $approx(op(x))$ is a function that returns an approximation to the actual result of $op(x)$.
- $est(op(x))$ is an estimate to the actual result of $op(x)$ and is generated by the corresponding $approx(op(x))$ operation.
- $refine(est(op(x)))$ is a function that refines $est(op(x))$.

The following example uses the 3DNow! instructions and it illustrates how to approximate a reciprocal calculation and how to further refine the result.

Example

Assume that register R_a contains the SP FP number a, and we want to calculate $1/a$ by first approximating the result and then by refining this estimate to achieve the desired SP accuracy. The first step requires the use of the PFRCP instruction:

   PFRCP R_b, R_a

Table 12 Approximation Operations

FP Operations	IA-64	MAX-2	MMX	SSE-2	3DNow!	AltiVec
$c_i = approx(1/a)$					√	
$c_i = approx(1/\sqrt{a})$					√	
$c_i = approx(1/a_i)$	√			√ [a]		√
$c_i = approx(1/\sqrt{a_i})$	√			√ [a]		√
$c_i = approx(\log_2 a_i)$						√
$c_i = approx(2a_i)$						√
$c_i = refine(est(1/a))$					√	
$c_i = refine(est(1/\sqrt{a}))$					√	

[a] Scalar versions of these operations also exist.

This gives a low-precision estimate of $1/a$. The result, which gets written into register R_b, is accurate to 14 bits in its significand field, compared to the 24-bit accuracy of a SP value. Increased accuracy requires the use of two additional instructions:

> PFRCPIT1 R_c, R_b, R_a
> PFRCPIT2 R_d, R_c, R_b

The first stage of refinement uses the PFRCPIT1 instruction with inputs to the instruction as R_a and R_b. PFRCPIT1 performs the first intermediate step of the Newton–Raphson algorithm to refine the approximation produced by PFRCP. The result of PFRCPIT1 instruction is written to R_c. The second and final step of the Newton–Raphson algorithm is performed by the PFRCPIT2 instruction. The inputs to the PFRCPIT2 are R_b and R_c. The result of the PFRCPIT2 instruction has the accuracy of a SP FP number, which is 24 bits in the significand field. Hence, at the end of three instructions, the value of $1/a$ is calculated to SP accuracy.

There is another instruction in the SSE-2 architecture that can be included in this section (Table 13). This is the *packed square-root* instruction, which operates on packed SP/DP FP operands and computes their square roots to SP/DP accuracy. However, this instruction is not an approximation instruction and, therefore, it requires more execution cycles to complete compared to an approximation instruction.

9 SUMMARY

We have described the latest multimedia instructions that have been added to current microprocessor instruction set architectures (ISAs) for native signal processing, or, more generally, for multimedia processing. We described these instructions by broad classes: *packed add/subtract* operations, *packed special arithmetic* operations, *packed multiply* operations, *packed compare* operations, *packed shift* operations, *subword rearrangement* operations, and *approximation* operations. For each of these instruction classes, we compared the instructions

Table 13 SSE-2 Features a Packed Instruction for Computing the Square Root of a FP Register

FP operation	IA-64	MAX-2	MMX	SSE-2	3DNow!	AltiVec
$c_i = \sqrt{a_i}$				$\sqrt{}$[a], SP, DP		

Note: The values in the source register can be SP or DP FP numbers.
[a] Scalar versions of these operations also exist.

in IA-64 [3], MMX [4], and SSE-2 [5] from Intel, MAX-2 [6,7] from Hewlett-Packard, 3DNow! [8,9] from AMD, and AltiVec [10] from Motorola.

The common theme in all these multimedia instructions is the implementation of subword parallelism. This is implemented for packed integers or fixed-point numbers in the integer datapaths, and for packed single-precision floating-point numbers in the floating-point datapaths. Visual multimedia data like images, video, graphics rendering, and animation involve pixel processing, which can fully exploit subword parallelism on the integer datapath. Higher-fidelity audio processing and graphics geometry processing require single-precision floating-point computations, which exploit subword parallelism on the floating-point datapath. Typical DSP operations like *multiply and accumulate* have also been added to the multimedia repertoire of general-purpose microprocessors. These multimedia instructions have embedded DSP and visual processing capabilities into general-purpose microprocessors, providing native signal processing and media processing [1] capabilities. In fact, most DSPs and media processors have also adopted subword parallelism in their architectures, as well as subword permutation instructions and other features often first introduced in microprocessors for native signal and media processing.

We see two trends in these multimedia ISAs. The first, "less is more" trend, is represented by the minimalist architecture approach of MAX-2, which adheres to the RISC architectural principles of defining as few instructions as necessary for high performance, with each instruction executable in a single pipeline cycle. The second, "more is better" trend, is represented by AltiVec, where very complex sequences of operations are represented by a single multimedia instruction, with such an instruction taking many cycles for execution. An example is the matrix multiply instruction, VSUMMBM, in AltiVec (see Fig. 24). More experience and evaluation of these instructions in multimedia processing applications can shed light on the effectiveness of these instructions. In the end, these two trends represent different stylistic preferences, akin to RISC and CISC instruction set architectural preferences. In fact, sometimes, RISC-like multimedia instructions have been added to CISC processor ISAs, and CISC-like multimedia instructions to RISC processor ISAs. The remarkable fact is that subword parallel multimedia instructions have achieved such rapid and pervasive adoption in both RISC and CISC microprocessors, DSPs, and media processors, attesting to their undisputed cost-effectiveness in accelerating multimedia processing in software.

REFERENCES

1. RB Lee, M Smith. Media processing: A new design target. IEEE Micro 16(4):6–9, 1996.

2. RB Lee. Efficiency of microSIMD architectures and index-mapped data for media processors. Proceedings of Media Processors 1999, IS&T/SPIE Symposium on Electric Imaging: Science and Technology, 1999, pp 34–46.

3. Intel. IA-64 Architecture Software Developer's Manual, Volume 3: Instruction Set Reference, Revision 1.1. Santa Clara, CA: Intel, 2000.

4. Intel. Intel Architecture Software Developer's Manual, Volume 2: Instruction Set Reference. Santa Clara, CA: Intel, 1999.

5. Intel. IA-32 Intel Architecture Software Developer's Manual with Preliminary Wilamette Architecture Information, Volume 2: Instruction Set Reference. Santa Clara, CA: Intel, 2000.

6. G Kane. PA-RISC 2.0 Architecture. Engelwood Cliffs, NJ: Prentice-Hall, 1996.

7. RB Lee. Subword parallelism with MAX-2. IEEE Micro 16(4):51–59, 1996.

8. AMD. 3DNow! Technology Manual. Sunnyvale, CA: AMD, 2000.

9. AMD. AMD Extensions to the 3DNow! and MMX Instruction Sets Manual. Sunnyvale, CA: AMD, March 2000.

10. Motorola. AltiVec Technology Programming Environments Manual, Revision 0.1. Denver, CO: Motorola, 1998.

11. RB Lee. Multimedia extensions for general-purpose processors. Proceedings of IEEE Signal Processing Systems '97, 1997, pp 9–23.

12. RB Lee. Accelerating multimedia with enhanced microprocessors. IEEE Micro 15(2):22–32, 1995.

13. RB Lee. Precision architecture. IEEE Computer 22(1):78–91, 1989.

14. V Bhaskaran, K Konstantinides, RB Lee, JP Beck. Algorithmic and architectural enhancements for real-time MPEG-1 decoding on a general purpose RISC workstation. IEEE Trans Circuits Syst Video Technol 5(5):380–386, 1995.

15. RB Lee, JP Beck, J Lamb, KE Severson. Real-time software MPEG video decoder on multimedia-enhanced PA7100LC processors. Hewlett-Packard J 46(2):60–68, April 1995.

16. M Tremblay, JM O'Connor, V Narayanan, H Liang. VIS speeds new media processing. IEEE Micro 16(4):10–20, 1996.

4

Reconfigurable Computing and Digital Signal Processing: Past, Present, and Future

Russell Tessier and Wayne Burleson
University of Massachusetts, Amherst, Massachusetts

1 INTRODUCTION

Throughout the history of computing, digital signal processing (DSP) applications have pushed the limits of computer power, especially in terms of real-time computation. Although processed signals have broadly ranged from media-driven speech, audio, and video waveforms to specialized radar and sonar data, most calculations performed by signal processing systems have exhibited the same basic computational characteristics. The inherent data parallelism found in many DSP functions has made DSP algorithms ideal candidates for hardware implementation, leveraging expanding VLSI (very-large-scale integration) capabilities. Recently, DSP has received increased attention due to rapid advancements in multimedia computing and high-speed wired and wireless communications. In response to these advances, the search for novel implementations of arithmetic-intensive circuitry has intensified.

Although application areas span a broad spectrum, the basic computational parameters of most DSP operations remain the same: a need for real-time performance within the given operational parameters of a target system and, in most cases, a need to adapt to changing datasets and computing conditions. In general, the goal of high performance in systems ranging from low-cost embedded radio components to special-purpose ground-based radar centers has driven the development of application and domain-specific chip sets. The development and financial cost of this approach is often large, motivating the need for new approaches to computer architecture that offer the same computational attributes as fixed-functionality architectures in a package that can be customized in the field. The second goal of system adaptability is generally addressed through the use of software-programmable, commodity digital signal processors. Although these platforms enable flexible deployment due to software development tools and great economies of scale, application designers and compilers must customize their processing approach to available computing resources. This flexibility often comes at the cost of performance and power efficiency.

As shown in Figure 1, *reconfigurable* computers offer a compromise between the performance advantages of fixed-functionality hardware and the flexibility of software-programmable substrates. Like application-specific integrated circuits (ASICs), these systems are distinguished by their ability to directly implement specialized circuitry directly in hardware. Additionally, like programmable processors, reconfigurable computers contain functional resources that may be modified easily after field deployment in response to changing operational parameters and datasets. To date, the core processing element of most reconfigurable computers has been the field programmable gate array (FPGA). These bit-programmable computing devices offer ample quantities of logic and register resources that can easily be adapted to support the fine-grained parallelism of many pipelined DSP applications. With current logic capacities exceeding 1 million gates per device, substantial logic functionality can be implemented on each programmable device. Although appropriate for some classes of implementation,

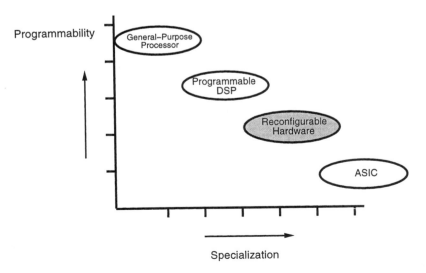

Figure 1 DSP implementation spectrum.

FPGAs represent only one possible implementation in a range of possible reconfigurable computing building blocks. A number of reconfigurable alternatives are presently under evaluation in academic and commercial environments.

In this survey, the evolution of reconfigurable computing with regard to digital signal processing is considered. This study includes an historical evaluation of reprogrammable architectures and programming environments used to support DSP applications. The chronology is supported with specific case studies which illustrate approaches used to address implementation constraints such as system cost, performance, and power consumption. It is seen that as technology has progressed, the richness of applications supported by reconfigurable computing and the performance of reconfigurable computing platforms have improved dramatically. Reconfigurable computing for DSP remains an active area of research as the need for integration with more traditional DSP technologies such as PDSPs becomes apparent and the goal of automated high-level compilation for DSP increases in importance.

The organization of this chapter is as follows. In Section 2, a brief history of the issues and techniques involved in the design and implementation of DSP systems is described. Section 3 presents a short history of reconfigurable computing. Section 4 describes why reconfigurable computing is a promising approach for DSP systems. Section 5 serves as the centerpiece of the chapter and provides a history of the application of various reconfigurable computing technologies to DSP systems and a discussion of the current state of the art. We conclude in Section 6 with some predictions about the future of reconfigurable computing for digital signal processing. These predictions are formulated by extrapolating the trends of reconfigurable technologies and describing future DSP applications which may be targeted to reconfigurable hardware.

1.1 Definitions

The following definitions are used to describe various attributes related to reconfigurable computing:

- *Reconfigurable or adaptive*: In the context of reconfigurable computing, this term indicates that the logic functionality and interconnect of a computing system or device can be customized to suit a specific application through postfabrication, user-defined programming.
- *Run-time (or dynamically) reconfigurable*: System logic functionality and/or interconnect connectivity can be modified during application execution. This modification may be either data driven or statically scheduled.
- *Fine-grained parallelism*: Logic functionality and interconnect connectivity is programmable at the bit level. Resources encompassing multiple logic bits may be combined to form parallel functional units.

- *Specialization*: Logic functionality can be customized to perform exactly the operation desired. An example is the synthesis of filtering hardware with a fixed constant value.

2   BACKGROUND IN DSP IMPLEMENTATION

2.1   DSP System Implementation Choices

Since the early 1960s, three goals have driven the development of DSP implementations: (1) data parallelism, (2) application-specific specialization, and (3) functional flexibility. In general, design decisions regarding DSP system implementation require trade-offs between these three system goals. As a result, a wide variety of specialized hardware implementations and associated design tools have been developed for DSP, including associative processing, bit-serial processing, on-line arithmetic, and systolic processing. As implementation technologies have become available, these basic approaches have matured to meet the needs of application designers.

As shown in Table 1, various cost metrics have been developed to compare the quality of different DSP implementations. Performance has frequently been the most critical system requirement because DSP systems often have demanding real-time constraints. In the past two decades, however, cost has become more significant as DSP has migrated from predominantly military and scientific applications into numerous low-cost consumer applications. Over the past 10 years, energy consumption has become an important measure as DSP techniques have been widely applied in portable, battery-operated systems such as cell phones, CD players, and laptops [1]. Finally, flexibility has emerged as one of the key differentiators in DSP implementations because it allows changes to system functionality at various points in the design life cycle. The results of

Table 1   DSP Implementation Comparison

	Performance	Cost	Power	Flexibility	Design effort (NRE)
ASIC	High	High	Low	Low	High
Programmable DSP	Medium	Medium	Medium	Medium	Medium
General-purpose processor	Low	Low	Medium	High	Low
Reconfigurable hardware	Medium	Medium	High	High	Medium

these cost trade-offs have resulted in four primary implementation options, including ASICs, programmable digital signal processors (PDSPs), general-purpose microprocessors, and reconfigurable hardware. Each implementation option presents different trade-offs in terms of performance, cost, power, and flexibility.

For many specialized DSP applications, system implementation must include one or more ASICs to meet performance and power constraints. Even though ASIC design cycles remain long, a trend toward automated synthesis and verification tools [2] is simplifying high-level ASIC design. Because most ASIC specification is done at the behavioral or register-transfer level, the functionality and performance of ASICs have become easier to represent and verify. Another, perhaps more important, trend has been the use of predesigned cores with well-defined functionality. Some of these cores are, in fact, PDSPs or reduced instruction set computer (RISC) microcontrollers, for which software has to be written and then stored on-chip. ASICs have a significant advantage in area and power, and for many high-volume designs, the cost-per-gate for a given performance level is less than that of high-speed commodity FPGAs. These characteristics are especially important for power-aware functions in mobile communication and remote sensing. Unfortunately, the fixed nature of ASICs limits their *reconfigurability*. For designs that must adapt to changing datasets and operating conditions, software-programmable components must be included in the target system, reducing available parallelism. Additionally, for low-volume or prototype implementations, the nonrecurring engineering (NRE) costs related to an ASIC may not justify its improved performance benefits.

The application domain of PDSPs can be identified by tracing their development lineage. Thorough summaries of programmable DSPs can be found in Refs. 3–5. In the 1980s, the first PDSPs were introduced by Texas Instruments. These initial processor architectures were primarily CISC (complex-instruction-set computer) pipelines augmented with a handful of special architectural features and instructions to support filtering and transform computations. One of the most significant changes to second-generation PDSPs was the adaptation of the Harvard architecture, effectively separating the program bus from the data bus. This optimization reduced the von Neumann bottleneck, thus providing an unimpeded path for data from local memory to the processor pipeline. Many early DSPs allowed programs to be stored in on-chip ROM and supported the ability to make off-chip accesses if instruction capacity was exceeded. Some DSPs also had coefficient ROMs, again recognizing the opportunity to exploit the relatively static nature of filter and transform coefficients.

Contemporary digital signal processors are highly programmable resources that offer the capability for in-field update as processing standards change. Parallelism in most PDSPs is not extensive but generally consists of overlapped data fetch, data operation, and address calculation. Some instruction set modifications are also used in PDSPs to specialize for signal processing. Addressing modes are provided to simplify the implementation of filters and transforms and, in general, control overhead for loops is minimized. Arithmetic instructions for fixed-point computation allow saturating arithmetic, which is important for avoiding overflow exceptions or oscillations. New hybrid DSPs contain a variety of processing and input/output (I/O) features, including parallel processing interfaces, very-long-instruction-word (VLIW) function unit scheduling, and flexible datapaths. Through the addition of numerous, special-purpose memories, on-chip DSPs can now achieve high-bandwidth and, to a moderate extent, reconfigurable interconnect. Due to the volume usage of these parts, costs are reduced and commonly used interfaces can be included. In addition to these benefits, the use of a DSP has specific limitations. In general, for optimal performance, applications must be written to utilize the resources available in the DSP. Although high-level compilation systems which perform this function are becoming available [6,7], often it is difficult to get exactly the mapping desired. Additionally, the interface to memory may not be appropriate for specific applications, creating a bandwidth bottleneck in getting data to functional units.

The 1990s have been characterized by the introduction of DSP to the mass commercial market. DSP has made the transition from a fairly academic acronym to one seen widely in advertisements for consumer electronics and software packages. A battle over the DSP market has ensued primarily between PDSP manufacturers, ASIC vendors, and developers of two types of general-purpose processor, desktop microprocessors and high-end microcontrollers. General-purpose processors, such as the Intel Pentium, can provide much of the signal processing needed for desktop applications such as audio and video processing, especially because the host microprocessor is already resident in the system and has highly optimized I/O and extensive software development tools. However, general-purpose desktop processors are not a realistic alternative for embedded systems due to their cost and lack of power efficiency in implementing DSP. Another category of general-purpose processors is the high-end microcontroller. These chips have also made inroads into DSP applications by presenting system designers with straightforward implementation solutions that have useful data interfaces and significant application-level flexibility.

One DSP hardware implementation compromise that has developed recently has been the development of domain-specific standard products in both programmable and ASIC formats. The PDSP community has determined that because certain applications have a high volume, it is worthwhile to tailor particular PDSPs to domain-specific markets. This has led to the availability of inexpensive, commodity silicon while allowing users to provide application differentiation in software. ASICs have also been developed for more general functions like MPEG decoding, in which standards have been set up to allow a large number of applications to use the same basic function.

Reconfigurable computing platforms for DSP offer an intermediate solution to ASICs, PDSPs, and general and domain-specific processors by allowing reconfigurable and specialized performance on a per-application basis. Although this emerging technology has primarily been applied to experimental rather than commercial systems, the application-level potential for these reconfigurable platforms is great. Following an examination of the needs of contemporary DSP applications, current trends in the application of reconfigurable computing to DSP are explored.

2.2 The Changing World of DSP Applications

Over the past 30 years, the application space of digital signal processing has changed substantially, motivating new systems in the area of reconfigurable computing. New applications over this time span have changed the definition of DSP and have created new and different requirements for implementation. For example, in today's market, DSP is often found in human–computer interfaces such as sound cards, video cards, and speech recognition system—application areas with limited practical significance just a decade ago. Because a human is an integral part of these systems, different processing requirements can be found, in contrast to communications front ends such as those found in DSL modems from Broadcom [8] or CDMA (code division multiple access) receiver chips from Qualcomm [9]. Another large recent application of DSP has been in the read circuitry of hard-drive and CD/DVD storage systems [10]. Although many of the DSP algorithms are the same as in modems, the system constraints are quite different.

Consumer products now make extensive use of DSP in low-cost and low-power implementations [11]. Both wireless and multimedia, two of the hottest topics in consumer electronics, rely heavily on DSP implementation. Cellular telephones, both GSM (global system for mobile communication) and CDMA, are currently largely enabled by custom silicon [12], although trends toward other implementation media such as PDSPs are growing. Modems for DSL, cable, local area networks (LANs), and, most recently, wireless all rely on sophisticated adaptive equalizers and receivers. Satellite set-top boxes rely on DSP for satellite reception using channel decoding as well as an MPEG decoder ASIC for video decompression. After the set-top box, the DVD player has now emerged as the fastest-growing consumer electronics product. The DVD player relies on DSP to avoid intersymbol interference, allowing more bits to be packed into a given area of disk. In the commercial video market, digital cameras and camcorders are rapidly becoming affordable alternatives to traditional analog cameras, largely supported by photo-editing, authoring software, and the Web.

Development of a large set of DSP systems has been driven indirectly by the growth of consumer electronics. These systems include switching stations for cellular, terrestrial, satellite and cable infrastructure as well as cameras, authoring studios, and encoders used for content production. New military and scientific applications applied to the digital battlefield, including advanced weapons systems and remote sensing equipment, all rely on DSP implementation that must operate reliably in adverse and resource-limited environments. Although existing DSP implementation choices are suitable for all of these consumer and military-driven applications, higher performance, efficiency, and flexibility will be needed in the future, driving current interest in reconfigurable solutions.

In all of these applications, data processing is considerably more sophisticated than the traditional filters and transforms which characterized DSP of the 1960s and 1970s. In general, performance has grown in importance as data rates have increased and algorithms have become more complex. Additionally, there is an increasing demand for flexible and diverse functionality based on environmental conditions and workloads. Power and cost are equally important because they are critical to overall system cost and performance.

Although new approaches to application-specific DSP implementation have been developed by the research community in recent years, their application in practice has been limited by the market domination of PDSPs and the reluctance of designers to expose schedule and risk-sensitive ASIC projects to nontraditional design approaches. Recently, however, the combination of new design tools and the increasing use of intellectual property cores [13] in DSP implementations have allowed some of these ideas to find wider use. These implementation choices include systolic architectures, alternative arithmetic (residue number system [RNS], logarithmic number system [LNS], digital-serial), word-length optimization, parallelizing transformations, memory partitioning, and power optimization techniques. Design tools have also been proposed which could close the gap between software development and hardware development for future hybrid DSP implementations. In subsequent sections, it will be seen that these tools will be helpful in defining the appropriate application of reconfigurable hardware to existing challenges in DSP. In many cases, basic design techniques used to develop ASICs or domain-specific devices can be reapplied to customize applications in programmable silicon by taking the limitations of the implementation technology into account.

3 A BRIEF HISTORY OF RECONFIGURABLE COMPUTING

Since their introduction in the mid-1980s, field programmable gate arrays (FPGAs) have been the subject of extensive research and experimentation. In this section, reconfigurable device architecture and system integration is investigated with an eye toward identifying trends likely to affect future development. Although this summary provides sufficient background to evaluate the impact of reconfigurable hardware on DSP, more thorough discussions of FPGAs and reconfigurable computing can be found in Refs. 14–17.

3.1 Field Programmable Devices

The concept of a digital hardware device which supports programmable logic was originated in the early 1960s with the introduction of cellular arrays. These devices contained built-in logic structures whose functionality could be set either in the final stages of production or in the field. Early cellular arrays, such as the Maitra cascade [18], contained extremely simple logic cells and supported linear, near-neighbor interblock connectivity. Each cell could generally perform a single-output Boolean function of two inputs which was determined through a programmable mask set late in the device fabrication process. Field programmable technology became a reality in the mid-1960s with the introduction of *cutpoint* cellular logic [19]. Like Maitra cascades, these devices contained a fixed interconnection between cells, but the logic functionality of each cell could be programmed in the field. Customization was typically accomplished by blowing programmable cell fuses through the use of programming currents or photoconductive exposure [19]. A direct forerunner of today's SRAM-based FPGA was a programmable array proposed and implemented by Wahlstrom [20] in 1967. Like today's FPGA devices, the operation of each logic cell was controlled by a user-defined bit stream which determined both internal logic functionality and connectivity to adjacent intercell wires and buses. The array could be reprogrammed to implement a variety of logic circuits and to accommodate in-field operational faults. Extensions and analysis of Wahlstrom's array were later documented in Ref. 21.

The modern era of reconfigurable computing was ushered in by the introduction of the first commercial SRAM-based FPGAs by Xilinx Corporation [22] in 1986. These early reprogrammable devices and subsequent offerings from both Xilinx and Altera Corporation [23] contain a collection of fine-grained programmable logic blocks interconnected via wires and programmable switches. Logic functionality for each block is specified via a small programmable memory, called a *look-up table*, driven by a limited number of inputs (typically less than five) which generates a single Boolean output. Additionally, each logic block typically contains one or more flip-flops for fine-grained storage. Although early FPGA architectures contained small numbers of logic blocks (typically less than 100), new device families have quickly grown to capacities of tens of thousands of look-up tables containing millions of gates of logic. As shown in Figure 2, fine-grained look-up table/flip-flop pairs are frequently grouped into tightly connected coarse-grained blocks to take advantage of circuit locality. Interconnection be-

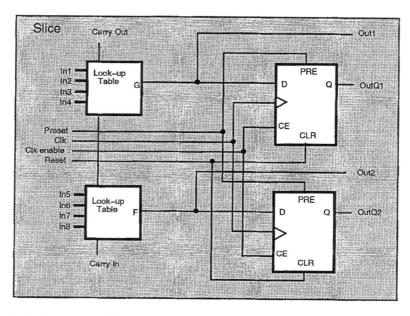

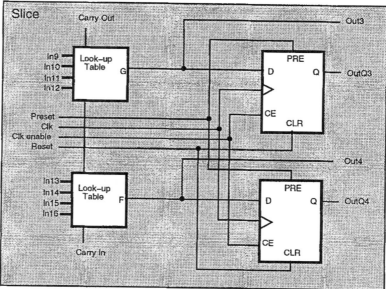

Figure 2   Simplified Xilinx Virtex logic block. Each logic block consists of two 2-LUT (look-up table) slices. (From Ref. 26.)

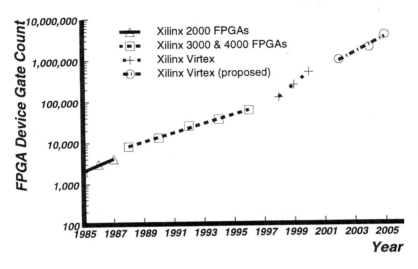

Figure 3 Growth of FPGA gate capacity.

tween logic blocks is provided via a series of wire segments located in channels between the blocks. Programmable pass transistors and multiplexers can be used to provide both block-to-segment connectivity and segment-to-segment connections.

Much of the recent interest in reconfigurable computing has been spurred by the development and maturation of field programmable gate arrays. The recent development of systems based on FPGAs has been greatly enhanced by an exponential growth rate in the gate capacity of reconfigurable devices and improved device performance due to shrinking die sizes and enhanced fabrication techniques. As shown in Figure 3, reported gate counts [24–26] for look-up table (LUT)-based FPGAs, from companies such as Xilinx Corporation, have roughly followed Moore's law over the past decade.* This increase in capacity has enabled complex structures such as multitap filters and small RISC processors to be implemented directly in a single FPGA chip. Over this same time period, the system performance of these devices has also improved exponentially. Whereas in the mid-1980s, system-level FPGA performance of 2–5 MHz was considered acceptable, today's LUT-based FPGA designs frequently approach performance

---

* In practice, usable gate counts for devices are often significantly lower than reported data book values (by about 20–40%). Generally, the proportion of per-device logic that is usable has remained roughly constant over the years, as indicated in Figure 3.

levels of 60 MHz and beyond. Given the programmable nature of reconfigurable devices, the performance penalty of a circuit implemented in reprogrammable technology versus a direct ASIC implementation is generally a factor on the order of 5 to 10.

3.2 Early Reprogrammable Systems

The concept of using reprogrammable logic to enhance the functional capabilities of a computing system is generally credited to Gerald Estrin [27]. In a feasibility study performed in the early 1960s, a digital system is described that contains both a sequential processor and a programmable logic core which can change logic functionality on a per-application basis. Even though a functioning hardware system based on the concept was not built, the study outlined the potential of application-level specialization of system hardware. Estrin's work motivated the later analysis of the use of cellular arrays for basic-block-level computation [28]. In this subsequent study, the potential of reconfigurability for use in design verification and algorithm development is addressed, setting the stage for contemporary multi-FPGA prototyping and development platforms.

Soon after the commercial introduction of the FPGA, computer architects began devising approaches for leveraging new programmable technology in computing systems. As summarized in Ref. 16, the evolution of reconfigurable computing was significantly shaped by two influential projects: Splash II [29] and Programmable Active Memories (PAM) [30]. Each of these projects addressed important programmable system issues regarding programming environment, user interface, and configuration management by applying pre-existing computational models in the areas of special-purpose coprocessing and statically scheduled communication to reconfigurable computing.

Splash II is a multi-FPGA parallel computer which uses orchestrated systolic communication to perform inter-FPGA data transfer. As shown in Figure 4, each board of multiboard Splash II systems contains 16 Xilinx XC4000 series FPGA processors (labeled with an X prefix), each with associated SRAM (labeled with an M prefix). Unlike its multi-FPGA predecessor, Splash [31], which was limited to strictly near-neighbor systolic communication, each Splash II board contains inter-FPGA crossbars for multihop data transfer and broadcast. Software development for the system typically involves the creation of VHDL (VHSIC hardware description language) circuit descriptions for individual systolic processors. These designs must meet size and performance constraints of the target FPGAs. Following processor creation, high-level inter-FPGA scheduling software is used to ensure that systemwide communication is synchronized. In general, the system is not dynamically reconfigured during operation. For applications with single instruction multiple data (SIMD) characteristics, a compiler [32] has been created to automatically partition processing across FPGAs and to synchronize interfaces to

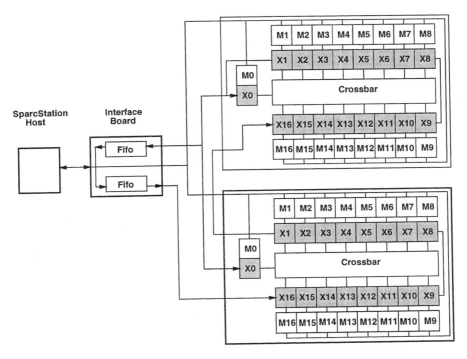

Figure 4   Two-board Splash II system. (From Ref. 29.)

local SRAMs. Numerous DSP applications have been mapped to Splash II, including audio and video algorithm implementations. These applications are described in greater detail in Section 5. Recently, FPGA-based systolic architectures based on the Splash II system have been developed by Annapolis Micro Systems [33]. The company's peripheral component interface (PCI) based Wildforce system contains five Xilinx XC4000XL devices aligned in a systolic chain. A similar, VME-based Wildstar board contains four Xilinx Virtex devices.

As shown in Figure 5, Programmable active memory DECPeRLe-1 system [30] contain arrangements of FPGA processors (labeled X) in a two-dimensional mesh with memory devices (labeled M) aligned along the array perimeter. PAMs were designed to create the architectural appearance of a functional memory for a host microprocessor and the PAM programming environment reflects this. From a programming standpoint, the multi-FPGA PAM can be accessed like a memory through an interface FPGA, XI, with written values treated as inputs and read values used as results. Designs are generally targeted to PAMs through hand-crafting of design subtasks, each appropriately sized to fit on an FPGA. The PAM

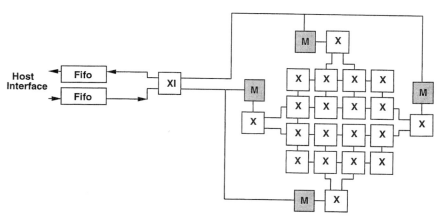

Figure 5   Programmable active memory DECPeRLe-1 system. (From Ref. 30.)

array and its successor, the Pamette [34], are interfaced to a host workstation through a backplane bus. Additional discussion of PAMs with regard to DSP applications appears in Section 5.

3.3   Reconfigurable Computing Research Directions

Over the past decade, interest in reconfigurable systems has progressed along four main paths [15]:

1. The proximity of reconfigurable hardware to a host CPU
2. The capability of hardware to support dynamic reconfiguration
3. Software support for high-level compilation and dynamic reconfiguration
4. The granularity of reconfigurable elements

Active research in these areas continues today in addition to a search for applications well-suited to the available architectural parameters.

As a result of the Prism I project [35], the first reconfigurable system which tightly coupled an off-the-shelf processor with an FPGA coprocessor was created. This project explored the possibility of augmenting the instruction set of a processor with special-purpose instructions that could be executed by an attached FPGA coprocessor in place of numerous processor instructions. For these instructions, the microprocessor would stall for several cycles while the FPGA-based coprocessor completed execution. More recently, the single-chip Napa [36] and OneChip [37] architectures have used similar approaches to synchronize processing between RISC processors and FPGA cores. As chip integration levels have increased, interest in tightly coupling both processor and reconfigurable resources at multiple architectural levels has grown. Single-chip architectures, such as Garp [38], now allow interfacing between processors and reconfigurable resources, both through coprocessor interfaces and through a shared data cache. A second approach to integrating reconfigurable logic and microprocessors has explored integrating reconfigurable logic *inside* the processor as special-purpose functional units. Although early approaches in this area attempted to keep reconfigurable functional unit timing consistent with other nonconfigurable resources [39], newer reconfigurable functional units [40] allow multicycle operation synchronized by the microprocessor control path.

An important aspect of reconfigurable devices is the ability to reconfigure functionality in response to changing operating conditions and application datasets. Although SRAM-based FPGAs have supported slow millisecond reconfiguration rates for some time, only recently have devices been created that allow for rapid device reconfiguration at run time. Dynamically reconfigurable FPGAs, or DPGAs [41,42], contain multiple interconnect and logic configurations for each programmable location in a reconfigurable device. Often these architectures are designed to allow configuration switching in a small number of system clock cycles, measuring nanoseconds rather than milliseconds. Although several DPGA devices have been developed in research environments, only one has been developed commercially. The Context Switching FPGA [43], developed commercially by Sanders Corporation, can simultaneously hold up to four complete configuration contexts. A context switch for the device can be performed in a single clock cycle. During the context switch, all internal data stored in registers are preserved. To promote reconfiguration at lower hardware cost, several commercial FPGA families [26,44] have been introduced that allow for fast, partial reconfiguration of FPGA functionality from off-chip memory resources. A significant challenge to the use of these reconfigurables is the development of compilation software which will partition and schedule the order in which computation will take place and will determine which circuitry must be changed. Although some preliminary work in this area has been completed [45,46], more advanced tools are needed to fully leverage the new hardware technology. Other software approaches that have been applied to dynamic reconfiguration include the definition of hardware subroutines [47] and the dynamic reconfiguration of instruction sets [48].

Although high-level compilation for microprocessors has been an active research area for decades, development of compilation technology for reconfigurable computing is still in its infancy. The compilation process for FPGA-based system is often complicated by a lack of identifiable coarse-grained structure in fine-grained FPGAs and the dispersal of logic resources across many pin-limited reconfigurable devices on a single computing platform. In particular, because most reconfigurable computers contain multiple programmable devices, design *partitioning* forms an important aspect of most compilation systems. Several compilation systems for reconfigurable hardware [49,50] have followed a traditional multidevice ASIC design flow involving pin-constrained device partitioning and individual device synthesis using RTL compilation. To overcome pin limitations and achieve full logic utilization on a per-device basis using this approach, either excessive internal device interconnect [49] or I/O counts [51] have been needed. In Ref. 52, a hardware virtualization approach is outlined that promotes high per-device logic utilization. Following design partitioning and placement, inter-FPGA wires are scheduled on interdevice wires at compiler-determined time slices, allowing pipelining of communication. Interdevice pipelining also forms the basis of several FPGA system compilation approaches that start at the behavioral level. A high-level synthesis technique described in Ref. 53 outlines inter-FPGA scheduling at the RTL level. In Refs. 54 and 55, functional allocation is performed that takes into account the amount of logic available in the target system and available interdevice interconnect. Combined communication *and* functional resource scheduling is then performed to fully utilize available logic and communication resources. In Ref. 56, inter-FPGA communication and FPGA-memory communication are virtualized because it is recognized that memory rather than inter-FPGA bandwidth is frequently the critical resource in reconfigurable systems. In Ref. 57, linear programming is used to partition MATLAB functions across sets of heterogeneous resources, including DSPs, RISC processors, and FPGAs. Scheduling, pipelining, and component-specific compilation are performed following partitioning to complete the mapping process.

4 THE PROMISE OF RECONFIGURABLE COMPUTING FOR DSP

Many of the motivations and goals of reconfigurable computing are consistent with the needs of signal processing applications. It will be seen in Section 5 that the deployment of DSP algorithms on reconfigurable hardware has aided in the advancement of both fields over the past 15 years. In general, the direct benefits of the reconfigurable approach for DSP can be summarized in three critical areas: functional specialization, platform reconfigurability, and fine-grained parallelism.

4.1 Specialization

As stated in Section 2.1, programmable digital signal processors are optimized to deliver efficient performance across a set of signal processing tasks. Although the specific implementation of tasks can be modified through instruction-configurable software, applications must frequently be customized to meet specific processor architectural aspects, often at the cost of performance. Currently, most DSPs remain inherently sequential machines, although some parallel VLIW and multifunction unit DSPs have recently been developed [58]. The use of reconfigurable hardware has numerous advantages for many signal processing systems. For many applications, such as digital filtering, it is possible to customize irregular datapath widths and specific constant values directly in hardware, reducing implementation area and power and improving algorithm performance. Additionally, if standards change, the modifications can quickly be reimplemented in hardware without expensive NRE costs. Because reconfigurable devices contain SRAM-controlled logic and interconnect switches, application programs in the form of device configuration data can be downloaded on a per-application basis. Effectively, this single, wide program instruction defines hardware behavior. Contemporary reconfigurable computing devices have little or no NRE cost because off-the-shelf development tools are used for design synthesis and layout. Although reconfigurable implementations may exhibit a 5–10 times performance reduction compared to the same circuit implemented in custom logic, limited manual intervention is generally needed to map a design to a reconfigurable device. In contrast, substantial NRE costs require ASIC designers to focus on high-speed physical implementation often involving hand-tuned physical layout and near-exhaustive design verification. Time-consuming ASIC implementation tasks can also lead to longer time-to-market windows and increased inventory, effectively becoming the critical path link in the system design chain.

4.2 Reconfigurability

Most reconfigurable devices and systems contain SRAM-programmable memory to allow full logic and interconnect reconfiguration in the field. Despite a wide range of system characteristics, most DSP systems have a need for configurability under a variety of constraints. These constraints include environmental factors such as changes in statistics of signals and noise, channel, weather, transmission rates, and communication standards. Although factors such as data traffic and interference often change quite rapidly, other factors such as location and weather change relatively slowly. Still other factors regarding communication standards vary infrequently across time and geography, limiting the need for rapid reconfiguration. Some specific ways that DSP can directly benefit from hardware reconfiguration to support these factors include the following:

- *Field customization*: The reconfigurability of programmable devices allows periodic updates of product functionality as advanced vendor firmware versions become available or product defects are detected. Field customization is particularly important in the face of changing standards and communication protocols. Unlike ASIC implementations, reconfigurable hardware solutions can generally be quickly updated

based on application demands without the need for manual field up-
grades or hardware swaps.

- *Slow adaptation*: Signal processing systems based on reconfigurable
  logic may need to be periodically updated in the course of daily operation based on a variety of constraints. These include issues such as
  variable weather and operating parameters for mobile communication
  and support for multiple, time-varying standards in stationary receivers.
- *Fast adaptation*: Many communication processing protocols [59] require nearly constant re-evaluation of operating parameters and can
  benefit from rapid adjustment of computing parameters. Some of these
  issues include adaptation to time-varying noise in communication channels, adaptation to network congestion in network configurations, and
  speculative computation based on changing datasets.

4.3 Parallelism

An abundance of programmable logic facilitates the creation of numerous functional units directly in hardware. Many characteristics of FPGA devices, in particular, make them especially attractive for use in digital signal processing systems.
The fine-grained parallelism found in these devices is well matched to the high
sample rates and distributed computation often required of signal processing applications in areas such as image, audio, and speech processing. Plentiful FPGA
flip-flops and a desire to achieve accelerated system clock rates have led designers
to focus on heavily pipelined implementations of functional blocks and interblock
communication. Given the highly pipelined and parallel nature of many DSP
tasks, such as image and speech processing, these implementations have exhibited
substantially better performance than standard PDSPs. In general, these systems
have been implemented using both task and functional unit pipelining. Many
DSP systems have featured bit-serial functional unit implementations [60] and
systolic interunit communication [29] that can take advantage of the synchronization resources of contemporary FPGAs without the need for software instruction
fetch and decode circuitry. As detailed in Section 5, bit-serial implementations
have been particularly attractive due to their reduced implementation area. However, as reconfigurable devices increase in size, more nibble-serial and parallel
implementations of functional units have emerged in an effort to take advantage
of data parallelism.

Recent additions to reconfigurable architectures have aided their suitability
for signal processing. Several recent architectures [26,61] have included 2–4-
kbit SRAM banks that can be used to store small amounts of intermediate data.
This allows for parallel access to data for distributed computation. Another important addition to reconfigurable architectures has been the capability to rapidly
change only small portions of device configuration without disturbing existing

device behavior. This feature has recently been leveraged to help adapt signal processing systems to reduce power [62]. The speed of adaptation may vary depending on the specific signal processing application area.

5 HISTORY OF RECONFIGURABLE COMPUTING AND DSP

Since the appearance of the first reconfigurable computing systems, DSP applications have served as important test cases in reconfigurable architecture and software development. In this section, a wide range of DSP design approaches and applications that have been mapped to functioning reconfigurable computing systems are considered. Unless otherwise stated, the design of complete DSP systems is stressed, including I/O, memory interfacing, high-level compilation, and real-time issues rather than the mapping of individual benchmark circuits. For this reason, a large number of FPGA implementations of basic DSP functions like filters and transforms that have not been implemented directly in system hardware have been omitted. Although our consideration of the history of DSP and reconfigurable computing is roughly chronological, some noted recent trends were initially investigated a number of years ago. To trace these trends, recent advancements are directly contrasted with early contributions.

5.1 FPGA Implementation of Arithmetic

Soon after the introduction of the FPGA in the mid-1980s, an interest developed in using the devices for DSP, especially for digital filtering which can take advantage of specialized constants embedded in hardware. Because a large portion of most filtering approaches involves the use of multiplication, efficient multiplier implementations in both fixed and floating points were of particular interest. Many early FPGA multiplier implementations used circuit structures adapted from the early days of large-scale integration (LSI) development and reflected the restricted circuit area available in initial FPGA devices [55]. As FPGA capacities have increased, the diversity of multiplier implementations has grown.

　　Since the introduction of the FPGA, bit-serial arithmetic has been used extensively to implement FPGA multiplication. As shown in Figure 6, taken from [Ref. 55], bit-serial multiplication is implemented using a linear systolic array that is well suited to the fine-grained nature of FPGAs. Two data values are input into the multiplier, including a parallel value in which all bits are input simultaneously and a sequential value in which values are input serially. In general, a data sampling rate of one value every M clock cycles can be supported, where M is the input word length. Each cell in the systolic array is typically implemented using one to four logic blocks similar to the one shown in

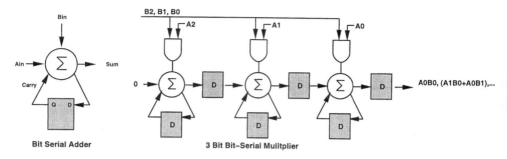

Figure 6   Bit-serial adder and multiplier. (From Ref. 55.)

Figure 2. Bit-serial approaches have the advantage that communication demands are independent of word length. As a result, low-capacity FPGAs can efficiently implement them. Given their pipelined nature, bit-serial multipliers implemented in FPGAs typically possess excellent area–time products. Many bit-serial formulations have been applied to finite impulse response filtering [63]. Special-purpose bit-serial implementations have included the canonic signed digit [64] and the power-of-2 sum or difference [65].

Given the dual use of look-up tables as small memories, distributed arithmetic (DA) has also been an effective implementation choice for LUT-based FPGAs. Because it is possible to group multiple LUTs together into a larger fanout memory, large LUTs for DA can easily be created. In general, distributed arithmetic requires the embedding of a fixed-input constant value in hardware, thus allowing the efficient precomputation of all possible dot-product outputs. An example of a distributed arithmetic multiplier, taken from Ref. 55, appears in Figure 7. It can be seen that a fast adder can be used to sum partial products based on nibble look-up. In some cases, it may be effective to implement the LUTs as RAMs so that new constants can be written during execution of the program.

To promote improved performance, several parallel arithmetic implementations on FPGAs have been formulated [55]. In general, parallel multipliers implemented in LUT-based FPGAs achieve a speedup of sixfold in performance when compared to their bit-serial counterparts with an area penalty of 2.5-fold. Specific parallel implementations of multipliers include a carry-save implementation [66], a systolic array with cordic arithmetic [67], and pipelined parallel [63,68,69].

As FPGA system development has intensified, more interest has been given to upgrading the accuracy of calculation performed in FPGAs, particularly through the use of floating-point arithmetic. In general, floating-point operations are difficult to implement in FPGAs due to the complexity of implementation

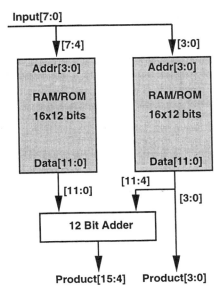

Figure 7 Distributed arithmetic multiplier. (From Ref. 55.)

and the amount of hardware needed to achieve desired results. For applications requiring extended precision, floating point is a necessity. In Ref. 70, an initial attempt was made to develop basic floating-point approaches for FPGAs that met IEEE-754 standards for addition and multiplication. Area and performance were considered for various FPGA implementations, including shift-and-add, carry-save, and combinational multiplier. Similar work was explored in Ref. 71, which applied 18-bit-wide floating-point adders/subtractors, multipliers, and dividers to 2D fast Fourier transform (FFT) and systolic FIR (finite impulse response) filters implemented on Splash II. This work was extended to a full 32-bit floating point in Ref. 72 for multipliers based on bit-parallel adders and digit-serial multipliers. More recent work [73] re-examines these issues with an eye toward greater area efficiency.

5.2 Reconfigurable DSP System Implementation

Although recent research in reconfigurable computing has been focused on advanced issues such as dynamic reconfiguration and special-purpose architecture, most work to date has been focused on the effective use of application parallelization and specialization. In general, a number of different DSP applications have been mapped to reconfigurable computing systems containing one, several, and many FPGA devices. In this subsection, a number of DSP projects that have been mapped to reconfigurable hardware are described. These implementations represent a broad set of DSP application areas and serve as a starting point for advanced research in years to come.

5.2.1   Image Processing Applications

The pipelined and fine-grained nature of reconfigurable hardware is a particularly good match for many image processing applications. Real-time image processing typically requires specialized datapaths and pipelining which can be implemented in FPGA logic. A number of projects have been focused in this application area. In Refs. 74 and 75, a set of image processing tasks mapped to the Splash II platform, described in Section 3.2, are outlined. Tasks such as Gaussian pyramid-based image compression, image filtering with 1D and 2D transforms, and image conversion using discrete fourier transform (DFT) operations are discussed. This work was subsequently extended to include the 2D discrete cosine transform (DCT) implemented on the Splash II platform in Ref. 76. The distributed construction of a stand-alone Splash II system containing numerous physical I/O ports is shown to be particularly useful in achieving high data rates. Because Splash II is effective in implementing systolic versions of algorithms that require repetitive tasks with data shifted in a linear array, image data can quickly be propagated in a processing pipeline. The targeted image processing applications are generally implemented as block-based systolic computations, with each FPGA operating as a systolic processor and groups of FPGAs performing specific tasks.

Additional reconfigurable computing platforms have also been used to perform image processing tasks. In Ref. 77, a commercial version of PAM, the turbochannel-based Pamette [34], is interfaced to a charge-coupled device (CCD) camera and a liquid-crystal polarizing filter is used to perform solar polarimetry. The activity of this application is effectively synchronized with software on an Alpha workstation. In Refs. 50 and 78, multi-FPGA systems are used to process 3D volume visualization data though ray casting. These implementations show favorable processing characteristics when compared to traditional microprocessor-based systems. In Ref. 79, a system is described in which a 2D DCT is implemented using a single FPGA device attached to a backplane bus-based processing card. This algorithm implementation uses distributed arithmetic and is initially coded in VHDL and subsequently compiled using RTL synthesis tools. In Ref. 80, a commercial multi-FPGA system is described that is applied to spatial median filtering. In Ref. 81, the application of a PCI-based FPGA board to 1D and 2D convolution is presented. Finally, in Ref. 82, a system implemented with a single-FPGA processing board is described that performs image interpolation. This system primarily uses bit-serial arithmetic and exploits dynamic reconfiguration to quickly swap portions of the computation located in the reconfigurable hardware. Each computational task has similar computational structure, so reconfiguration time of the FPGA is minimal.

5.2.2 Video Processing Applications

Like image processing, video processing requires substantial data bandwidth and processing capability to handle data obtained from analog video equipment. To support this need, several reconfigurable computing platforms have been adapted for video processing. The PAM system [30], described in Section 3.2, was the first platform used in video applications. A PAM system programmed to perform stereo vision was applied to applications requiring 3D elevation maps such as those needed for planetary exploration. A stereo-matching algorithm was implemented that was shown to be substantially faster than programmable DSP-based approaches. This implementation employed dynamic reconfiguration by requiring the reconfiguration of programmable hardware among three distinct processing tasks at run time. A much smaller single-FPGA system, described in Ref. 83, was focused primarily on block-based motion estimation. This system tightly coupled SRAM to a single FPGA device to allow for rapid data transfer.

An interesting application of FPGAs for video computation is described in Ref. 84. A stereo transform is implemented across 16 FPGA devices by aligning two images together to determine the depth between the images. Scan lines of data are streamed out of adjacent memories into processing FPGAs to perform the comparison. In an illustration of the benefit of a single-FPGA video system, in Ref. 85 a processing platform is described in which a T805 transputer is tightly coupled with an FPGA device to perform frame object tracking. In Ref. 86, a single-FPGA video coder, which is reconfigured dynamically among three different subfunctions (motion estimation, DCT, and quantization), is described. The key idea in this project is that the data located in hardware do not move, but rather the functions which operate on it are reconfigured in place.

5.2.3 Audio and Speech Processing

Whereas audio processing typically requires less bandwidth than video and image processing, audio applications can benefit from datapath specialization and pipelining. To illustrate this point, a sound synthesizer was implemented using the multi-FPGA PAM system [30], producing real-time audio of 256 different voices at up to 44.1 kHz. Primarily designed for the use of additive synthesis techniques based on look-up tables, this implementation included features to allow frequency modulation synthesis and/or nonlinear distortion and was also used as a sampling machine. The physical implementation of PAM as a stand-alone processing system facilitated interfacing to tape recorders and audio amplifiers. The system setup was shown to be an order-of-magnitude faster than a contemporary off-the-shelf DSP.

Other smaller projects have also made contributions in the audio and speech processing areas. In Ref. 87, a methodology is described to perform audio processing using a dynamically reconfigurable FPGA. Audio echo production is facilitated by dynamically swapping filter coefficients and parameters into the device from an adjacent SRAM. Third-party DSP tools are used to generate the coefficients. In Ref. 69, an inventive FPGA-based cross-correlator for radio astronomy is described. This system achieves high processing rates of 250 MHz inside the FPGA by heavily pipelining each aspect of the data computation. To support speech processing, a bus-based multi-FPGA board, Tabula Rasa [88], was programmed to perform Markov searches of speech phenomes. This system is particularly interesting because it allowed the use of behavioral partitioning and contained a codesign environment for specification, synthesis, simulation, and evaluation design phases.

5.2.4   Target Recognition

Another important DSP application that has been applied to Splash II is target recognition [89]. To support this application, images are broken into columns and compared to precomputed templates stored in local memory along with pipelined video data. As described in Section 3.2, near-neighbor communication is used with Splash II to compare pass-through pixels with stored templates in the form of partial sums. After an image is broken into pieces, the Splash II implementation performs second-level detection by roughly identifying sections of sub-images that conform to objects through the use of templates. In general, the use of FPGAs provides a unique opportunity to quickly adapt target recognition to new algorithms, something not possible with ASICs. In another FPGA implementation of target recognition, researchers [90] broke images into pieces called chips and analyzed them using a single FPGA device. By swapping target templates dynamically, a range of targets may be considered. To achieve high-performance design, templates were customized to meet the details of the target technology. In Ref. 91, a description is given of a novel software system that is used to map a high-level description of a target recognition algorithm to a multi-FPGA system. This software tool set converts algorithmic descriptions previously targeted to the Khoros [92] design environment into a format which can be loaded into a Wildforce system from Annapolis Micro Systems [33].

5.2.5   Communication Coding

In modern communication systems, signal-to-noise ratios make data coding an important aspect of communication. As a result, convolutional coding can be used to improve signal-to-noise ratios based on the constraint length of codes without increasing the power budget. Several reconfigurable computing systems have been configured to aid in the transmission and receipt of data. One of the first applications of reconfigurable hardware to communications involved the PAM project [30]. On-board PAM system RAM was used to trace through 2^{14} possible states of a Viterbi encoder, allowing for the computation of 4 states per clock cycle. The flexibility of the system allowed for quick evaluation of new encoding algorithms. A run-length Viterbi decoder, described in Ref. 93, was created and implemented using a large reconfigurable system containing 36 FPGA devices. This constraint length 14 decoder was able to achieve decode rates of up to 1 Mbit/sec. In Ref. 94, a single-FPGA system is described that supports variable-length code detection at video transfer rates.

5.3 Reconfigurable Computing Architecture and Compiler Trends for DSP

Over the past decade, the large majority of reconfigurable computing systems targeted to DSP have been based on commercial FPGA devices and have been programmed using RTL and structural hardware description languages. Although these architectural and programming methodologies have been sufficient for initial prototyping, more advanced architectures and programming languages will be needed in the future. These advancements will especially be needed to support advanced features such as dynamic reconfiguration and high-level compilation over the next few years. In this subsection, recent trends in reconfigurable computing-based DSP with regard to architecture and compilation are explored. Through near-term research advancement in these important areas, the breadth of DSP applications that are appropriate for reconfigurable computing is likely to increase.

5.3.1 Architectural Trends

Most commercial FPGA architectures have been optimized to perform efficiently across a broad range of circuit domains. Recently, these architectures have been changed to better suit specific application areas.

Specialized FPGA Architectures for DSP. Several FPGA architectures specifically designed for DSP have been proposed over the past decade. In Ref. 95, a fine-grained programmable architecture is considered that uses a customized LUT-based logic cell. The cell is optimized to efficiently perform addition and multiplication through the inclusion of XOR gates within LUT-based logic blocks. Additionally, device intercell wire lengths are customized to accommodate both local and global signal interconnections. In Ref. 96, a specialized DSP operator array is detailed. This architecture contains a linear array of adders and shifters connected to a programmable bus and is shown to efficiently implement

FIR filters. In Ref. 97, the basic cell of a LUT-based FPGA is augmented to include additional flip-flops and multiplexers. This combination allows for tight interblock communication required in bit-serial DSP processing. External routing was not augmented for this architecture due to the limited connectivity required by bit-serial operation.

Whereas fine-grained look-up table FPGAs are effective for bit-level computations, many DSP applications benefit from modular arithmetic operations. This need has led to an interest in reconfigurables with coarse-grained functional units. One such device, Paddi [98], is a DSP-optimized parallel computing architecture that includes eight ALUs and localized memories. As part of the architecture, a global instruction address is distributed to all processors, and instructions are fetched from a local instruction store. This organization allows for high instruction and I/O bandwidth. Communication paths between processors are configured through a communication switch and can be changed on a per-cycle basis. The Paddi architecture was motivated by a need for high data throughput and flexible datapath control in real-time image, audio, and video processing applications. The coarse-grained Matrix architecture [99] is similar to Paddi in terms of block structure, but it exhibits more localized control. Whereas Paddi has a VLIW-like control word which is distributed to all processors, Matrix exhibits more multiple instruction multiple data (MIMD) characteristics. Each Matrix tile contains a small processor, including a small SRAM and an ALU which can perform 8 bit data operations. Both near-neighbor and length-4 wires are used to interconnect individual processors. Interprocessor data ports can be configured to support either static or data-dependent dynamic communication.

The ReMarc architecture [100], targeted to multimedia applications, was designed to perform a SIMD-like computation with a single control word distributed to all processors. A 2D grid of 16-bit processors is globally controlled with a SIMD-like instruction sequencer. Interprocessor communication takes place either through near-neighbor interconnect or through horizontal and vertical buses. The MorphoSys architecture [101] was also designed for SIMD operation, but, unlike ReMarc, it offers support for efficient dynamic reconfiguration. Functional blocks in this architecture can perform either 8- or 16-bit ALU operations. A three-level hierarchy of interconnect provides for flexible interblock communication. The Chess architecture [102] is based on 4-bit ALUs and contains pipelined near-neighbor interconnect. Each computational tile in the architecture contains memory which can either store local processor instructions or local data memory. The Colt architecture [103] was specially designed as an adaptable architecture for DSP that allows interconnect reconfiguration. This coarse-grained architecture allows run-time data to steer programming information to dynamically determined points in the architecture. A mixture of both 1-bit and 16-bit functional units allows both bit and word-based processing.

Whereas coarse-grained architectures organized in a 2D array offer significant interconnect flexibility, often signal processing applications, such as filtering, can be accommodated with a linear computational pipeline. Several coarse-grained reconfigurable architectures have been created to address this class of applications. PipeRench [104] is a pipelined, linear computing architecture that consists of a sequence of computational *stripes*, each containing look-up tables and data registers. The modular nature of PipeRench makes dynamic reconfiguration on a per-stripe basis straightforward. Rapid [105] is a reconfigurable device based on both linear data and control paths. The coarse-grained architecture for this datapath includes multipliers, adders, and pipeline registers. Unlike PipeRench, the interconnect bus for this architecture is segmented to allow for nonlocal data transfer. In general, communication patterns built using Rapid interconnect are static, although some dynamic operation is possible. A pipelined control bus that runs in parallel to the pipelined data can be used to control computation.

DSP Compilation Software for Reconfigurable Computing. Although some high-level compilation systems designed to target DSP algorithms to reconfigurable platforms have been outlined and partially developed, few complete synthesis systems have been constructed. In Ref. 106, a high-level synthesis system is described for reconfigurable systems that promotes high-level synthesis from a behavioral synthesis language. For this system, DSP designs are represented as a high-level flowgraph and user-specified performance parameters in terms of a maximum and minimum execution schedule are used to guide the synthesis process. In Ref. 60, a compilation system is described that converts a standard ANSI C representation of filter and FFT operations into a bit-serial circuit that can be applied to an FPGA or to a field programmable multichip module. In Ref. 107, a compiler, debugger, and linker targeted to DSP data acquisition is described. This work uses a high-level model of communicating processes to specify computation and communication in a multi-FPGA system. By integrating digital-to-analog (D/A) and A/D converters into the configurable platform, a primitive digital oscilloscope is created.

The use of dynamic reconfiguration to reduce area overhead in computing systems has recently motivated renewed interest in reconfigurable computing. Although a large amount of work remains to be completed in this area, some preliminary work in the development of software to manage dynamic reconfiguration for DSP has been accomplished. In Ref. 108, a method of specifying and optimizing designs for dynamic reconfiguration is described. Through selective configuration scheduling, portions of an application used for 2D image processing is dynamically reconfigured based on need. Later work [46] outlined techniques based on bipartite matching to evaluate which portions of an dynamic application should be reconfigured. The technique is demonstrated using an image filtering example.

Several recent DSP projects address the need for both compile-time and run-time management of dynamic reconfiguration. In Ref. 109, a run-time manager is described for a single-chip reconfigurable computing system with a large FIR filter used as a test case. In Ref. 45, a compile-time analysis approach to aid reconfiguration is described. In this work, all reconfiguration times are statically determined in advance and the compilation system determines the minimum circuit change needed at each run-time point to allow for reconfiguration. Benchmark examples which use this approach include arithmetic units for FIR filters which contain embedded constants. Finally, in Ref. 62, algorithms are described that perform dynamic reconfiguration to save DSP system power in time-varying applications such as motion estimation. The software tool created for this work dynamically alters the search space of motion vectors in response to changing images. Because power in the motion estimation implementation is roughly correlated with search space, a reduced search proves to be beneficial for applications such as mobile communications. Additionally, unused computational resources can be scheduled for use as memory or rescheduled for use as computing elements as computing demands require.

Although the integration of DSP and reconfigurable hardware is just now being considered for single-chip implementation, several board-level systems have been constructed. GigaOps provided the first commercially available DSP and FPGA board in 1994 containing an Analog Devices 2101 PDSP, 2 Xilinx XC4010s, 256KB of SRAM, and 4MB of DRAM. This PC-based system was used to implement several DSP applications, including image processing [110]. Another board-based DSP/FPGA product line is the Arix-C67 currently available from MiroTech Corporation [111]. This system couples a Xilinx Virtex FPGA with a TMS320C6701 DSP. In addition to supporting several PC-bus interfaces, this system has an operating system, a compiler, and a suite of debugging software.

6 THE FUTURE OF RECONFIGURABLE COMPUTING
AND DSP

The future of reconfigurable computing for DSP systems will be determined by the same trends that affect the development of these systems today: system integration, dynamic reconfiguration, and high-level compilation. DSP applications are increasingly demanding in terms of computational load, memory requirements, and flexibility. Traditionally, DSP has not involved significant run-time adaptivity, although this characteristic is rapidly changing. The recent emergence of new applications that require sophisticated, adaptive, statistical algorithms to extract optimum performance has drawn renewed attention to run-time reconfigurability. Major applications driving the move toward adaptive computation in-

clude wireless communications with DSP in hand-sets, base stations and satel-
lites, multimedia signal processing [112], embedded communications systems
found in disk drive electronics [10] and high-speed wired interconnects [113],
and remote sensing for both environmental and military applications [114]. Many
of these applications have strict constraints on cost and development time due
to market forces.

The primary trend impacting the implementation of many contemporary
DSP systems is Moore's law, resulting in consistent exponential improvement
in integrated circuit device capacity and circuit speeds. According to the National
Technology Roadmap for Semiconductors, growth rates based on Moore's law
are expected to continue until at least the year 2015 [115]. As a result, some of
the corollaries of Moore's law will require new architectural approaches to deal
with the speed of global interconnect, increased power consumption and power
density, and system and chip-level defect tolerance. Several architectural approaches have been suggested to allow reconfigurable DSP systems to make
the best use of large amounts of VLSI resources. All of these architectures are
characterized by heterogeneous resources and novel approaches to interconnection. The term *system-on-a-chip* is now being used to describe the level of
complexity and heterogeneity available with future VLSI technologies. Figures
8 and 9 illustrate various characteristics of future reconfigurable DSP systems.
These are not mutually exclusive and some combination of these features will
probably emerge based on driving application domains such as wireless hand-
sets, wireless base stations, and multimedia platforms. Figure 8, taken from
Ref. 116, shows an architecture containing an array of DSP cores, a RISC microprocessor, large amounts of uncommitted SRAM, a reconfigurable FPGA fabric,
and a reconfigurable interconnection network. Research efforts to condense
DSPs, FPGA logic, and memory on a single substrate in this fashion are being
pursued in the Pleiades project [116,117]. This work focuses on selecting the
correct collection of functional units to perform an operation and then intercon-

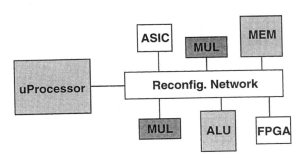

Figure 8   Architectural template for a single-chip Pleiades device. (From Ref. 116.)

necting them for low power. An experimental compiler has been created for this system [116] and testing has been performed to determine appropriate techniques for building a low-power interconnect. An alternate, adaptive approach [118] that takes a more distributed view of interconnection appears in Figure 9. This figure shows how a regular tiled interconnect architecture can be overlaid on a set of heterogeneous resources. Each tile contains a *communication switch* which allows for statically scheduled communication between adjacent tiles. Cycle-by-cycle communications information is held in embedded communication switch SRAM (SMEM).

The increased complexity of VLSI systems enabled by Moore's law presents substantial challenges in design productivity and verification. To support the continued advancement of reconfigurable computing, additional advances will be needed in hardware synthesis, high-level compilation, and design verification. Compilers have recently been developed which allow software development to be done at a high level, enabling the construction of complex systems including significant amounts of design reuse. Additional advancements in multicompilers [119] will be needed to partition designs, generate code, and synchronize interfaces for a variety of heterogeneous computational units. VLIW compilers [120] will be needed to find substantial amounts of instruction-level parallelism in DSP code, thereby avoiding the overhead of run-time parallelism extraction. Finally, compilers that target the codesign of hardware and software and leverage techniques such as static interprocessor scheduling [56] will allow truly reconfigurable systems to be specialized to specific DSP computations.

A critical aspect of high-quality DSP system design is the effective integration of reusable components or cores. These cores range from generic blocks like RAMs and RISC microprocessors to more specific blocks like MPEG decoders

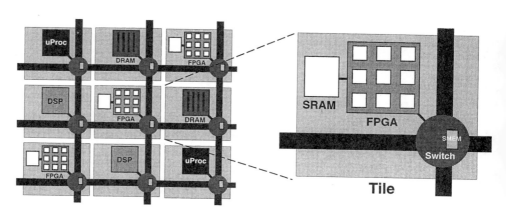

Figure 9   Distributed single-chip DSP interconnection network. (From Ref. 118.)

and PCI bus interfaces. Trends involving core development and integration will continue and tools to support core-based design will emerge, allowing significant user interaction for both design-time and run-time specialization and reconfiguration. Specialized synthesis tools will be refined to leverage core-based design and to extract optimum efficiency for DSP kernels while using conventional synthesis approaches for the surrounding circuitry [1,121].

Verification of complex and adaptive DSP systems will require a combination of simulation and emulation. Simulation tools like Ptolemy [122] have already made significant progress in supporting heterogeneity at a high level and will continue to evolve in the near future. Newer verification techniques based on logic emulation will emerge as effective mechanisms for using reconfigurable multi-FPGA platforms to verify DSP systems are developed. Through the use of new generations of FPGAs and advanced emulation software [123], new emulation systems will provide the capability to verify complex systems at near real-time rates.

Power consumption in DSP systems will be increasingly important in coming years due to expanding silicon substrates and their application to battery-powered and power-limited DSP platforms. The use of dynamic reconfiguration has been shown to be one approach that can be used to allow a system to adapt its power consumption to changing environments and computational loads [62]. Low-power core designs will allow systems to be assembled without requiring detailed power optimizations at the circuit level. Domain-specific processors [116] and loop transformations [124] have been proposed as techniques for avoiding the inherent power inefficiency of von Neumann architectures [125]. Additional computer-aided design tools will be needed to allow high-level estimation and optimization of power across heterogeneous architectures for dynamically varying workloads.

The use of DSP in fields such as avionics and medicine have created high-reliability requirements that must be addressed through available fault tolerance. Reliability is a larger system goal, of which power is only one component. As DSP becomes more deeply embedded in systems, reliability becomes even more critical. The increasing complexity of devices, systems, and software all introduce numerous failure points which need to be thoroughly verified. New techniques must especially be developed to allow defect tolerance and fault tolerance in the reconfigurable components of DSP systems. One promising technique which takes advantage of FPGA reconfiguration at various grain sizes is described in Ref. 126.

Reconfiguration for DSP systems is driven by many different goals: performance, power, reliability, cost, and development time. Different applications will require reconfiguration at different granularities and at different rates. DSP systems that require rapid reconfiguration may be able to exploit regularity in their algorithms and architectures to reduce reconfiguration time and power consumption. An approach called *dynamic algorithm transforms* (DAT) [127,128] is based on the philosophy of moving away from designing algorithms and architectures for worst-case operating conditions in favor of real-time reconfiguration to support the current situational case. This is the basis for reconfigurable ASICs (RAS-ICs) [129], where just the amount of flexibility demanded by the application is introduced. Configuration cloning [130], caching, and compression [131] are other approaches to address the need for dynamic reconfiguration. Techniques from computer architecture regarding instruction fetch and decode need to be modified to deal with the same tasks applied to configuration data.

In conclusion, reconfiguration is a promising technique for the implementation of future DSP systems. Current research in this area leverages contemporary semiconductors, architectures, computer-aided design tools, and methodologies in an effort to support the ever-increasing demands of a wide range of DSP applications. There is much work still to be done, however, because reconfigurable computing presents a very different computational paradigm for DSP system designers as well as DSP algorithm developers.

REFERENCES

1. D Singh, J Rabaey, M Pedram, F Catthor, S Rajgopal, N Sehgal, T Mozdzen. Power-conscious CAD tools and methodologies: A perspective. Proc IEEE 83(4): 570–594, 1995.
2. J Rabaey, R Broderson, T Nishitani. VLSI design and implementation fuels the signal-processing revolution. IEEE Signal Process Mag 5:22–38, January 1998.
3. E Lee. Programmable DSP architectures, Part I. IEEE Signal Process Mag 5:4–19, October 1988.
4. E Lee. Programmable DSP architectures, Part II. IEEE Signal Process Mag 6:4–14, January 1989.
5. J Eyre, J Bier. The evolution of DSP processors: From early architecture to the latest developments. IEEE Signal Process Mag 17:44–51, March 2000.
6. A Kalavade, J Othmer, B Ackland, K Singh. Software environment for a multiprocessor DSP. Proceedings of the 36th Design Automation Conference, 1999.
7. P Schaumont, S Vernalde, L Rijnders, M Engels I Bolsens. A programming environment for the design of complex high speed ASICs. Proceedings of the 35th Design Automation Conference, June 1998, pp 315–320.
8. Broadcom Corporation, www.broadcom.com, 2000.
9. Qualcomm Corporation, www.qualcomm.com, 2000.
10. N Nazari. A 500 Mb/s disk drive read channel in .25 μm CMOS incorporating programmable noise predictive Viterbi detection and Trellis coding. Proceedings, IEEE International Solid State Circuits Conference, 2000.
11. A Bell. The dynamic digital disk. IEEE Spectrum 36:28–35, October 1999.
12. G Weinberger. The new millennium: Wireless technologies for a truly mobile society. Proceedings, IEEE International Solid State Circuits Conference, 2000.

13. W Strauss. Digital signal processing: The new semiconductor industry technology driver. IEEE Signal Process Mag 17:52–56, March 2000.

14. S Hauck. The role of FPGAs in reprogrammable systems. Proc IEEE 86:615–638, April 1998.

15. W Mangione-Smith, B Hutchings, D Andrews, A Dehon, C Ebeling, R Hartenstein, O Mencer, J Morris, K Palem, V Prasanna, H Spaanenberg. Seeking solutions in configurable computing. IEEE Computer 30:38–43, December 1997.

16. J Villasenor, B Hutchings. The flexibility of configurable computing. IEEE Signal Process Mag 15:67–84, September 1998.

17. J Villasenor, W Mangione-Smith. Configurable computing. Sci Am 276:66–71, June 1997.

18. KK Maitra. Cascaded switching networks of two-input flexible cells. IEEE Trans Electron Computing EC-11:136–143, April 1962.

19. RC Minnick. A survey of microcellular research. J Assoc Computing Mach 14: 203–241, April 1967.

20. SE Wahlstrom. Programmable arrays and networks. Electronics 40:90–95, December 1967.

21. R Shoup. Programmable cellular logic arrays. PhD thesis, Carnegie Mellon University, 1970.

22. Xilinx Corporation, www.xilinx.com, 2000.

23. Altera Corporation, www.altera.com, 2000.

24. Xilinx Corporation. The Programmable Logic Data Book. San Jose, CA: Xilinx Corporation, 1994.

25. Xilinx Corporation. The Programmable Logic Data Book. San Jose, CA: Xilinx Corporation, 1998.

26. Xilinx Corporation. Virtex Data Sheet. San Jose, CA: Xilinx Corporation, 2000.

27. G Estrin. Parallel processing in a restructurable computing system. IEEE Trans Electron Computers 747–755, December 1963.

28. FP Manning. Automatic test, configuration, and repair of cellular arrays. PhD thesis, Massachusetts Institute of Technology, 1975.

29. J Arnold, D Buell, E Davis. Splash II. Proceedings, 4th ACM Symposium of Parallel Algorithms and Architectures, 1992, pp 316–322.

30. J Vuillemin, P Bertin, D Roncin, M Shand, H Touati, P Boucard. Programmable active memories: reconfigurable systems come of age. IEEE Trans VLSI Syst 4: 56–69, March 1996.

31. M Gokhale, W Holmes, A Kopser, S Lucas, R Minnich, D Sweeney, D Lopresti. Building and using a highly parallel programmable logic array. Computer 24:81–89, January 1991.

32. M Gokhale, R Minnich. FPGA computing in a data parallel C. Proceedings, IEEE Workshop on FPGAs for Custom Computing Machines, 1993, pp 94–101.

33. Annapolis Micro Systems, www.annapmicro.com, 2000.

34. M Shand. Flexible image acquisition using reconfigurable hardware. Proceedings, IEEE Workshop on FPGAs for Custom Computing Machines, 1995, pp 125–134.

35. P Athanas, H Silverman. Processor reconfiguration through instruction set metamorphosis: Architecture and compiler. Computer 26:11–18, March 1993.

36. National Semiconductor Corporation. NAPA 1000 Adaptive Processor. Santa Clara, CA: National Semiconductor Corporation, 1998.

37. R Wittig, P Chow. OneChip: An FPGA processor with reconfigurable logic. Proceedings, IEEE Workshop on FPGAs for Custom Computing Machines, 1996, pp 126–135.

38. J Hauser, J Wawrzynek. Garp: A MIPS processor with a reconfigurable coprocessor. Proceedings, IEEE Symposium on Field-Programmable Custom Computing Machines, 1997, pp 24–33.

39. R Razdin, MD Smith. A high-performance microarchitecture with hardware-programmable functional units. Proceedings, International Symposium on Microarchitecture, 1994, pp 172–180.

40. S Hauck, T Fry, M Hosler, J Kao. The Chimaera reconfigurable functional unit. Proceedings, IEEE Symposium on Field-Programmable Custom Computing Machines, 1997, pp 87–97.

41. XP Ling, H Amano. WASMII: A data driven computer on a virtual hardware. Proceedings, IEEE Workshop on FPGAs for Custom Computing Machines, 1993, pp 33–42.

42. A Dehon. DPGA-coupled microprocessors: Commodity ICs for the 21st century. Proceedings, IEEE Workshop on FPGAs for Custom Computing Machines, 1994, pp 31–39.

43. S Scalera, J Vazquez. The design and implementation of a context switching FPGA. Proceedings, IEEE Symposium on Field-Programmable Custom Computing Machines, 1998, pp 78–85.

44. Atmel Corporation. AT6000 Data Sheet. San Jose, CA: Amtel Corporation, 1999.

45. JP Heron, R Woods, S Sezer, RH Turner. Development of a run-time reconfiguration system with low reconfiguration overhead. J VLSI Signal Process 28(1):97–113, 2001.

46. N Shirazi, W Luk, PY Cheung. Automating production of run-time reconfigurable designs. Proceedings, IEEE Symposium on Field-Programmable Custom Computing Machines, 1998, pp 147–156.

47. N Hastie, R Cliff. The implementation of hardware subroutines on field programmable gate arrays. Proceedings, IEEE Custom Integrated Circuits Conference, 1990.

48. M Wirthlin, B Hutchings. A dynamic instruction set computer. Proceedings, IEEE Workshop on FPGAs for Custom Computing Machines, 1995, pp 99–107.

49. R Amerson, R Carter, WB Culbertson, P Kuekes, G Snider. Teramac—Configurable custom computing. Proceedings, IEEE Workshop on FPGAs for Custom Computing Machines, 1995, pp 32–38.

50. WB Culbertson, R Amerson, R Carter, P Kuekes, G Snider. Exploring architectures for volume visualization on the Teramac computer. Proceedings, IEEE Workshop on FPGAs for Custom Computing Machines, 1996, pp 80–88.

51. J Varghese, M Butts, J Batcheller. An efficient logic emulation system. IEEE Trans VLSI Syst 1:171–174, June 1993.

52. J Babb, R Tessier, M Dahl, S Hanono, D Hoki, A Agarwal. Logic emulation with virtual wires. IEEE Trans Computer-Aided Design Integrated Circuits Syst 10:609–626, June 1997.

53. H Schmit, L Arnstein, D Thomas, E Lagnese. Behavioral synthesis for FPGA-based computing. Proceedings, IEEE Workshop on FPGAs for Custom Computing Machines, 1994, pp 125–132.

54. A Duncan, D Hendry, P Gray. An overview of the COBRA–ABS high level synthesis system. Proceedings, IEEE Symposium on Field-Programmable Custom Computing Machines, 1998, pp 106–115.

55. RJ Peterson. An assessment of the suitability of reconfigurable systems for digital signal processing master's thesis, Brigham Young University, 1995.

56. J Babb, M Rinard, CA Moritz, W Lee, M Frank, R Barua, S Amarasinghe. Parallelizing applications to silicon. Proceedings, IEEE Symposium on Field-Programmable Custom Computing Machines, 1999.

57. P Banerjee, N Shenoy, A Choudary, S Hauck, C Bachmann, M Haldar, P Joisha, A Jones, A Kanhare, A Nayak, S Periyacheri, M Walkden, D Zaretsky. A MATLAB compiler for distributed, heterogeneous, reconfigurable computing systems. Proceedings, IEEE Symposium on Field-Programmable Custom Computing Machines, 2000.

58. Texas Instruments Corporation. TMS320C6201 DSP Data Sheet. Dallas, TX: Texas Instruments Corporation, 2000.

59. D Goeckel. Robust adaptive coded modulation for time-varying channels with delayed feedback. Proceedings of the Thirty-Fifth Annual Allerton Conference on Communication, Control, and Computing, 1997, pp 370–379.

60. T Isshiki, WWM Dai. Bit-serial pipeline synthesis for multi-FPGA systems with C++ design capture. Proceedings, IEEE Workshop on FPGAs for Custom Computing Machines, 1996, pp 38–47.

61. Altera Corporation. Flex10K Data Sheet. San Jose, CA: Altera Corporation, 1999.

62. SR Park, W Burleson. Reconfiguration for power savings in real-time motion estimation. Proceedings, International Conference on Acoustics, Speech, and Signal Processing, 1997, pp 3037–3040.

63. GR Goslin. A guide to using field programmable gate arrays for application-specific digital signal processing performance. Xilinx Application Note. San Jose, CA: Xilinx Corporation, 1998.

64. S He, M Torkelson. FPGA implementation of FIR filters using pipelined bit-serial canonical signed digit multipliers. Custom Integrated Circuits Conference, 1994, pp 81–84.

65. YC Lim, JB Evans, B Liu. An efficient bit-serial FIR filter architecture. Circuits, Systems, and Signal Processing 14(5):639–650, 1995.

66. JB Evans. Efficient FIR filter architectures suitable for FPGA implementation. IEEE Trans. Circuits Syst 41:490–493, July 1994.

67. CH Dick. FPGA based systolic array architectures for computing the discrete Fourier transform. Proceedings, International Symposium on Circuits and Systems, 1995, pp 465–468.

68. P Kollig, BM Al-Hashimi, KM Abbott. FPGA implementation of high performance FIR filters. Proceedings, International Symposium on Circuits and Systems, 1997, pp 2240–2243.

69. BV Herzen. Signal processing at 250 MHz using high performance FPGAs. Pro-

ceedings, International Symposium on Field Programmable Gate Arrays, 1997, pp 62–68.

70. B Fagin, C Renard. Field programmable gate arrays and floating point arithmetic. IEEE Trans VLSI Syst 2:365–367, September 1994.

71. N Shirazi, A Walters, P Athanas. Quantitative analysis of floating point arithmetic on FPGA-based custom computing machines. Proceedings, IEEE Workshop on FPGAs for Custom Computing Machines, 1995, pp 155–162.

72. L Louca, WH Johnson, TA Cook. Implementation of IEEE single precision floating point addition and multiplication on FPGAs. Proceedings, IEEE Workshop on FPGAs for Custom Computing Machines, 1996, pp 107–116.

73. WB Ligon, S McMillan, G Monn, F Stivers, KD Underwood. A re-evaluation of the practicality of floating-point operations on FPGAs. Proceedings, IEEE Symposium on Field-Programmable Custom Computing Machines, 1998.

74. AL Abbott, P Athanas, L Chen, R Elliott. Finding lines and building pyramids with Splash 2, Proceedings, IEEE Workshop on FPGAs for Custom Computing Machines, 1994, pp 155–161.

75. P Athanas, AL Abbott. Real-time image processing on a custom computing platform. IEEE Computer 28:16–24, February 1995.

76. N Ratha, A Jain, D Rover. Convolution on Splash 2. Proceedings, IEEE Workshop on FPGAs for Custom Computing Machines, 1995, pp 204–213.

77. M Shand, L Moll. Hardware/software integration in solar polarimetry. Proceedings, IEEE Symposium on Field-Programmable Custom Computing Machines, 1998, pp 18–26.

78. M Dao, TA Cook, D Silver, PS D'Urbano. Acceleration of template-based ray casting for volume visualization using FPGAs. Proceedings, IEEE Workshop on FPGAs for Custom Computing Machines, 1995.

79. R Woods, D Trainer, J-P Heron. Applying an XC6200 to real-time image processing. IEEE Design Test Computers 15:30–37, January 1998.

80. B Box. Field programmable gate array based reconfigurable preprocessor. Proceedings, IEEE Workshop on FPGAs for Custom Computing Machines, 1994, pp 40–48.

81. S Singh, R Slous. Accelerating Adobe photoshop with reconfigurable logic. Proceedings, IEEE Symposium on Field-Programmable Custom Computing Machines, 1998, pp 18–26.

82. RD Hudson, DI Lehn, PM Athanas. A run-time reconfigurable engine for image interpolation. Proceedings, IEEE Symposium on Field-Programmable Custom Computing Machines, 1998, pp 88–95.

83. J Greenbaum, M Baxter. Increased FPGA capacity enables scalable, flexible CCMs: An example from image processing. Proceedings, IEEE Symposium on Field-Programmable Custom Computing Machines, 1997.

84. J Woodfill, BV Herzen. Real-time stereo vision on the PARTS reconfigurable computer. Proceedings, IEEE Symposium on Field-Programmable Custom Computing Machines, 1997, pp 242–250.

85. I Page. Constructing hardware–software systems from a single description. J VLSI Signal Process 12(1):87–107, 1996.

86. J Villasenor, B Schoner, C Jones. Video communications using rapidly reconfigur-

able hardware. IEEE Trans Circuits Syst Video Technol 5:565–567, December 1995.

87. L Ferguson. Generating audio effects using dynamic FPGA reconfiguration. Computer Design, February 1997, p 50.

88. DE Thomas, JK Adams, H Schmit. A model and methodology for hardware–software codesign. IEEE Design Test Computers 10:6–15, September 1993.

89. M Rencher, BL Hutchings. Automated target recognition on Splash II. Proceedings, IEEE Symposium on Field-Programmable Custom Computing Machines, 1997, pp 192–200.

90. J Villasenor, B Schoner, K-N Chia, C Zapata. Configurable computing solutions for automated target recognition. Proceedings, IEEE Workshop on FPGAs for Custom Computing Machines, 1996, pp 70–79.

91. S Natarajan, B Levine, C Tan, D Newport, D Bouldin. Automatic mapping of Khoros-based applications to adaptive computing systems. Proceedings, 1999 Military and Aerospace Applications of Programmable Devices and Technologies International Conference (MAPLD), 1999, pp 101–107.

92. JR Rasure, S Kubica. The Khoros application development environment. Khoros Research Technical Memo, 2000; www.khoral.com.

93. D Yeh, G Feygin, P Chow. RACER: A reconfigurable constraint-length 14 Viterbi decoder. Proceedings, IEEE Workshop on FPGAs for Custom Computing Machines, 1996.

94. G Brebner, J Gray. Use of reconfigurability in variable-length code detection at video rates. Proceedings, Field Programmable Logic and Applications (FPL'95), 1995, pp 429–438.

95. M Agarwala, PT Balsara. An architecture for a DSP field-programmable gate array. IEEE Trans VLSI Syst 3:136–141, March 1995.

96. T Arslan, HI Eskikurt, DH Horrocks. High level performance estimation for a primitive operator filter FPGA. Proceedings, International Symposium on Circuits and Systems, 1998, pp V237–V240.

97. A Ohta, T Isshiki, H Kunieda. New FPGA architecture for bit-serial pipeline datapath. Proceedings, IEEE Symposium on Field-Programmable Custom Computing Machines, 1998.

98. DC Chen, J Rabaey. A reconfigurable multiprocessor IC for rapid prototyping of algorithmic-specific high speed DSP data paths. IEEE J Solid-State Circuits 27: 1895–1904, December 1992.

99. E Mirsky, A Dehon. MATRIX: A reconfigurable computing architecture with configurable instruction distribution and deployable resources. Proceedings, IEEE Workshop on FPGAs for Custom Computing Machines, 1996, pp 157–166.

100. T Miyamori, K Olukotun. A quantitative analysis of reconfigurable coprocessors for multimedia applications. Proceedings, IEEE Symposium on Field-Programmable Custom Computing Machines, 1998.

101. F Kurdahi, E Filho. Design and implementation of the MorphoSys reconfigurable computing processor. J VLSI Signal Process 24(2):147–164, 2000.

102. A Marshall, T Stansfield, I Kostarnov, J Vuillemin, B Hutchings. A reconfigurable arithmetic array for multimedia applications. Proceedings, International Symposium on Field Programmable Gate Arrays, 1999, pp 135–143.

103. R Bittner, P Athanas. Wormhole run-time reconfiguration. Proceedings, International Symposium on Field Programmable Gate Arrays, 1997, pp 79–85.

104. SC Goldstein, H Schmit, M Moe, M Budiu, S Cadambi, RR Taylor, R Laufer. PipeRench: A coprocessor for streaming multimedia acceleration. Proceedings, International Symposium on Computer Architecture, 1999, pp 28–39.

105. C Ebeling, D Cronquist, P Franklin, J Secosky, SG Berg. Mapping applications to the RaPiD configurable architecture. Proceedings, IEEE Symposium on Field-Programmable Custom Computing Machines, 1997, pp 106–115.

106. M Leeser, R Chapman, M Aagaard, M Linderman, S Meier. High level synthesis and generating FPGAs with the BEDROC system. J VLSI Signal Process 6(2): 191–213, 1993.

107. A Wenban, G Brown. A software development system for FPGA-based data acquisition systems. Proceedings, IEEE Workshop on FPGAs for Custom Computing Machines, 1996, pp 28–37.

108. W Luk, N Shirazi, PY Cheung. Modelling and optimising run-time reconfigurable systems. Proceedings, IEEE Workshop on FPGAs for Custom Computing Machines, 1996, pp 167–176.

109. J Burns, A Donlin, J Hogg, S Singh, M de Wit. A dynamic reconfiguration run-time system. Proceedings, IEEE Symposium on Field-Programmable Custom Computing Machines, 1997, pp 66–75.

110. P Athanas, R Hudson. Using rapid prototyping to teach the design of complete computing solutions. Proceedings, IEEE Workshop on FPGAs for Custom Computing Machines, 1996.

111. Mirotech Corporation, www.mirotech.com, 1999.

112. P Pirsch, A Freimann, M Berekovic. Architectural approaches for multimedia processors. Proc Multimedia Hardware Architect SPIE. 3021, 2–13, 1997.

113. W Dally, J Poulton. Digital Systems Engineering. Cambridge: Cambridge University Press, 1999.

114. M Petronino, R Bambha, J Carswell, W Burleson. An FPGA-based data acquisition system for a 95 GHz W-band radar. Proceedings, International Conference on Acoustics, Speech, and Signal Processing, 1997, pp 4105–4108.

115. D Sylvester, K Keutzer. Getting to the bottom of deep submicron. Proceedings, International Conference on Computer-Aided Design, 1998, pp 203–211.

116. M Wan, H Zhang, V George, M Benes, A Abnous, V Prabhu, J Rabaey, Design methodology of a low-energy reconfigurable single-chip DSP system. VLSI Signal Process 28(1):47–61, 2001.

117. H Zhang, V Prabhu, V George, M Wan, M Benes, A Abnous, JM Rabaey. A 1V heterogeneous reconfigurable processor IC for baseband wireless applications. Proceedings, IEEE International Solid State Circuits Conference, 2000.

118. J Liang, S Swaminathan, R Tessier. aSOC: A scalable, single-chip communication architecture. Proceedings, International Conference on Parallel Architectures and Compilation Techniques, 2000, pp 37–46.

119. K McKinley, SK Singhai, GE Weaver, CC Weems. Compiler architectures for heterogeneous processing. Languages and Compilers for Parallel Processing. Lecture Notes in Computer Science. Berlin: Springer-Verlag, 1995, pp 434–449.

120. K Konstantinides. VLIW architectures for media processing. IEEE Signal Process Mag 15:16–19, March 1998.

121. Synopsys Corporation, www.synopsys.com, 2000.
122. JT Buck, S Ha, EA Lee, DG Messerschmitt. Ptolemy: A framework for simulating and prototyping heterogeneous systems. Int J Computer Simul 4:155–182, April 1994.
123. R Tessier. Incremental compilation for logic emulation. Proceedings, IEEE Tenth International Workshop on Rapid System Prototyping, 1999, pp 236–241.
124. H DeMan, J Rabaey, J Vanhoof, G Goosens, P Six, L Claesen. CATHEDRAL-II—A computer-aided synthesis system for digital signal processing VLSI systems. Computer-Aided Eng J 5:55–66, April 1988.
125. M Horowitz, R Gonzalez. Energy dissipation in general purpose processors. J Solid State Circuits 31:1277–1284, November 1996.
126. V Lakamraju, R Tessier. Tolerating operational faults in cluster-based FPGAs. Proceedings, International Symposium on Field Programmable Gate Arrays, 2000, pp 187–194.
127. M Goel, NR Shanbhag. Dynamic algorithm transforms for low-power adaptive equalizers. IEEE Trans Signal Process 47:2821–2832, October 1999.
128. M Goel, NR Shanbhag. Dynamic algorithm transforms (DAT): A systematic approach to low-power reconfigurable signal processing. IEEE Trans VLSI Syst 7:463–476, December 1999.
129. J Tschanz, NR Shanbhag. A low-power reconfigurable adaptive equalizer architecture. Proceedings of the Asilomar Conference on Signals, Systems, and Computers, 1999.
130. SR Park, W Burleson. Configuration cloning: Exploiting regularity in dynamic DSP architectures. Proceedings, International Symposium on Field Programmable Gate Arrays, 1999.
131. S Hauck, Z Li, E Schwabe. Configuration compression for the Xilinx XC6200 FPGA. Proceedings, IEEE Symposium on Field-Programmable Custom Computing Machines, 1998, pp 138–146.

5

Parallel Architectures for Programmable Video Signal Processing

Zhao Wu and Wayne Wolf
Princeton University, Princeton, New Jersey

1 INTRODUCTION

Modern digital video applications, ranging from video compression to content analysis, require both high computation rates and the ability to run a variety of complex algorithms. As a result, many groups have developed programmable architectures tuned for video applications. There have been four solutions to this problem so far: modifications of existing microprocessor architectures, application-specific architectures, fully programmable video signal processors (VSPs), and hybrid systems with reconfigurable hardware. Each approach has both advantages and disadvantages. They target the market from different perspectives. Instruction set extensions are motivated by the desire to speed up video signal processing (and other multimedia applications) by software solely rather than by special-purpose hardware. Application-specific architectures are designed to implement one or a few applications (e.g., MPEG-2 decoding). Programmable VSPs are architectures designed from the ground up for multiple video applications and may not perform well on traditional computer applications. Finally, reconfigurable systems intend to achieve high performance while maintaining flexibility.

Generally speaking, video signal processing covers a wide range of applications from simple digital filtering through complex algorithms such as object recognition. In this survey, we focus on advanced digital architectures, which are intended for higher-end video applications. Although we cannot address every possible video-related design, we cover major examples of video architectures that illustrate the major axes of the design space. We try to enumerate all the cutting-edge companies and their products, but some companies did not provide much detail (e.g., chip architecture, performance, etc.) about their products,so we do not have complete knowledge about some Integrated circuits (ICs) and systems. Originally, we intended to study only the IC chips for video signal processing, but reconfigurable systems also emerge as a unique solution, so we think it is worth mentioning these systems as well.

The next section introduces some basic concepts in video processing algorithms, followed by an early history of VSPs in Section 3. This is just to serve as a brief introduction of the rapidly evolving industry. Beginning in Section 6, we discuss instruction set extensions of modern microprocessors. In Section 5, we compare the existing architectures of some dedicated video codecs. Then, in Section 6, we contrast in detail and analyze the pros and cons of several programmable VSPs. In Section 7, we introduce systems based on reconfigurable computing, which is another interesting approach for video signal processing. Finally, conclusions are drawn in Section 8.

2 BACKGROUND

Although we cannot provide a comprehensive introduction to video processing algorithms here, we can introduce a few terms and concepts to motivate the architectural features found in video processing chips. Video compression was an early motivating application for video processing; today, there is increased interest in video analysis.

The Motion Pictures Experts Group (MPEG) (www.cselt.it) has been continuously developing standards for video compression. MPEG-1, -2, and -4 are complete, and at this writing, work on MPEG-7 is underway. We refer the reader to the MPEG website for details on MPEG-1 and -2 and to a special issue of *IEEE Transactions on Circuits and Systems for Video Technology* for a special issue on MPEG-4. The MPEG standards apply several different techniques for video compression. One technique, which was also used for image compression in the JPEG standard (JPEG book) is coding using the discrete cosine transform (DCT). The DCT is a frequency transform which is used to transform an array of pixels (an 8×8 array in MPEG and JPEG) into a spatial frequency spectrum; the two-dimensional DCT for the 2D array can be found by computing two 1D DCTs on the blocks. Specialized algorithms have been developed for computing the DCT efficiently. Once the DCT is computed, lossy compression algorithms will throw away coefficients which represent high-spatial frequencies, because those represent fine details which are harder to resolve by the human eye, particularly in moving objects. DCT is one of the two most computation-intensive operations in MPEG.

The other expensive operation in MPEG-style compression is block motion estimation. Motion estimation is used to encode one frame in terms of another (DCT is used to compress data within a single frame). As shown in Figure 1, in MPEG-1 and -2, a macroblock (a 16×16 array of pixels composed of four blocks) taken from one frame is correlated within a distance p of the macroblock's current position (giving a total search window of size $2p + 1 \times 2p + 1$). The reference macroblock is compared to the selected macroblock by two-dimensional correlation: Corresponding pixels are compared and the sum of the magnitudes of the differences is computed. If the selected macroblock can be matched within a given tolerance, in the other frame, then the macroblock need be sent only once for both frames. A region around the macroblock's original position is chosen as the search area in the other frame; several algorithms exist which avoid performing the correlation at every offset within the search region. The macroblock is given a motion vector that describes its position in the new frame relative to its original position. Because matches are not, in general, exact, a difference pattern is sent to describe the corrections made after applying the macroblock in the new context.

MPEG-1 and -2 provide three major types of frames. The I-frame is coded without motion estimation. DCT is used to compress blocks, but a lossily compressed version of the entire frame is encoded in the MPEG bit stream. A P-

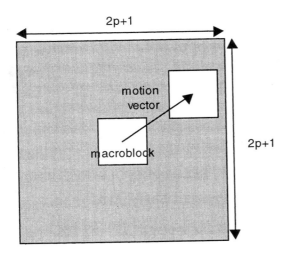

Figure 1   Block motion estimation.

frame is predicted using motion estimation. A P-frame is encoded relative to an earlier I-frame. If a sufficiently good macroblock can be found from the I-frame, then a motion vector is sent rather than the macroblock itself; if no match is found, the DCT-compressed macroblock is sent. A B-frame is bidirectionally encoded using motion estimation from frames both before and after the frame in time (frames are buffered in memory to allow bidirectional motion prediction). MPEG-4 introduces methods for describing and working with objects in the video stream. Other detailed information about the compression algorithm can be found in the MPEG standard [1].

Wavelet-based algorithms have been advocated as an alternative to block-based motion estimation. Wavelet analysis uses filter banks to perform a hierarchical frequency decomposition of the entire image. As a result, wavelet-based programs have somewhat different characteristics than block-based algorithms.

Content analysis of video tries to extract useful information from video frames. The results of content analysis can be used either to search a video database or to provide summaries that can be viewed by humans. Applications include video libraries and surveillance. For example, algorithms may be used to extract key frames from videos. The May and June 1998 issues of the *Proceedings of the IEEE* and the March 1998 issue of *IEEE Signal Processing Magazine* survey multimedia computing and signal processing algorithms.

3   EARLY HISTORY OF VLSI VIDEO PROCESSING

An early programmable VSP was the Texas Instruments TMS34010 graphics system processor (GSP) [2]. This chip was released in 1986. It is a 32-bit microprocessor optimized for graphics display systems. It supports various pixel formats (1-, 2-, 4-, 8-, and 16-bit) and operations and can accelerate graphics interface efficiently. The processor operates at a clock speed from 40 to 60 MHz, achieving a peak performance of 7.6 million instructions per second (MIPS).

Philips Semiconductors developed early dedicated video chips for specialized video processors. Philips announced two digital multistandard color decoders at almost the same time. Both the SAA9051 [3] and the SAA7151 [4] integrate a luminance processor and chrominance processor on-chip and are able to separate 8-bit luminance and 8-bit chrominance from digitized S-Video or composite video sources as well as generate all the synchronization and control signals. Both VSPs support PAL, NTSC, and SECAM standards.

In the early days of JPEG development, its computational kernels could not be implemented in real time on typical CPUs, so dedicated DCT/IDCT (discrete cosine transform–inverse DCT) units, Huffman encoder/decoder, were built to form a multichip JPEG codec [another solution was multiple digital signal processors (DSPs)]. Soon, the multiple modules could be integrated onto a single chip. Then, people began to think about real-time MPEG. Although MPEG-1 decoders were only a little more complicated than JPEG decoders, MPEG-1 encoders were much more difficult. At the beginning, encoders that are fully compliant to MPEG-1 standards could not be built. Instead, people had to come up with some compromise solutions. First, motion-JPEG or I-frame-only (where the motion estimation part of the standard is completely dropped) encoders were designed. Later, forward prediction frames were added in IP-frame encoders. Finally, bidirectional prediction frames were implemented. The development also went through a whole procedure from multichip to singlechip. Meanwhile, the microprocessors became so powerful that some software MPEG-1 players could support real-time playback of small images. The story of MPEG-2 was very similar to MPEG-1 and began as soon as the first single-chip MPEG-1 decoder was born. Like MPEG-1, it also experienced asymptotic approaches from simplified standards to fully compliant versions, and from multichip solutions to single chip solutions.

The late 1980s and early 1990s saw the announcement of several complex, programmable VSPs. Important examples include chips from Matsushita [5], NTT [6], Philips [7], and NEC [8]. All of these processors were high-performance parallel processors architected from the ground up for real-time video signal processing. In some cases, these chips were designed as showcase chips to display the capabilities of submicron very-large-scale integration (VLSI) fabrication processes. As a result, their architectural features were, in some cases, chosen for their ability to demonstrate a high clock rate rather than their effectiveness for video processing. The Philips VSP-1 and NEC processor were probably the most heavily used of these chips.

The software (compression standards, algorithms, etc.) and hardware (instruction set extensions, dedicated codecs, programmable VSPs) developments of video signal processing are in parallel and rely heavily on each other. On one hand, no algorithms could be realized without hardware support; on the other hand, it is the software that makes a processor useful. Modern VLSI technology not only makes possible but also encourages the development of coding algorithms—had developers not been able to implement MPEG-1 in hardware, it may not have become popular enough to inspire the creation of MPEG-2.

4 INSTRUCTION SET EXTENSIONS FOR VIDEO SIGNAL PROCESSING

The idea of providing special instructions for graphics rendering in a general-purpose processor is not new; it appeared as early as 1989 when Intel introduced i860, which has instructions for Z-buffer checks [9]. Motorola's 88110 is another example of using special parallel instructions to handle multiple pixel data simultaneously [10]. To accommodate the architectural inefficiency for multimedia applications, many modern general-purpose processors have extended their instruction set. This kind of patch is relatively inexpensive as compared to designing a VSP from the very beginning, but the performance gain is also limited. Almost all of the patches adopt single instruction multiple data (SIMD) model, which operates on several data units at a time. Apparently, the supporting facts behind this idea are as follows: First, there is a large amount of parallelism in video applications; second, video algorithms seldom require large data sizes. The best part of this approach is that few modifications need to be done on existing architectures. In fact, the area overhead is only 0.1% (HP PA-RISC MAX2) to 3% (Sun UltraSparc) of the original die in most processors. Already having a 64-bit datapath in the architecture, it takes only a few extra transistors to provide pixel-level parallelism on the wide datapath. Instead of working on one 64-bit word, the new instructions can operate on 8 bytes, four 16-bit words, or two 32-bit words simultaneously (with the same execution time), octupling, quadrupling, or doubling the performance, respectively. Figure 2 shows the parallel operations on four pairs of 16-bit words.

In addition to the parallel arithmetic, shift, and logical instructions, the new instruction set must also include data transfer instructions that pack and unpack data units into and out of a 64-bit word. In addition, some processors (e.g., HP PA-RISC MAX2) provide special data alignment and rearrangement instructions to accelerate algorithms that have irregular data access patterns (e.g., zigzag scan in discrete cosine transform). Most instruction set extensions provide three ways to handle overflow. The default mode is modular, nonsaturating arithmetic, where any overflow is discarded. The other two modes apply saturating arithmetic. In signed saturation, an overflow causes the result to be clamped to its maximum or minimum signed value, depending on the direction of the overflow. Similarly, in unsigned saturation, an overflow sets the result to its maximum or minimum unsigned value.

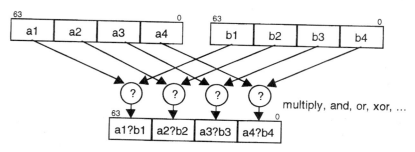

Figure 2   Examples of SIMD operations.

Table 1   Instruction Set Extensions for Multimedia Applications

Vendor	Microprocessor	Extension	Release date	Ref.
Hewlett-Packard	HP PA-RISC 1.0 and 2.0	MAX1	Jan. 1994	11
		MAX2	Feb. 1996	12
Intel	Pentium and Pentium Pro	MMX	March 1996	14
Sun	UltraSPARC-I, -II, and -III	VIS	Dec. 1994	15
DEC	Alpha 21264	MVI	Oct. 1996	17
MIPS	MIPS R10000	MDMX	March 1997	18

An important issue for instruction set extension is compatibility. Multimedia extensions allow programmers to mix multimedia-enhanced code with existing applications. Table 1 shows that all the modern microprocessors have added multimedia instructions to their basic architecture. We will discuss the first three microprocessors in detail.

4.1   Hewlett-Packard MAX2 (Multimedia Acceleration eXtensions)

Hewlett-Packard was the first CPU vendor to introduce multimedia extensions for general-purpose processors in a product [11]. MAX1 and MAX2 were released in 1994 and 1996, respectively, for 32-bit PA-RISC and 64-bit PA-RISC processors.

Table 2 lists the MAX2 instructions in PA-RISC 2.0 [12]. Having observed a large portion of constant multiplies in multimedia processing, HP added *hshladd* and *hshradd* to speed up this kind of operation. The mix and permute instructions are useful for subword data formatting and rearrangement operations. For example, the mix instructions can be used to expand 16-bit subwords into 32-bit subwords and vice versa.

Another example is matrix transpose, where only eight mix instructions are required for a 4 × 4 matrix. The permute instruction takes 1 source register and produces all the 256 possible permutations of the 16-bit subwords in that register, with or without repetitions.

From Table 3 we can see that MAX2 not only reduces the execution time significantly but also requires fewer registers. This is because the data rearrangement instructions need fewer temporary registers and saturation arithmetic saves registers that hold the constant clamping value.

4.2   Intel MMX (Multi Media eXtensions)

Table 4 lists all the 57 MMX instructions, which, according to Intel's simulations of the P55C processor, can improve performance on most multimedia applica-

Table 2 MAX2 Instructions in PA-RISC 2.0

Group	Mnemonic	Description
Parallel add	Hadd	Add 4 pairs of 16-bit operands, with modulo arithmetic
	Hadd,ss	Add 4 pairs of 16-bit operands, with signed saturation
	Hadd,us	Add 4 pairs of 16-bit operands, with unsigned saturation
Parallel subtract	Hsub	Subtract 4 pairs of 16-bit operands, with modulo arithmetic
	Hsub,ss	Subtract 4 pairs of 16-bit operands, with signed saturation
	Hsub,us	Subtract 4 pairs of 16-bit operands, with unsigned saturation
Parallel shift and add	Hshladd	Multiply 4 first operands by 2, 4, or 8 and add corresponding second operands
	hshradd	Divide 4 first operands by 2, 4, or 8 and add corresponding second operands
Parallel average	havg	Arithmetic mean of 4 pairs of operands
Parallel shift	hshr	Shift right by 0 to 15 bits, with sign extension on the left
	hshr,u	Shift right by 0 to 15 bits, with zero extension on the right
	hshl	Shift left by 0 to 15 bits, with zeros shifted in on the right
Mix	mixh,L mixh,R mixw,L mixw,R	Interleave alternate 16-bit [h] or 32-bit [w] subwords from two source registers, starting from leftmost [L] subword or ending with rightmost [R] subword
Permute	permh	Rearrange subwords from one source register, with or without repetition

Source: Ref. 13.

tions by 50–100%. Compared to HP's MAX2, the MMX multimedia instruction set is more flexible on the format of the operand. It not only works on four 16-bit words but also supports 8 bytes and two 32-bit words. In addition, it provides packed multiply and packed compare instructions. Using packed multiply, it requires only 6 cycles to calculate four products of 16×16 multiplication on a P55C, whereas on a non-MMX Pentium, it takes 10 cycles for a single 16×16 multiplication. The behavior of pack and unpack instructions is very similar to that of the mix instructions in MAX2. Figure 3 illustrates the function of two MMX instructions. The DSP-like PMADDWD multiplies two pairs of 16-bit

Table 3 Performance of Multimedia Kernels With (and Without) MAX2 Instructions

Kernel algorithm	16 × 16 Block match	8 × 8 Matrix transpose	3 × 3 Box filter	8 × 8 IDCT
Cycles	160 (426)	16 (42)	548 (2324)	173 (716)
Registers	14 (12)	18 (22)	15 (18)	17 (20)
Speedup	2.66	2.63	4.24	4.14

Source: Ref. 13.

words and then sums each pair to produce two 32-bit results. On a P55C, the execution takes three cycles when fully pipelined. Because multiply-add operations are critical in many video signal processing algorithms such as DCT, this feature can improve the performance of some video applications (e.g., JPEG and MPEG) greatly. The motivation behind the packed compare instructions is a common video technique known as chroma key, which is used to overlay an object on another image (e.g., weather person on weather map). In a digital implementation with MMX, this can be done easily by applying packed logical operations after packed compare. Up to eight pixels can be processed at a time.

Unlike MAX2, MMX instructions do not use general-purpose registers; all the operations are done in eight new registers (MM0–MM7). This explains why the four packed logical instructions are needed in the instruction set. The MMX registers are mapped to the floating-point registers (FP0–FP7) in order to avoid introducing a new state. Because of this, floating-point and MMX instructions cannot be executed at the same time. To prevent floating-point instructions from corrupting MMX data, loading any MMX register will trigger the busy bit of all the FP registers, causing any subsequent floating-point instructions to trap. Consequently, an EMMS instruction must be used at the end of any MMX routine to resume the status of all the FP registers. In spite of the awkwardness, MMX has been implemented in several Pentium models and also inherited in Pentium II and Pentium III.

4.3 Sun VIS

Sun UltraSparc is probably today's most powerful microprocessor in terms of video signal processing ability. It is the only off-the-shelf microprocessor that supports real-time MPEG-1 encoding and real-time MPEG-2 decoding [15]. The horsepower comes from a specially designed engine: VIS, which accelerates multimedia applications by twofold to sevenfold, executing up to 10 operations per cycle [16].

Table 4 MMX Instructions

Group	Mnemonic	Description
Data transfer, pack and unpack	MOV[D,Q][a]	Move [double,quad] to/from MM register
	PACKUSWB	Pack words into bytes with unsigned saturation
	PACKSS[WB,DW]	Pack [words into bytes, doubles into words] with signed saturation
	PUNPCKH[BW,WD,DQ]	Unpack (interleave) high-order [bytes, words, doubles] from MM register
	PUNPCKL[BW,WD,DQ]	Unpack (interleave) low-order [bytes, words, doubles] from MM register
Arithmetic	PADD[B,W,D]	Packed add on [byte, word, double]
	PADDS[B,W]	Saturating add on [byte, word]
	PADDUS[B,W]	Unsigned saturating add on [byte, word]
	PSUB[B,W,D]	Packed subtract on [byte, word, double]
	PSUBS[B,W]	Saturating subtract on [byte, word]
	PSUBUS[B,W]	Unsigned saturating subtract on [byte, word]
	PMULHW	Multiply packed words to get high bits of product
	PMULLW	Multiply packed words to get low bits of product
	PMADDWD	Multiply packed words, add pairs of products
Shift	PSLL[W,D,Q]	Packed shift left logical [word, double, quad]
	PSRL[W,D,Q]	Packed shift right logical [word, double, quad]
	PSRA[W,D]	Packed shift right arithmetic [word, double]
Logical	PAND	Bit-wise logical AND
	PANDN	Bit-wise logical AND NOT
	POR	Bit-wise logical OR
	PXOR	Bit-wise logical XOR
Compare	PCMPEQ[B,W,D]	Packed compare "if equal" [byte, word, double]
	PCMPGT[B,W,D]	Packed compare "if greater than" [byte, word, double]
Misc	EMMS	Empty MMX state

[a] Intel's definitions of word, double word, and quad word are, respectively, 16-bit, 32-bit, and 64-bit.
Source: Ref. 14.

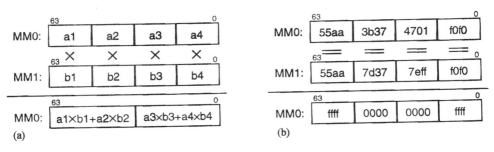

Figure 3 Operations of (a) packed multiply-add (PMADDWD) and (b) packed compare-if-equal (PCMPEQW). (From Ref. 14.)

For a number of reasons, the visual instruction set (VIS) instructions are implemented in the floating-point unit rather than integer unit. First, some VIS instructions (e.g., partitioned multiply and pack) take multiple cycles to execute, so it is better to send them to the floating-point unit (FPU) which handles multiple-cycle instructions like floating-point add and multiply. Second, video applications are register-hungry; hence, using FP registers can save integer registers for address calculation, loop counts, and so forth. Third, the UltraSparc pipeline only allows up to three integer instructions per cycle to be issued; therefore, using FPU again saves integer instruction slots for address generation, memory load/store, and loop control. The drawback of this is that the logical unit has to be duplicated in the floating-point unit, because VIS data are kept in the FP registers.

The VIS instructions (listed in Table 5) support the following data types: pixel format for true-color graphics and images, fixed16 format for 8-bit data, and fixed32 format for 8-, 12-, or 16-bit data. The partitioned add, subtract, and multiply instructions in VIS function very similar to those in MAX2 and MMX. In each cycle, the UltraSparc can carry out four 16×8 or two 16×16 multiplications. Moreover, the instruction set has quite a few highly specialized instructions. For example, *EDGE* instructions compare the address of the edge with that of the current pixel block, and then generate a mask, which later can be used by partial store (*PST*) to store any appropriate bytes back into the memory without using a sequence of read–modify–write operations. The *ARRAY* instructions are specially designed for three-dimensional (3D) visualization. When the 3D dataset is stored linearly, a 2D slice with arbitrary orientation could yield very poor locality in cache. The *ARRAY* instructions convert the 3D fixed-point addresses into a blocked-byte address, making it possible to move along any line or plane with good spatial locality. The same operation would require 24 RISC-equivalent instructions. Another outstanding instruction is *PDIST*, which calculates the SAD (sum of absolute difference) of two sets of eight pixels in parallel. This is the

Table 5 Summary of VIS Instructions

Opcode[a]	Operands	Description
FPADD16/32(S)	Fsrc1, fsrc2, fdest	Four 16-bit or two 32-bit partitioned add or
FPSUB16/32(S)	Fsrc1, fsrc2, fdest	subtract
FPACK16	Fsrc2, fdest	Pack four 16-bit pixels into fdest
FPACK32	Fsrc1, fsrc2, fdest	Add two 32-bit pixels into fdest
FPACKFIX	Fsrc2, fdest	Pack two 32-bit pixels into fdest
FEXPAND	Fsrc2, fdest	Expand four 8-bit pixels into fdest
FPMERGE	Fsrc1, fsrc2, fdest	Merge two sets of four 8-bit pixels
FMUL8 $\times$ 16(opt)	Fsrc1, fsrc2, fdest	Multiply four 8-bit pixels by four 16-bit constants
ALIGNADDR(L)	src1, src2, dest	Set up for unaligned access
FALIGNDATA	fsrc1, fsrc2, fdest	Align data from unaligned access
FZERO(S)	Fdest	Fill fdest with zeroes
FONE(S)	Fdest	Fill fdest with ones
FSRC(S)	fsrc, fdest	Copy fsrc to fdest
FNOT(S)	fsrc, fdest	Negate fsrc in fdest
Flogical(S)	fsrc1, fsrc2, fdest	Perform one of 10 logical operations (AND, OR, etc.)
FCMPcc16/32	fsrc1, fsrc2, dest	Perform four 16-bit or two 32-bit compares with results in dest
EDGE8/16/32(L)	src1, src2, dest	Edge boundary processing
PDIST	fsrc1, fsrc2, dest	Pixel distance calculation
ARRAY8/16/32	src1, src2, dest	Convert 3D address to blocked byte address
PST	fsrc, [address]	Partial store
FLD, STF	[address], fdest	8- or 16-Bit load/store to FP register
QLDA	[address], dest	128-Bit atomic load
BLD, BST	[address], dest	64-Bit block load/store

[a] S = single-precision option; L = little-endian option.
Source: Ref. 15.

most time-consuming part in MPEG-1 and MPEG-2 encoders, which normally needs more than 1500 conventional instructions for a 16 $\times$ 16 block search; however, the same job can be done with only 32 *PDIST* instructions on Ultra-Sparc. Needless to say, VIS has vastly enhanced the capability and role of UltraSparc in high-end graphics and video systems.

4.4 Commentary

Instruction set extensions increase the processing power of general-purpose processors by adding new functional units dedicated for video processing and/or modifying existing architecture. All of the extensions take advantage of subword parallelism. The new instruction set not only accelerates video applications greatly but also can benefit other applications that bear the same kind of subword parallelism. The extended instruction sets get the processors more involved in video signal processing and lengthens the lifetime of those general-purpose processors.

5 APPLICATION-SPECIFIC PROCESSORS

Although some of today's modern microprocessors are powerful enough to support computation intensive video applications such as MPEG-1 and MPEG-2, it is still worthwhile to design dedicated VSPs that are tailored for a specific applications. Many dedicated VSPs are available now (see Table 6). They display a variety of architectures, including the array processor [19], pipelined architecture [20], and the application-specific processor (ASIC) [21]. Application-specific processors are often used in cost-sensitive applications, such as digital cable boxes and DVD players. Because these processors are highly optimized for limited functionality, they usually achieve better performance/cost ratio for application-specific systems than multimedia-enhanced microprocessors or programmable VSPs; hence, they will continue to exist in some cost-sensitive environments.

Most dedicated VSPs have been designed for MPEG-1 and MPEG-2 encoding and decoding. By adopting special-purpose components (e.g., DCT/IDCT unit, motion estimation unit, run-length encoder/decoder, Huffman encoder/decoder, etc.) in a heterogeneous solution, dedicated VSPs can achieve very high performance at a relatively inexpensive cost.

5.1 8 × 8 VCP and LVP

The 8 × 8 ("8 × 8" is a product name) 3104 video codec processor (VCP) and 3404 low bit-rate video Processor (LVP) have the same architecture, which is shown in Figure 4. They can be used to build videophones capable of executing all the components of the ITU H.324 specification. Both chips are members from 8 × 8's multimedia processor architecture (MPA) family. The RISC IIT is a 32-bit pipelined microprocessor running at 33 MHz. Instead of using an instruction cache, it has a 32-bit interface to external SRAM for fast access. The RISC processor also supervises the two direct memory access (DMA) controllers, which provide 32-bit multichannel data passage for the entire chip. The embedded vision processor (VPe) carries out all the compression and decompression operations as well as preprocessing and postprocessing functions required by various applications. The chips can also be programmed for other applications, such as I-frame encoding, video decoding, and audio encoding/decoding for MPEG-1. The

Table 6 Summary of Some Dedicated VSPs

Vendor	Product(s)	Application(s)	Architecture	Peak perform.	Technology
8 × 8	3104 (VCP) 3404 (LVP)	H.324 videophone MPEG-1 I-frame encoder MPEG-1 decoder	Multimedia processor architecture (MPA), DSP-like engine	33 MHz	240-PQFP, 225-BGA, 5 V, 2 W
Analog Devices	ADV 601 ADV 601LC	4:1 to 350:1 real-time wavelet compression	Wavelet kernel, adaptive quantizer, and coder	27–29.5 MHz	120-PQFP, 5 V, low cost
	ADV 611 ADV 612	Real-time compression/decompression of CCIR-601 video at up to 7500:1	Wavelet kernel plus precise compressed bit rate control	27 MHz	120-LQFP
C-Cube	DV^x 5110 DV^x 6210	MPEG-2 main profile at main level encoder	DV^x multimedia architecture	100 MHz >10 BOPS	352-BGA, 3.3 V
	CLM 4440	MPEG2 authoring encoder	CL4040 Video RISC Processor 3 (VRP-3) loaded with different microcode	60 MHz 5.7 BOPS	240-MQUAD, 3 W
	CLM 4725 CLM 4740	MPEG-2 storage encoder MPEG2 broadcast encoder			
ESS Technology	AViA 500 AViA 502	MPEG-2 audio/video decoder	Video RISC processor-based architecture	80 MHz	160-PQFP, 3.3 V, 1.6 W
	ES3308	MPEG-2 audio/video decoder	RISC processor and MPEG processor		208-PQFP, 3.3 V, <1 W
IBM	MPEGME31 MPEGME30 (chipset)	MPEG-2 main profile at main level encoder	RISC-based architecture loaded with different microcode	54 MHz	304-CQFP, 0.5 µm, 3.3 V, 3.0–4.8 W
	MPEGCS22 MPEGCD21	MPEG-2 audio/video decoder	RISC-based architecture loaded with different microcode		160-PQFP, 0.4/0.5 µm, 3.3 V, 1.4 W

Company	Product	Application	Architecture	Clock	Package/Process
InnovaCom	DV Impact	MPEG-2 main profile at main level encoder		54 MHz	304-BGA, 4.5 W
LSI Logic	VISC (chipset)	MPEG-2 main profile at main level encoder	MIPS-compatible RISC core	54 MHz	208-QFP, 0.5 μm, 3.3 V
	L64002 L64005 L64020	MPEG-2 audio/video decoder / DVD decoder	Customized RISC engine	27 MHz	160-PQFP, 3.3 V
Matsushita	VDSP2 COMET	MPEG-2 main profile at main level encoder	SIMD DSP-core and motion-estimation processor	100 MHz / 80 MHz	261-PGA / 144-PGA
Mitsubishi	DISP II (chipset)	MPEG-2 main profile at main level encoder	RISC-processor-based architecture		393-PGA, 257-PGA, 152-QFP
NTT	ENC-C ENC-M	MPEG-2 simple profile at main level encoder		81 MHz	208-CQFP / 304-CQFP
Philips	SAA6750H	Single-chip low-cost MPEG-2 encoder	Motion estimator plus preprocessing	27 MHz	0.5 μm, 198 mm²
	SAA7201	MPEG-2 audio/video/graphics decoder	Video decoder, audio decoder, and graphics unit	27 MHz	160-PQFP, 3.3 V
	SAA4991	Motion-compensated field-rate conversion	Top-level processor and coprocessors for interpolation, motion estimation and vector	33 MHz / 10 BOPS	0.8 μm, 1 million transistor, 84-PLCC, 5 V, 1.8 W
Sony	CXD1922Q	MPEG-2 main profile at main level encoder	DSP controller and coprocessors	27 MHz (DSP)	208-PQFP, 0.4 μm, 4.5 million transistor
	CXD1930Q	MPEG-2 audio/video decoder	RISC, audio DSP and video processor	27 MHz	208-PQFP, 0.4 μm, 3.3 V
Vision Tech	MVision 10	MPEG-2 main profile at main level encoder	MIMD massively parallel scalable processor	40.5 MHz	304-CQFP, 0.5 μm, 5.2 million transistor

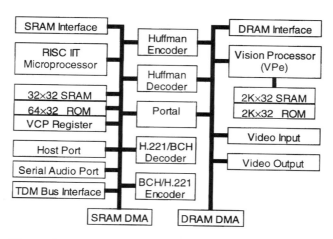

Figure 4   Architecture of 8 × 8 VCP and LVP. (From Ref. 22.)

microprogram is stored in the 2K × 32 on-chip ROM; the 2K × 32 SRAM provides alternatives to download new code.

   The RISC processor can be programmed using an enhanced optimizing C compiler, but further information about the software developing tools is not available. Targeting at low bit-rate video applications, both VCP and LVP are low-end VSPs which do not support real-time applications such as MPEG-1 encoding.

5.2   Analog Devices ADV601 and ADV601LC

Unlike other VSPs which target DCT, the ADV601 and ADV601LC [23] target wavelet-based schemes, which have been advocated for having advantages over classical DCT compression. Wavelet-basis functions are considered to have a better correlation to the broad-band nature of images than the sinusoidal waves used in DCT approaches. One specific advantage of wavelet-based compression is that its entire image filtering eliminates the block artifacts seen in DCT-based schemes. This not only offers more graceful image degradation at high compression ratios but also preserves high image quality in spatial scaling, even up to a zoom factor of 16. Furthermore, because the subband data of the entire image is available, a number of image processing functions such as scaling can be done with little computational overhead. Because of these reasons, both JPEG 2000 and the upcoming MPEG-4 incorporate wavelet schemes in their definition.

   Both the ADV601 and ADV601LC are low-cost (the 120-pin TQFP ADV601LC is, at this writing, $14.95 each, in quantities of 10,000 units) real-time video codecs that are capable of supporting all common video formats, in-

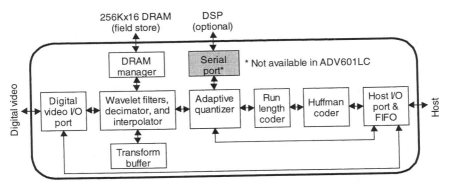

Figure 5 Block diagram of Analog Devices ADV601 (ADV601LC). (From Ref. 23.)

cluding CCIR-656. It has precise compressed bit-rate control, with a wide range of compression ratios from visually lossless (4:1) to 350:1. The glueless video and host interfaces greatly reduce system cost while yielding high-quality images. As shown in Figure 5, the ADV601 consists of four interface blocks and five processing blocks. The wavelet kernel contains a set of filters and decimators that process the image in both horizontal and vertical directions. It performs forward and backward biorthogonal 2D separable wavelet transforms on the image. The transform buffer provides delay line storage, which significantly reduces bandwidth when calculating wavelet transforms on horizontally scanned images. Under the control of an external host or digital signal processor (DSP), the adaptive quantizer generates quantized wavelet coefficients at a near-constant bit-rate regardless of scene changes.

5.3 C-Cube DV^x and Other MPEG-2 Codecs

The C-Cube DV^x 5110 and DV^x 6210 [24] were designed to provide single-chip solutions to MPEG-2 video encoding at both main- and high-level MPEG-2 profiles (see Table 7) at up to 50 Mbit/sec. Main profile at mail level (MP@ML) is one of the MPEG-2 specifications used in digital satellite broadcasting and digital video disks (DVD). SP@ML is a simplified specification, which uses only I-frames and P-frames in order to reduce the complexity of compression algorithms.

The DV^x architecture (Fig. 6), which is an extension of the C-Cube Video RISC Processor (VRP) architecture, extends the VRP instruction set for efficient MPEG compression/decompression and special video effects. The chip includes two programmable coprocessors. A motion estimation coprocessor can perform hierarchical motion estimation on designated frames with a horizontal search

Table 7 Profiles and Levels for MPEG-2 Bit Stream

Profile	Parameter	Level			
		Low (CIF)	Main (CCIR 601)	High 1440 (HDTV 4:3)	High (HDTV 16:9)
Simple (I- and P-frames only, 4:2:0)	Image size		720 × 576 (480)		
	Frame rate		25 (30) Hz		
	Bit rate		15 Mbit/sec		
Main (4:2:0)	Image size	352 × 288 (240)	720 × 576 (480)	1440 × 1152 (960)	1920 × 1152 (960)
	Frame rate	25 (30) Hz	25 (30) Hz	50 (60) Hz	50 (60) Hz
	Bit rate	4 Mbit/sec	15 Mbit/sec	60 Mbit/sec	80 Mbit/sec
SNR scalable (4:2:0)	Image size	352 × 288 (240)	720 × 576 (480)		
	Frame rate	25 (30) Hz	25 (30) Hz		
	Bit rate	3 Mbit/sec	10 Mbit/sec		
	Image size	352 × 288 (240)	720 × 576 (480)		
	Frame rate	25 (30) Hz	25 (30) Hz		
	Bit rate	4 Mbit/sec	15 Mbit/sec		

Profile	Image size	Frame rate	Bit rate
Spatially scalable (4:2:0)	720 × 576 (480)	25 (30) Hz	15 Mbit/sec
	1440 × 1152 (960)	50 (60) Hz	40 Mbit/sec
	1440 × 1152 (960)	50 (60) Hz	60 Mbit/sec
High (4:2:2, 4:2:0)	352 × 288 (240)	25 (30) Hz	4 Mbit/sec
	720 × 576 (480)	25 (30) Hz	15 Mbit/sec
	720 × 576 (480)	25 (30) Hz	20 Mbit/sec
	1440 × 1152 (960)	50 (60) Hz	20 Mbit/sec
	1440 × 1152 (960)	50 (60) Hz	60 Mbit/sec
	1440 × 1152 (960)	50 (60) Hz	80 Mbit/sec
	960 × 576 (480)	25 (30) Hz	25 Mbit/sec
	1920 × 1152 (960)	50 (60) Hz	80 Mbit/sec
	1920 × 1152 (960)	50 (60) Hz	100 Mbit/sec

Note: No shading = base layer; light shading = enhancement layer 1; dark shading = enhancement layer 2.
Source: Ref. 25.

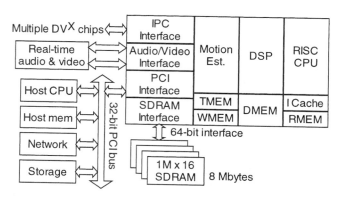

Figure 6 C-Cube DV^X platform architecture block diagram. (From Ref. 24.)

range of ± 202 pixels and vertical range of ± 124 pixels. A DSP coprocessor can execute up to 1.6 billion arithmetic pixel-level operations per second. The IPC interface coordinates multiple DV^x chips (at the speed of 80 Mbyte/sec) to support higher quality and resolution. The video interface is a programmable high-speed input/output (I/O) port which transfers video streams into and out of the processor. MPEG audio is implemented in a separate processor.

Both the AViA500 and AVia502 support the full MPEG-2 video main profile at the main level and two channels of layer-I and layer-II MPEG-2 audio, with all the synchronization done automatically on-chip. Their architectures are shown in Figure 7. In addition, the AViA502 supports Dolby Digital AC-3 sur-

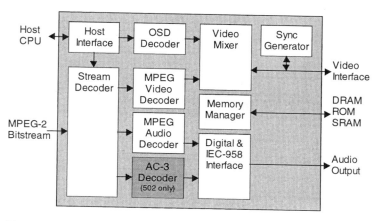

Figure 7 Architecture of AviA500 and Avia502. (From Ref. 24.)

round-sound decoding. The two MPEG-2 audio/video decoders each require 16 Mbit external DRAM.

These processors are sold under a business model which is becoming increasingly common in the multimedia hardware industry but may be unfamiliar to workstation users. C-Cube develops code for common applications for its processors and licenses the code chip customers. However, C-Cube does not provide tools for customers to write their own programs.

5.4 ESS Technology ES3308

As we can see from Figure 8, the ES3308 MPEG-2 audio, video, and transport-layer decoder [26] from ESS Technology has a very similar architecture to 8 × 8's VCP or LVP. Both chips have a 32-bit pipelined RISC processor, a microcode programmable low-level video signal processor, a DRAM DMA controller, a Huffman decoder, a small amount of on-chip memory, and interfaces to various devices. The RISC processor of ES3308 is an enhanced version of MIPS-X prototype, which can be programmed using optimizing C compilers. In an embedded system, the RISC processor can be used to provide all the system controls and user features such as volume control, contrast adjustment, and so forth.

5.5 IBM MPEG-2 Encoder Chipset

The IBM chipset for MPEG-2 encoding [27] consists of three chips: an I-frame (MPEGSE10 in chipset MPEGME30, MPEGSE11 in chipset MPEGME31) chip, a Refine (MPEGSE20/21) chip, and a Search (MPEGSE30/31) chip. These chips can operated in one-, two-, or three-chip configurations, supporting a wide range

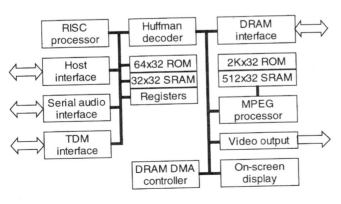

Figure 8 ES3308 block diagram.

of applications economically. In a one-chip configuration, a single "I" chip produces I-frame-only-encoded pictures. In a two-chip configuration, the "I" and "R" chips work together to produce IP-encoded pictures. Finally, in a three-chip configuration, B-frames are generated for IPB-encoded pictures. The chipset offers expandable solutions for different needs. For example, I-frame-only bit streams are good enough for video editing, IP-encoded bit streams can reduce coding delay in video conferencing, and IPB-encoded bit streams offer a good compression ratio for applications like DVD. Furthermore, the chipset is also able to generate a $4:2:2$ MPEG-2 profile at the main level. The encoder chipset has an internal RISC processor powered by a different microcode. IBM is releasing the microcode for the variable bit-rate (VBR) encoder. Little information is available from IBM about the architecture of the internal RISC processor and they do not offer tools for microcode-level development.

5.6   Philips SAA6750H, SAA7201, and SAA4991

The Philips SAA6750H [28] is a single-chip, low-cost MPEG-2 encoder which requires only 2 Mbytes of external DRAM. The chip includes a special-purpose motion estimation unit. It is able to generate bit streams that contain I-frames and P-frames. The designers claimed that "the disadvantage of omitting the B-frames can almost completely be eliminated using sophisticated on-chip pre-processing" and "at 10 Mbit/s, the CCIR picture quality is comparable with DV coding, while at 2.5 Mbit/s the SIF picture quality is comparable with Video CD" [28].

The SAA7201 [29] is an integrated MPEG-2 audio and video decoder. In addition, it incorporates a graphics decoder on-chip, which enhances region-based graphics and facilitates on-screen display. Using an optimized architecture, the AVG (audio, video, and graphics) decoder only requires $1M \times 16$ SDRAM, yet more than 1.2 Mbits (2.0 Mbits for a 60-Hz system) is available for graphics. The internal video decoder can handle all the MPEG-compliant streams up to the main profile at the main level, and the layer-1 and layer-2 MPEG audio decoder supports mono, stereo, surround sound, and dual-channel modes. The on-chip graphics unit and display unit allow multiple graphics boxes with background loading, fast switching, scrolling, and fading. Featuring a fast CPU access, the full bit-map can be updated within a display field period.

The Philips SAA4991 WP (MELZONIC) [30] is a motion-compensation chip, designed using Phideo, a special architecture synthesis tool for video applications developed by Philips Research [31]. This chip can automatically identify the original frame transition and correctly interpolate the motion up to a field rate of 100 Hz. In addition, it also performs noise reduction, vertical zoom functions, and $4:3$ to $16:9$ conversion. Four different types of SRAM and DRAM totaling 160 Kbits are embedded on-chip in order to deliver an overall memory bandwidth of 25 Gbit/sec.

5.7 Sony Semiconductor CXD1922Q and CXD1930Q

The CXD1922Q [32] is a low-cost MPEG-2 video encoder for a real-time main profile at the main level. The on-chip encoding controller supports variable bit-rate encoding, group-of-pictures (GOP) structure, adaptive frame/field MC/DCT (motion compensation–DCT) coding and programmable quantization matrix tables, and so forth. The chip uses multiple clocks for different modules (a 67.5-MHz clock for SRAM control; a 45-MHz clock for motion estimation and motion compensation, which has a wide search range of -288 to $+287.5$ pixels in horizontal and -96 to $+95.5$ pixels in vertical; a 22.5-MHz clock for variable-length encoding block; a 13.5-MHz clock for front-end filters; and a 27-MHz clock for the DSP core), yet it only consumes 1.2 W.

The CXD1930Q [33] is another member of Sony Semiconductor's Virtuoso family. It incorporates the MPEG-1/MPEG-2 (main profile at main level) video decoder, MPEG-1/MPEG-2/Dolby Digital AC-3 audio decoder, programmable preparser for system streams, programmable display controller, subpicture decoder for DVD and letter box, and some other programmable modules. The chip targets low-cost consumer applications such as DVD players. The embedded RISC processor in the CXD1930Q is able to support real-time multitasking through Sony's proprietary nano-OS operating system.

5.8 Other Dedicated Codecs

InnovaCom DVImpact [34] is a single-chip MPEG-2 encoder that supports main profile at main level. This chip has been designed from the perspective of the systems engineer; a multiplexing function has been built in so as to relieve the customer's task of writing interfacing code. Although the detailed architecture is not available, it is not difficult to infer that the kernel must be a RISC processor plus a powerful motion estimator, like the ones used in C-Cube's DV^x architecture.

The LSI Logic Video Instruction Set Computing (VISC) encoder chipset [35] consists of three ICs: the L64110 video input processor (VIP) for image preprocessing, the L64120 advanced motion estimation processor (AMEP) for computation-intensive motion search, and the L64130 advanced video signal processor (AVSP) for coding operations such as DCT, zigzag ordering, quantization, and bit-rate control. Although the VIP and AVSP are required in all the configurations, users can choose one to three AMEPs, depending on the desired image quality. The AMEP performs a wide search range of ±128 pixels in both horizontal and vertical directions. All the three chips need external VRAMs to achieve a high bandwidth. Featuring the CW4001 32-bit RISC (which has a compatible instruction set with MIPS) core, the VIP, AMEP, and AVSP can be programmed using the C/C++ compilers for MIPS, which greatly simplifies the development of the firmware.

Mitsushita Electric Industrial's MPEG-2 encoder chipset [36] consists of a video digital signal processor (VDSP2) and a motion estimation processor (COMET). To support MPEG-2 main profile encoding at main level, two of each are required; but an MPEG-2 decoder can be implemented with just one VDSP2. Inside the VDSP2, there are a DRAM controller, a DCT/IDCT unti, a variable-length-code encoder/decoder, a source data input interface, a communication interface, and a DSP core which further include four identical vector processing units (VPU) and one scalar unit. Each VPU has its own ALU, multiplier, accumulator, shifters and memories based on the vector-pipelined architecture [37]. Therefore the entire DSP core is like a VLIW engine.

Mitsubishi's DISP II chipset [38] includes three chips: a controller (M65721), a pixel processor (M65722) and a motion estimation processor (M65727). In a minimum MPEG-2 encoder system, a controller, a pixel processor, and four motion-estimation processors are required to provide a search range of 31.5×15.5. Like some other chipsets, the DISP II is also expandable. By adding four more motion-estimation processors, the search range can be enlarged to 63.5×15.5.

The MVision 10 from VisionTech [39] is yet another real-time MPEG-2 encoder for the main profile at the main level. It is a single-chip Multiple Instruction Multiple Data (MIMD) processor, which requires eight $1M \times 16$ extended data out (EDO) DRAMs and four $256K \times 8$ DRAM FIFOs. Detailed information about the internal architecture is not available.

5.9 Summary of MPEG-2 Encoders

Digital satellite broadcasting and DVD have been offering great market opportunities for MPEG-2. MPEG-2 encoders are important for broadcast companies, DVD producers, nonlinear editing, and so forth, and it will be widely used in tomorrow's video creation and recording products (e.g., camcorders, VCRs, and PCs). Because they reflect the processing ability and represent the most advanced stage of dedicated VSPs, we summarize them in Table 8.

5.10 Commentary

Dedicated video codecs, which are optimized for one or more video-compression standards, achieve high performance in the application domain. Due to the complexity of the video standards, all of the VSPs have to use microprogrammable

Table 8 Summary of MPEG-2 Encoders

Vendor	Product(s)	Chip count	Supported profile at main level	Encoding bit rate (Mbit/sec)	Var. bit rate	Group of pictures	Search range (pixels) H	V	External memory for MP@ML
C-Cube	DV^X 5110	1	SP, MP	2–15	Yes	I, IP, IBP, IBBP	±202	±124	8 MB DRAM
	DV^X 6210	2	SP, MP, 4:2:2	2–50	Yes	I, IP, IBP, IBBP	±202	±124	16 MB DRAM
	CLM 4725	7	SP, MP	3–15	Yes		±100^P ±52^B	±28^P ±24^B	14 MB DRAM
IBM	MPEGME 30/31	3	SP, MP, 4:2:2	1.5–40	Yes	I, IP, IBP	±64	±56	5–6 MB DRAM 256 KB SRAM
Innova Com	DV Impact	1	SP, MP			I, IP, IBP, IBBP, IBBBP	±64	±64	9 MB DRAM
LSI Logic	VISC	5	SP, MP	2–15			±128	±128	10–14 MB VRAM
Matsushita	VDSP2 & COMET	4–6	SP, MP	Up to 15			±48^P ±32^B	±48^P ±32^B	14 MB DRAM
Mitsubishi	DISP II	6 or 10	SP, MP	1–20			±64^P ±32^B	±16	5.5 MB SDRAM
NTT	ENC-C & ENC-M	2	SP	1.8–15			±48.5	±24.5	512 KB VRAM 2 MB SDRAM
Sony	CXD1922Q	1	SP, MP	Up to 25	Yes	I, IP, IBP, IBBP	±288	±96	8 MB SDRAM
Vision Technology	Mvision10	1	SP, MP	1–24			±50	±34	16 MB DRAM 1 MB FIFO

RISC cores. It is important for the dedicated VSPs to be configurable or programmable to accept different compression standards and/or parameters.

6 PROGRAMMABLE VSP

Although MPEG is an important application, it is only one of many video applications. Therefore, it would be extremely helpful to develop highly programmable VSPs that can support a whole range of applications. Video applications are continuously becoming more complex and diverse. While still working hard on MPEG-2, people have already proposed MPEG-4, and MPEG-7 is on the schedule. Apparently, dedicated VSPs cannot keep pace with the rapid evolution of new and some existing video applications.

Although usually less expensive, dedicated VSPs may not be the overall winner in a comprehensive system. We might need quite a few different dedicated VSPs in a complicated system which must support several different multimedia applications such as video compression/decompression, graphics acceleration, and audio processing. Furthermore, the development cost of dedicated VSPs is not inexpensive, as the designers must hand-tune many parts to achieve the best performance/cost ratio. Because of the large potential market demand, programmable VSPs seem to be relatively inexpensive.

Consequently, the need for greater functionality, as well as increased cost and time-to-market pressures, will push the video industry toward programmable VSPs. The industry has already seen a similar trend in modem and audio codecs. More and more, new systems incorporate DSPs instead of dedicated controllers.

Generally speaking, all of the VSPs are programmable to some degree; some of them have multiple powerful microprocessors, some have a RISC core and several coprocessors, and others only have programmable registers for system configuration. Our definition of ''programmable'' excludes the last category. Due to the complexity of video encoding algorithms, dedicated encoders have to use a processor core and many special-purpose functional units optimized for various parts of the algorithm, such as a motion-estimation unit, a vector quantization unit, a variable-length code encoder, and so on. By loading different microcodes into the core processor, the chip is able to generate different data formats for different standards. As a matter of fact, most of the dedicated video encoders support several standards. A demonstrative example is the DV^x architecture developed by C-Cube mentioned earlier. The kernel of this architecture is a 32-bit embedded RISC CPU; it also contains two programmable coprocessors. However, the architecture is designed and optimized for MPEG-2 encoding and decoding, not for general video applications.

What we are interested in is highly programmable VSPs that are more flexible and adaptable to new applications. These chips would be somewhat similar to general-purpose processors in terms of functionality and programmability.

However, the difference is that the VSPs are dedicated to video signal processing. Many video applications belong to the category of scientific calculation and have some good properties such as regular control flow and symmetry in data processing. For a variety of video applications, including H.263, MPEG-1, MPEG-2, and MPEG-4, the whole video frame is divided into several blocks and then the data processing procedure for each block is exactly the same. This kind of symmetry carries a huge amount of parallelism. A high-performance programmable VSP must have a well-defined parallel architecture that is capable of exploring the potential parallelism.

All of the VSPs in Tables 9 and 10 are programmable to some extent. They all have at least one internal microprocessor, on which programs or microcodes are running. However, not all of the manufacturers provide developing tools for users to implement and test their own applications. Some only provide firmware necessary to support end applications.

6.1 Chromatic Research Mpact2 Media Processor

Media processors are different from traditional VSPs in that they are not only dedicated to accelerate video processing but are also capable of improving other multimedia functions (e.g., audio, graphics). The Mpact media processor [40] is a low-cost, multitasking, supercomputerlike chip that works in conjunction with an x86/MMX processor to provide a wide range of digital multimedia functions. In an Mpact-based multimedia system, specialized software called Mpact mediaware runs on both the Mpact chip and x86 processor, delivering multimedia functions such as DVD, videophone, video editing, 2D/3D graphics, audio, fax/modem, and telephony.

The block diagram of the data path of Mpact2 R/6000 is shown in Figure 9. Two major enhancements were made to Mpact2 over the initial Mpact architecture to support 3D graphics: a pipelined floating-point unit and a 3D-graphics rendering unit (ALU group 6) and its associated texture cache. In each cycle, the Mpact2 can start a pair of floating-point add operations and a pair of floating-point multiply operations, yielding a peak performance of 500 MFLOPS (mega floating-point operations per second) at 125 MHz. The dedicated 3D-graphics rendering unit is a 35-stage scan-conversion pipeline which can render 1 million triangles per second. The Mpact2 also has a VLIW core, for which each instruction word is 81 bits long and contains two instructions which may cause eight single-byte operations to be executed on each of the four ALU groups (groups 1–4). The Mpact2 data paths are all 72 bits wide. Data exchanges are done via a crossbar on a 792-bit interconnection bus, which can transfer eleven 72-bit results simultaneously at an aggregate bandwidth of 18 Gbyte/sec.

In multimedia applications, external bandwidth is as critical as internal throughput. Chromatic chose Rambus RDRAMs because they provide a high bandwidth at a very low pin count. The Mpact2 memory controller supports two

Table 9 Summary of Some Programmable VSPs by Vendor

Vendor	Processor(s)	Application(s)	Architecture	Peak perform.	Technology
Chromatic Research	Mpact2 R/6000	MPEG-1 encoding MPEG-1 & -2 decoding Windows GUI acceleration 2D/3D graphics H.320/H.324 videophone Audio, FAX/modem	VLIW (SIMD) 6 ALU groups, 2 Rambus RAC channels, and 5 DMA bus controllers	125 MHz 6 BOPS	0.35 µm, 352-BGA, 3.3 V
MicroUnity	Cronus	Communicating and processing at broad-band rates	Single instruction group data, 5 threads	300 MHz	0.6 µm, 441-BGA, 3.3 V
Mitsubishi	D30V	MPEG-1 & -2 decoding Dolby AC-3 decoding H.263 codec 2D/3D graphics; modem	Two-way VLIW RISC core	250 MHz 1 BOPS	0.3 µm, 135-PGA, 2V
NEC	V830R/AV	MPEG-1 encoding MPEG-1 & -2 decoding	Superscalar RISC core and SIMD multimedia extension	200 MHz 2 BOPS	0.25 µm, 208-PQFP, 2.5 V, 2 W

Philips	TriMedia TM-1000	MPEG-1 encoding MPEG-1 & -2 decoding 2D/3D graphics H.32x videophone V.34 modem	VLIW core with 27 function units and 5 issue slots; coprocessors: video I/O, audio I/O, timer, image coprocessor, etc.	100 MHz 4 BOPS	0.35 μm, 240-BGA, 3.3 V, 4 W
Samsung	MSP-1	MPEG-1 & -2 decoding H.324 codec 2D/3D graphics AC-3 decoding, wave-table Modem, telephony	ARM7 RISC core and vector processor in shared memory	100 MHz 6.4 BOPS 8-bit int. 1.6 GFLOPS	0.5/0.35 μm, 128/256 pin, 3.3 V, 4 W
Texas Instruments	TMS320-C80 (MVP)	MPEG-1 encoding/decoding H.261 codec JPEG; 2D/3D graphics Image and vector processing	MIMD architecture with 4 integer DSPs and 1 floating-point RISC	50 MHz 2 BOPS 100 MFLOPS	305-CPGA, 3.3 V
	TMS320-C6201	Modems; servers Wireless based stations Multi-channel telephony	VLIW DSP core with 8 functional units and dual data-path	200 MHz 1.6 BOPS	0.25 μm, 352-BGA, 2.5 V, 4.2 W
	TMS320-C6701	High-performance fixed- and floating-point digital signal processing	TMS320C6201 added floating-point units	167 MHz 1 GFLOPS	0.18 μm, 352-BGA, 1.8 V

Table 10 Summary of Some Programmable VSPs by Processor

Processor(s)	Func. units	Issue slots	Data width (bit)	Register file	Inst. width (bit)	Inst. $/RAM	Data $/RAM	Float. point	Develop. tools
Mpact2	8	2	72	512×72	81	20 Kb	18-Kb texture cache	Yes	No
Cronus	5	1	128	$5 \times 64 \times 64$	32	256 Kb	256-Kb cache/buffer	No	C/C++, assembly
D30V	7	2	32	64×32	64	256 Kb	256 Kb RAM	No	
V830R/AV	V 3	1	32	32×32	16/32	128 Kb	128-Kb cache	No	
	M 6		64	32×64	32				
TriMedia TM-1000	27	5	32	128×32	220	256 Kb	128-Kb cache	Yes	C/C++, assembly
MSP-1	2	2	32	31×32	32	16 Kb	40-Kb cache	Yes	MSP tools
TMS320-C80	5	1	32	31×32	32	32 Kb	32-Kb cache	Yes	C, assembly
(MVP)	4×5	4×4	32	$4 \times 44 \times 32$	64	4×16 Kb	272-Kb RAM		
TMS320-C6201	8	8	32	32×32	256	512 Kb	512-Kb data RAM	No	C, assembly
TMS320-C6701	8	8	32	32×32	256	512 Kb	512-Kb data RAM	Yes	C, assembly

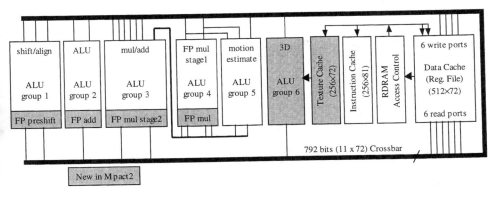

Figure 9 Architecture of Mpact2 datapath. (From Ref. 41.)

9-bit 300-MHz Rambus channels of RDRAM media memory (1.2 Gbyte/sec), which store both instructions and data. All of the other Mpact2 I/O ports have also been improved from the first generation: The PCI interface now operates at 66 MHz, enabling a peak transfer rate of 264 Mbyte/sec; the display interface incorporates a 220-MHz RAMDAC on-chip; and the digital video interface becomes fully duplex.

The Mpact2 media processor runs an optimized real-time multitasking kernel to simultaneously execute several Mpact mediaware modules. This kernel, along with the programming model and instruction set, is completely proprietary and developed in-house only. Besides the drivers for accelerating multimedia applications, Chromatic does not provide any tools for users to write their own microcode.

6.2 MicroUnity's Broad-Band Media Processor

''A broadband media processor is a general-purpose processor system with sufficient computing resources to communicate and process digital, audio, data, and radio frequency signals at broadband rates (more than 1.5 Mbit/s)'' [42]. The MicroUnity broad-band media processor is intended to be the sole processor in client or terminal systems. As such, it is a general-purpose microprocessor, including a memory management unit. Figure 10 shows the architecture of the media processor. In the first BiCMOS implementation, the 1-GHz clock drives a 512-Gbit/sec computing bandwidth and a 128-Gbit/sec memory bandwidth. An integrated SDRAM interface supports a peak bandwidth of 3.2 Gbit/sec. The media channel also offers a great communication bandwidth as high as 32 Gbit/sec, which can be used to construct a multiprocessor system. A very simple packet

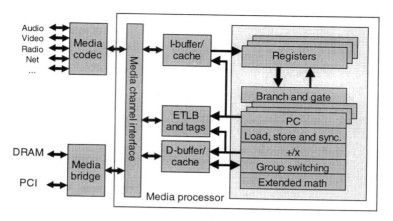

Figure 10   Structure of MicroUnity broad-band media processor. (From Ref. 43.)

control protocol is used in the multiprocessor system, with several types of hardware support such as low-latency memory-mapped I/O.

The MicroUnity media processor exploits parallelism from two perspectives: group instructions and multithreading. The group instructions specify operations on four 128-bit register pairs, totaling a bandwidth of 512 bits/instruction. This architecture, referred to as Single Instruction Group Data (SIGD), is almost exactly the same as the instruction set extensions introduced in Section 6, and a very similar concept can also be found in Texas Instruments' MVP, where the splittable ALUs can be reconfigured dynamically. There is some difference, however, in the size and number of operands. Although most other general-purpose microprocessors work on two 64-bit source operands, MicroUnity's media processor can take up to three source register pairs, each 128 bits long (a register pair consists of two 64-bit registers and can be used to represent different data granularities from two 64-bit words to 128 single bits). In order to deal with unaligned or mixed-precision data, the broad-band media processor also provides switching instructions which can shift, shuffle, extract, expand, and swizzle operands as well as other kinds of manipulation. These switching instructions are much more powerful than any other processors. Other instructions include control instructions, which can effectively reduce branch overhead. For example, branch-gateway instruction fetches 128 bits from memory into a pair of registers (code and data pointers, respectively) while checking translation lookaside buffer (TLB) for access control; then it jumps to the code pointer, storing a result link in its space. This is extremely helpful for active message in message passing based multiprocessor systems. In addition, MicroUnity's media processor provides extended math operations such as multiply over 8-bit Galois fields [i.e.

GF(256)]. This makes the processor not only suitable for source coding (e.g., MPEG-2) but also useful for channel coding (e.g., Reed–Solomon error-correcting code). All of the instructions are summarized in Table 11.

Another key feature of the architecture is that it supports multithreading. Like superscalar, multithreading is another technique to exploit instruction-level parallelism. A thread is an encapsulation of the control flow in a program. Multithreaded programs have multiple threads running through different code paths concurrently, so multithread processors interleave a number of threads in a single execution pipeline. Although in a fine-grain view, only one instruction from one thread is running in a single pipeline, this technique can hide dependency. Instead of waiting for a stalled instruction, which would cause a pipeline bubble, multithreaded processors switch to another thread, which is independent of the previous one. By the time the processor is back on the thread that had dependency constraints, the next instruction in the thread can be issued immediately as the dependency would have gone, maximizing pipeline utilization. In MicroUnity's media processor, five hardware-based threads of control share a common datapath, instruction cache, data cache, DRAM interface, and a set of I/O interfaces.

MicroUnity provides an open programming platform. Their software development environment includes assembler, C/C++ compiler, source-code debugger, profiler, and media and communications software libraries for various standards. In addition, the company also has a real-time micro kernel for client devices and a 64-bit Open Software Foundation UNIX for server applications [44].

6.3 Mitsubishi D30V

The D30V [45] is not really a VSP but an optimized processor core for video signal processing. By integrating a few other components, an MPEG-2 decoder can be implemented on a single chip. The architecture of the D30V core is shown in Figure 11.

There are three execution units in the D30V processor core: a memory unit, an integer unit, and a branch unit (inside the instruction decode unit). In addition to program sequencing control, the memory unit is also able to crunch data in its ALU and shifter. It supports several data types from byte (signed and unsigned) to 64-bit word. Its load and store instructions can operate on multiple operands using packing and unpacking. Other features of the memory unit include postincrement/decrement and modulo addressing, which ease programming. The integer unit contains a 32-bit ALU, a 32-bit shifter, a 32 × 32 multiplier, and two 64-bit accumulators. The 64-bit accumulators support 64-bit arithmetic, and all of the other functional units operate on both 32-bit and 16-bit operands. The integer unit has been optimized for video applications. It has subword and half-word operations to further exploit parallelism, and a few added video operations

Table 11 Media Processor Instruction Set Summary

Category	Instructions	Optional features	Interval-issue-latency (cycles)
Storage (8, 16, 32, 64, or 128 bits) and synchronization (64 bits)	Load 8, 16, 32, 64, or 128 bits, little- or big-endian	Unsigned, aligned, immediate	2-1-2
	Store 8, 16, 32, 64, or 128 bits, little- or big-endian	Aligned, immediate	4-1-0
	Store add, compare, or multiplex 64 bits	Immediate, -and-swap	8-7-7
Branch (64 bits)	Branch and-equal, and-not-equal, less, or less-equal-zero		2-1-1 for pipelined, 2-1-4 for unpipelined
	Branch equal, not-equal, less, or greater-equal		2-1-1
	Branch floating-point equal, not-equal, less, or greater-equal (16, 32, 64, or 128 bits)		
	Branch	Immediate, -and-link	2-1-1
	Branch gateway	Immediate	2-2-1
	Branch down or back		2-1-1
Fixed point (64 bits) and group (128 × 1, 64 × 2, 32 × 4, 16 × 8, 8 × 16, 4 × 32, or 2 × 64 bits)	Add or subtract	Immediate, overflow	1-1-1
	Multiply	Unsigned, -and-add	1-5-7 for 32-bit, 1-20-22 for 64-bit multiply, 1-23-25 for 64-bit multiply-and-add, 1-2-4 for others
	Divide	Unsigned	

Operation	Modes	Code
AND, OR, AND-NOT, OR-NOT, XOR, XNOR, NOR, or NAND	Immediate	1-1-1
Shuffle, deal, or swizzle		1-1-2
Compress or expand	Unsigned, immediate	1-1-2
Extract	Unsigned, immediate	1-2-3
Deposit or withdraw immediate	Unsigned, merge	1-1-2
Shift or rotate right or left	Unsigned, immediate, overflow	1-1-2
4- or 8-Way multiplex	Shuffle, transpose	1-1-2
Select bytes		1-1-2
Set or sub, equal, not-equal, less, or greater-equal	Unsigned, immediate	
Multiplex		1-1-1
AND sum of bits		1-1-3
Log most significant bit		1-1-2
Galois-field multiply, polynomial multiply-divide, 8 or 64 bits		1-4-5
Add, subtract, multiply, or divide	Near, truncate, floor, ceiling, or exact	
Multiply-and-add or -subtract	Near, truncate, floor, ceiling, or exact	
Square-root, sink, float, or deflate	Near, truncate, floor, ceiling, or exact	
Absolute, negate, inflate	Exception	
Set equal, not-equal, less, greater-equal	Exception	

Floating-point scalar (16, 32, 64, or 128 bits) and group (8 × 16, 4 × 32, or 2 × 64 bits)

Source: Ref. 42.

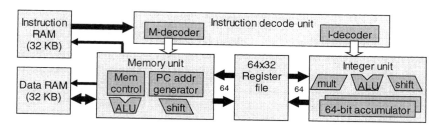

Figure 11 Mitsubishi D30V processor core. (From Ref. 46.)

such as variable length saturation instruction, join instruction, add sign instruction, and so on. The branch unit has a variable number of delay slots and additional conditional branches (e.g., test zero and branch, test notzero and branch), which enable zero-delay branches and zero-overhead loops. All of the functional units in the D30V core are fully pipelined using a four-stage pipeline; they are controlled by a 64-bit VLIW instruction, which contains two short or one long RISC subinstructions. The dual-issue processor has used some advanced techniques to improve the performance. These techniques include predicated execution and speculative execution.

The D30V processor core was designed not only to meet computational requirements and cost but also to provide the flexibility of programmable processors. However, the dual-issue processor is not powerful enough for computation-intensive video applications like MPEG-2 encoding. In an implementation of an MPEG-2 decoder, two D30V cores and several small processing units are required in addition to a dedicated motion-estimation processor.

6.4 Philips TriMedia

The Philips TriMedia [47] is also a general-purpose microprocessor enhanced for multimedia processing. As shown in Figure 12, at the center of the chip is a 400-Mbyte/sec high-speed backbone that connects autonomous modules and provides accesses to internal control registers. The data highway, consisting of a 32-bit data bus and a 32-bit address bus, uses a block transfer protocol, which can transfer 64 bytes at a burst.

TM-1000 incorporates independent DMA-driven peripheral units and coprocessors to streamline data throughput. These on-chip processing units can be masters or slaves on the data highway, and they manage input, output, and formatting of multimedia data streams as well as perform specific functions. While sending an image from SDRAM to the video frame buffer (could be in the SDRAM or the host's PCI-based graphics system), the image coprocessor

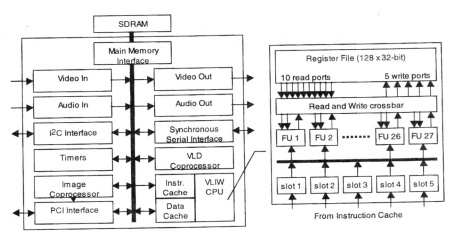

Figure 12   Architecture of TriMedia. (From Ref. 48.)

can perform horizontal or vertical image filtering and scaling, YUV to RGB color space conversion, as well as overlay live video on background image. The variable-length decoder (VLD) coprocessor offloads the VLIW CPU by decoding Huffman-encoded video bit streams. Due to the characteristics of the algorithm, this task has little inherent parallelism and, hence, is not suited for VLIW processing. The two coprocessors are microprogrammable. They are independent of the VLIW CPU and are synchronized with it using an interrupt mechanism.

The VLIW processor has a rich instructions set (197 instructions), including many extensions for handling multimedia data types. Parallelism is achieved by incorporating 27 functional units in the VLIW engine and feeding them with five instruction issue slots. The type and number of functional units are listed in Table 12. All of the functional units are pipelined, with a depth ranging from 1 to 17 stages. The five constant units do not perform any calculation except providing ports for accessing immediate values stored in the instruction word. Like many other processors, TM-1000 also provides pack/unpack and group instructions, which can manipulate 4 bytes or two 16-bit words at one time, exploiting subword parallelism. Other special instructions include *me8* for motion estimation, which is similar to the *PDIST* instruction in UltraSparc's visual instruction set. Most instructions accept a guard register for predicated execution. Although the TM-1000 processor has 27 functional units, in each cycle it can issue only up to 5 instructions.

The TM-1000 has a dedicated instruction cache and a data cache on-chip, both of which are eight-way set-associative with LRU (least-recently used) replacement and locking mechanism to improve performance. The 16 KB dual-

Table 12 TM-1000 Functional Units

Name	Quantity	Latency (cycles)	Recovery (cycles)
Constant	5	1	1
Integer ALU	5	1	1
Memory load/store	2	3	1
Shift	2	1	1
DSP ALU	2	2	1
DSP multiply	2	3	1
Branch	3	3	1
Floating-point ALU	2	3	1
Integer/floating-point multiply	2	3	1
Floating-point compare	1	1	1
Floating-point sqrt/divide	1	17	16

Source: Refs. 47 and 48.

ported data cache allows two simultaneous nonblocking accesses. To save bandwidth and storage space, the VLIW instructions are stored and cached using a 2–23-byte compressed format until they are fetched from the instruction cache. The chip also has a glueless interface to support four banks of external SDRAMs.

The TM-1000 development environment includes a VLIW compiler, a C/C++ software development environment, and the pSOS+ real-time operating system.

The TriMedia CPU64 [49] is a successor to the TM-1000. The CPU64 has a 64-bit word and uses subword parallelism within the VLIW CPU to increase parallelism on small data words.

6.5 Samsung Media Signal Processor

The Samsung media signal processor MSP-1 [50] is a cache-based dual-processor architecture (Fig. 13). The architecture consists of a floating-point vector processor for digital signal processing, an ARM7 RISC CPU for system control and management, a bit-stream processor for parsing the video stream, I/O interfaces, and 10K unused gates for optimal customization. The vector processor, running at 100 MHz, supports various type of integer (from bytes to 32-bit words) and 32-bit IEEE 754 floating-point numbers, with a peak performance of 6.4 billion operations per second (BOPS) 8-bit integers and 1.6 BOPS 32-bit floating points. The ARM7 CPU, running at 50 MHz, is responsible for general functions such as real-time scheduling. Both processors share the same cache subsystem and can operate simultaneously. In this dual-processor architecture, the bulk of the

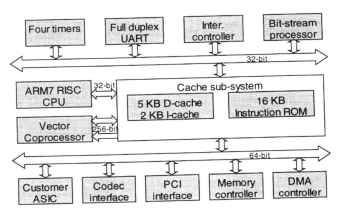

Figure 13   MSP-1 microarchitecture. (From Ref. 50.)

video processing is performed by the vector processor because the RISC CPU only deals with general system functions.

The MSP is a fully programmable media processor with a rich instruction set, including standard ARM RISC instructions for scalar processing, high-performance SIMD instructions for vector processing, I/O instructions for block load/store, and special instructions for filtering and MPEG applications. The programming model also has macro library instructions such as *DCT, CONV*, and *MULM*. The software development tools include MSP-oriented assembler, compiler, linker, debugger, and simulator.

6.6   Texas Instruments' TMS320C8x Multimedia Video Processor

Texas Instruments' TMS320C8x (MVP) [51] family has four members and two architectures. The TMS320C80 is a highly integrated multiprocessor and was an early single-chip MPEG-1 encoder. As shown in Figure 14, the C80 includes four advanced integer DSPs (ADSPs) and a floating-point RISC master processor (MP), which are integrated with a transfer controller (TC), a video controller (VC), and five memory banks. The TMS320C80 allows 5 instruction fetches and 10 parallel data accesses in each cycle, allowing a transfer rate as high as 1.8 Gbyte/sec for instruction and 2.4 Gbyte/sec for data. The younger member, TMS320C82, is a scaled down version of the TMS320C80. It provides better cost/performance ratio for some cost-sensitive applications by removing two integer DSPs, the video controller, and some on-chip memory.

The master processor of C80 is a general-purpose RISC processor with an IEEE 754 compatible three-stage floating-point pipeline. The RISC processor is

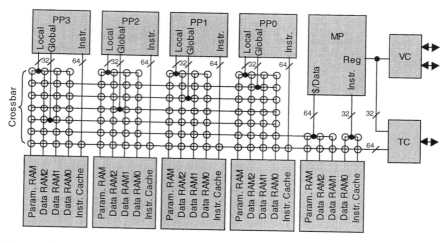

Figure 14 Architecture of Texas Instruments' TMS320C80 (MVP). (From Ref. 51.)

powered by 32-bit instructions and can issue, in each cycle, a parallel multiply, an add, and a 64-bit load/store, yielding 100 MFLOPS at 50 MHz. The floating-point unit can perform both single- and double-precision arithmetic.

Each of the four ADSPs is a 32-bit integer DSP optimized for bit- and pixel-oriented imaging and graphics applications. Each parallel processor can issue, in each cycle, a multiply, an ALU operation, and two memory accesses within a single 64-bit instruction. The parallelism comes from two independent datapaths. The multiplier datapath includes a three-stage 16×16 multiplier, a half-word swapper, and rounding hardware. The ALU datapath includes a 32-bit three-input ALU, a barrel rotator, a mask generator, a 1-bit to n-bit expander, a left/rightmost and left/rightmost bit-change logic, and several multipliers. The 32-bit three-input ALU can perform all of the 256 three-input Boolean combinations as well as many other mixed logical and arithmetic operations. Both the multiplier and the ALU are splittable. Although the 16×16 multiplier can be split into two 8×8 multipliers, the 32-bit ALU can be divided into two 16-bit ALUs or four 8-bit ALUs. The big register file contains 8 data registers, 10 address registers, 6 index registers, and 20 other user-visible registers. Three hardware loop controllers enable zero-overhead looping/branching and multiple-loop end points. The ADSPs provide conditional operation (also referred to as predicated execution).

The video controller handles both video input and output and can simultaneously support two independent capture or display systems. The transfer controller combines a memory interface and a DMA engine, handling data movement within the MVP system as requested by the master processor, parallel processors, video controller, and external devices. In addition to handling internal block transfers, it also supports direct interface to SRAM, DRAM, SDRAM, and VRAM, with a peak external bandwidth of 400 Mbyte/sec.

Processing elements in the MVP architecture communicate by a global crossbar (actually an incomplete crossbar because it is missing some interconnect paths), which provides internal fast data communication on a cycle-by-cycle basis. The crossbar automatically sets up the connection between a module and a memory bank every cycle without special configuration instructions. It allows 5 instruction fetches and 10 parallel data accesses at a time, totaling 1.8 Gbyte/sec instruction transfer and 2.4 Gbyte/sec data transfer.

The C80 comes with an open programming environment which includes assembler, C compiler, linker, simulator, debugger, and emulator.

6.7   Texas Instruments' VelociTI

Texas Instruments' VelociTI [52] is an advanced VLIW architecture for digital signal processing. Although it is not designed for video applications and contains no functional units dedicated to video operation, the potential performance can greatly accelerate some multimedia applications. Therefore, we highlight its features here.

Announced in February 1997, TMS320C6201 is the first member of the VelociTI architecture and it offers a breakthrough for digital signal processing by enabling a sustained throughput of up to eight 32-bit instructions every cycle, achieving 1600 MIPS at 200 MHz. As shown in Figure 15, the C6201 has two independent 32-bit datapaths, each having four functional units (.L, .S,. M, and .D; see descriptions in Table 13) and a register file containing 16, 32-bit registers. A data bus joins the two register files and provides fast data exchanges between them, so logically there is only one unified register file. The instruction set is in favor of digital signal processing: It supports saturation, bit-field set/clear and extract, bit counting, normalization, and various data types from a 1-bit to a 32-bit word. All instructions are conditional (predicated) to provide more parallelism. The 4-Gbyte space is byte-addressable with dual-endian support and a variety of addressing modes, including circular addressing with a 5–15-bit offset. The C6201 incorporates large amount of memory on-chip. The 1-Mbit RAM is split between data and instruction memory. The external memory interface (EMIF), capable of operating at 200 MHz, supports both SRAM and synchronous DRAM.

Using the same VelociTI architecture, the C67x extends the C6201 by adding floating-point capability to six of the eight functional units. The new instruction set is a superset of the integer DSP, and old binary programs can be run without any modification. The new processor can execute two floating-point arith-

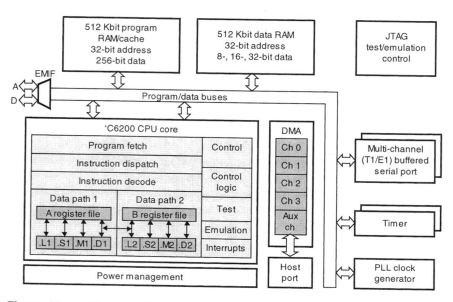

Figure 15   Architecture of Texas Instruments' TMS320C6201. (From Ref. 52.)

Table 13   Functional Units and Descriptions

Functional unit	Description
.L (.L1 and .L2)	32/40-Bit arithmetic and compare operations
	Finds leftmost 1 or 0 bit for 32-bit register
	Normalization count for 32 and 40 bits
	32-Bit logical operations
.S (.S1 and .S2)	32-Bit arithmetic operations
	32/40-Bit shifts and 32-bit bit-field operations
	32-Bit logical operations
	Branching
	Constant generation
	Register transfers to/from the control register file
.M (.M1 and .M2)	16×16-Bit multiplies
.D (.D1 and .D2)	32-Bit add, subtract, linear, and circular address calculation

Source: Ref. 52.

metic operations, two floating-point reciprocal/absolute value/square root operations, and two floating-point multiply operations per cycle, resulting in 1 GFLOPS at 167 MHz.

This architecture also supports an open programming environment. The C compiler performs a variety of optimizations, including software pipelining, which can effectively improve code performance on VLIW machines.

6.8 Commentary

Programmable VSPs represent a new trend in multimedia systems. They tend to be more versatile in dealing with multimedia applications, including video, audio, and graphics. VSPs have to be very powerful because the amount of computation required by video compression is enormous. To meet the performance demands, all of the VSPs employ parallel processing techniques to some degree: VLIW (SIMD), multiprocessor-based MIMD, or the concept from vector processing (SIGD). However, none of these programmable VSPs are able to compete with dedicated state-of-the-art VSPs—none of them could support real-time MPEG-2 encoding yet.

It is not surprising to see that many programmable VSPs adopt VLIW architecture. There are basically two reasons for doing this. First, there is much parallelism in video applications [53]. Second, in VLIW machines, a high degree of parallelism and high clock rates are made possible by shifting part of the hardware workload to software. This kind of shift once happened in the microprocessor evolution from CISC to RISC. By relieving the hardware burden, RISC achieved a new level that CISC was unable to compete with and the revolution has been a milestone in microprocessor history. Analogously, we would expect VLIW to outperform other architectures. Unlike their superscalar counterparts, VLIW processors rely on the compilers entirely to exploit the parallelism; static scheduling is performed by sophisticated optimizing compilers. All of this raises challenges for next-generation compilers. More discussions on the VLIW architecture as well as its associated compiler and coding techniques can be found in Fisher et al.'s review [54]. Although offering architectural advantages for general-purpose computing (where unpredictability and irregularity are high), multithreading architectures are not as optimal for video processing where regularity and predictability are much higher.

7 RECONFIGURABLE SYSTEMS

Reconfigurable computing is yet another approach to balancing performance and flexibility. In contrast with VSPs where the programmability relies in the instruction set architecture (ISA), the flexibility of reconfigurable systems comes from

Table 14 Comparison of Different Solutions for Video Signal Processing

Solution	Performance	Flexibility	Power	Cost	Density
Multimedia instruction extension	Low	High	Medium	High	High
Application-specific codec	High	Low	Low	Low	Medium
Programmable VSP	Medium	High	Medium	High	Medium
Reconfigurable computing	Medium	Medium	High	Medium	Low

a much lower level—logic gate arrays. Table 14 compares different solutions for video signal processing from several perspectives.

Reconfigurable computing has evolved from the original field programmable gate array (FPGA), which was invented in the early 1980s and has been undergoing vast improvements ever since. Traditionally, FPGAs were only used as a replacement of glue-logic and fast prototyping, but their applications have been widened in the past decade. The introduction of SRAM-based FPGAs by Xilinx Corporation [55] in 1986 opened a new era. SRAM-based FPGAs use SRAM cells to store logic functionality and wiring information and thus can be programmed an infinite number of times. Almost all of the modern FGPAs choose look-up-table (LUT)-based design, where the each logic cell consists of one or two LUT units, each driven by a limited number of inputs. The LUT units can be configured to implement any multiple-input (usually less than five) single-output Boolean function, providing fine-grained parallelism. With technology advances, the density (reported in equivalent gate counts) of state-of-the-art FPGAs is approaching 1 million gates. Further discussion on FPGA technologies is beyond the scope of this survey, so we refer interested readers to other literature [56]. In the following subsections, we will focus on using reconfigurable computing for video signal processing.

7.1 Implementation Choices

Unlike the other three approaches we have discussed previously, we have not yet seen single-chip or even chipset solutions for reconfigurable video signal processing. However, there are systems existing for this application. Reconfigurable systems typically consist of a general-purpose microprocessor and some reconfigurable logic such as FPGAs. Although the computational cores are mapped to the reconfigurable logic, the microprocessor takes care of everything else that cannot be implemented efficiently with reconfigurable hardware, including branches, loops, and other expensive operations. A natural question then arises: Where does one draw the boundary between the CPU and the FPGA?

Table 15 Implementation Choices for Combining CPU and Reconfigurable Logic

Location	Role	Bus	Bandwidth
Inside CPU	Functional unit	Internal data bus	5–10 Gbyte/sec
CPU–L1 cache ⎱ L1–L2 cache ⎰	Closely coupled coprocessor	CPU bus	2–5 Gbyte/sec 1–2 Gbyte/sec
L2–main memory	Loosely coupled coprocessor	Memory bus	500–1000 Mbyte/sec
Beyond main memory	Standalone processor	I/O bus	66–200 Mbyte/sec

There are many implementation choices for how to combine general-purpose microprocessor with reconfigurable logic in the CPU-memory hierarchy. As can be seen in Table 15, the closer the configurable logic sits to the CPU, the higher the bandwidth will be. It is difficult to compare in general which one is better, because different applications have different needs and they yield different results on different systems.

Reconfigurable logic supports implementing different logic functions in hardware. This has two implications. First, it means that reconfigurable computing has the potential to offer massive parallelism. Numerous studies have shown that video applications bear a huge amount of parallelism, so, theoretically, reconfigurable computing is a sound solution for video signal processing. Second, LUT-based FPGAs exploit parallelism at a very fine granularity. When dealing with fine-grained parallelism, it is desirable for the reconfigurable logic to sit closer to the CPU. This is because fine-grained parallelism will yield many intermediate results, which requires a high bandwidth to exchange. In this approach, the reconfigurable logic can be viewed as a functional unit, providing functions that can be altered every once a while, depending on the need for reconfiguration. This flexibility can even be built into the instruction set architecture, generating an application-specific instruction set on a general-purpose microprocessor, which can speed up many different applications potentially. Although this idea is very attractive, it comes with a significant cost. In order to be flexible, reconfigurable hardware has a large overhead in the wiring structure as well as inside the LUT-based logic cell. As the silicon resource becomes more and more precious inside microprocessors, it is probably not worth using it for reconfigurable logic; putting some memory or fixed functional units in the same area is likely to yield better performance.

By moving the reconfigurable logic outside a CPU, we may achieve a better utilization of the microprocessor real estate, but we will have to sacrifice some bandwidth. In this approach, reconfigurable resources are used to speed up certain operations which cannot be done efficiently on the microprocessor (e.g., bit-serial operations, deeply pipelined procedures, application-specific computations, etc.). Depending on the application, if we can map a coarser level of parallelism to the FPGA coprocessor, we may still benefit considerably from this kind of reconfigurable computing.

For coarse-grained parallelism or complex functions, we may need to seek heterogeneous system solutions. In this approach, commercial FPGAs are used to form a stand-alone processing unit which performs some complicated and reconfigurable functions. Because the FPGAs are loosely coupled with the microprocessor, this kind of system is often implemented as an add-on module to an existing platform. Like other alternatives, this one has both advantages and disadvantages. Although it can implement very complicated functions and be CPU-agnostic, it introduces a large communication delay between the reconfigurable processing unit and the main CPU. If not carefully designed, the I/O bus in between can become a bottleneck, severely hampering the throughput.

In addition to the whereabouts of reconfigurable logic in a hybrid system, there are many other hardware issues, such as how to interconnect multiple FPGAs, how to reconfigure quickly, how to change part of a configuration, and so forth. Due to limited space, we refer users to some good surveys on FPGA [56,57].

7.2 Implementation Examples

The fine granularity of parallelism and pipelined nature of reconfigurable computing make it a particularly good match for many video processing algorithms. Among several implementation options for reconfigurable computing system, it is not clear which one is the winner. In the following paragraphs, we will enumerate a few reconfigurable computing systems, with emphasis on the hardware architecture instead of the application software. Note that our interests are in multimedia applications, so we will not address every important reconfigurable system.

Splash II [58], a systolic array processor based on multiple FPGAs, is one of a few influential projects in the history of reconfigurable computing. The 16 Xilinx FPGAs on each board in Splash II are connected in a mesh structure. In addition, a global crossbar is provided to facilitate multihop data transfer and broadcast. To synchronize communication at the system level, a high-level inter-FPGA scheduler as well as a compiler are developed to coordinate the FPGAs and the associated SRAMs. Among many DSP applications that have been mapped to the Splash II architecture, there are various image filtering [59], 2D DCT [60], target recognition [61], and so forth.

Another important milestone during the evolution of reconfigurable computing is the Programmable Active Memory (PAM) [62] developed by DEC (now Compaq). PAM also consists of an array of FPGAs arranged in a two-dimensional mesh. With the interface FPGA, PAM looks like a memory module except that the data written in may be read out differently after being massaged by the reconfigurable logic. PAM has also been used in numerous image processing and video applications, including image filtering, 1D and 2D transforms, stereo vision, and so forth.

The Dynamic Instruction Set Computer (DISC) [63] is an example of integrating FPGA resources into a CPU. Due to limited silicon real estate, DISC treats instructions as swappable modules and pages them in and out continuously during program execution through partial reconfiguration. In some sense, the on-chip FPGA resources function like an instruction cache in microprocessors. Each instruction in DISC is implemented as an independent module, and different modules are paged in and out based on the program needs. The advantage with this approach is that limited resources can be fully utilized, but the downside is that the context switching can cause delay, conflict, and complexity. DISC adopts a linear, one-dimensional hardware structure to simplify routing and reallocation of the modules. Although the logic cells are organized in an array, only adjacent rows can be used for one instruction. The width of each instruction module is fixed, but the height (number of rows) is allowed to vary (Fig. 16). In an experiment of mean image filtering, the authors reported a speedup of 23.5 over a general-purpose microprocessor setup.

The Garp architecture combines reconfigurable hardware with a standard MIPS processor on the same die to achieve high performance [64]. The top-level block diagram of the integration is shown in Figure 17. The internal architecture

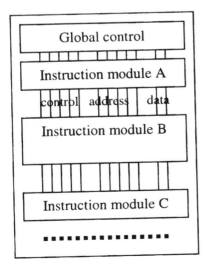

Figure 16   Linear reconfigurable instruction modules.

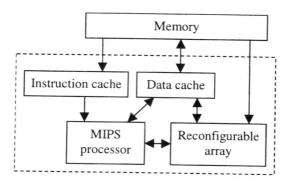

Figure 17 Block diagram of Garp.

of the reconfigurable array is very similar to DISC (Fig. 16). Each row of the array contains 23 logic blocks, each capable of handling 2 bits. In addition, there is a distributed cache built in the reconfigurable array. Similar to an instruction cache, the configuration cache holds the most recent configuration so as to expedite dynamic reconfiguration. Simulation results show speedups ranging from 2 to 24 against a 167-MHz Sun UltraSPARC 1/170.

The REMARC reconfigurable array processor [65] also couples some reconfigurable hardware with a MIPS microprocessor. It consists of a global control unit and 8 × 8 programmable logic blocks called nanoprocessors (Fig. 18). Each nanoprocessor is a small 16-bit processor: It has a 32-entry instruction RAM, a 16-entry data RAM 1 ALU, 1 instruction register, 8, 16-bit data registers, 4 data input registers, and 1 data output register. The nanoprocessor are interconnected in a mesh structure. In addition, there are eight horizontal buses and eight vertical busses for global communication. All of the 64 nanoprocessors are controlled by the same program counter, so the array processor is very much like a VLIW processor. REMARC is not based on FPGA technology, but the authors compared it with an FPGA-based reconfigurable coprocessor (which is about 10 times larger than REMARC) and found that both have similar performance, which is 2.3–7.3 times as fast as the MIPS R3000 microprocessor. Simulation results also show that both reconfigurable systems outperform Intel MMX instruction set extensions.

Used as an attached coprocessor, PipeRench [66] explores parallelism at a coarser granularity. It employs a pipelined, linear reconfiguration to solve the problems of compilability, configuration time, and forward compatibility. Targeting at stream-based functions such as finite impulse response (FIR) filtering, PipeRench consists of a sequence of stripes, which are equivalent to pipeline stages. However, one physical stripe can function as several pipeline stages in a

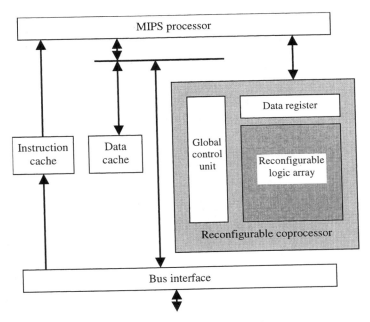

Figure 18   Block diagram of REMARC with a microprocessor.

time-sharing fashion. For example, a five-stage pipeline can be implemented on three stripes with slight reconfiguration overhead. As shown in Figure 19, each stripe has an interconnect network and some processing elements (PEs), which are composed of ALUs and registers. The ALUs are implemented using look-up tables and some extra logic for carry chains, zero detection, and so forth. In addition to the local interconnect network, there are also four global buses for forwarding data between stripes that are not next to each other. Evaluation of certain multimedia computing kernels shows a speedup factor of 11–190 over a 330-MHz UltraSPARC-II.

The Cheops imaging system is a stand-alone unit for acquisition, processing, and display of digital video sequences and model-based representations of moving scenes [67]. Instead of using a number of general-purpose microprocessors and DSPs to achieve the computation power for video applications, Cheops abstracts out a set of basic, computationally intensive stream operations required for real-time performance of a variety of applications and embodies them in a compact, modular platform. The Cheops system uses stream processors to handle video data like a data flow machine. It can support up to four processor modules. The block diagram of the overall architecture is depicted in Figure 20.

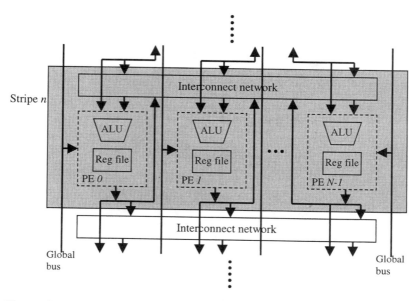

Figure 19 Stripe architecture.

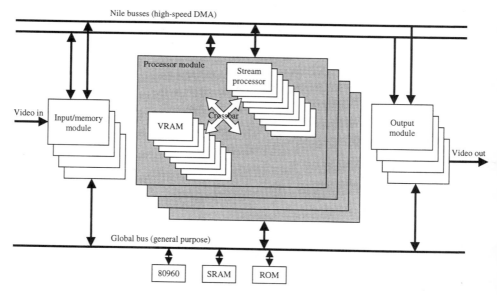

Figure 20 Block diagram of the Cheops system.

Each processor module consists of eight memory units and eight stream processors which are connected together through a cross-point switch. The DMA controllers on each VRAM bank are capable of handling one- and two-dimensional arrays and they can pad or decimate depending on the direction of data flow. The Cheops is a rather complete system which even includes a multitasking operating system with a dynamic scheduler.

7.3 Commentary

Reconfigurable computing systems to some degree combine the speed of ASIC and the flexibility of software together. They emerge as a unique approach to high-performance video signal processing. Quite a few systems have been built to speed up video applications and they have proven to be more efficient than systems based on general-purpose microprocessors.

This approach also opens a new research area and raises many challenges for both hardware and software development. On the hardware side, how to couple reconfigurable components with microprocessors still remains open, and the granularity, speed, and portion of reconfiguration as well as routing structures are also subjects of active research. On the software side, CAD tools need great improvement to automate or accelerate the process of mapping applications to reconfigurable systems.

Although reconfigurable systems have shown the ability to speed up video signal processing as well as many other types of applications, they have not met the requirement of the marketplace; most of their applications are limited to individual research groups and institutions.

8 CONCLUSIONS

In this chapter, we have discussed four major approaches to digital video signal processing architectures: Instruction set extensions try to improve the performance of modern microprocessors; dedicated codecs seem to offer the most cost-effective solutions for some specific video applications such as MPEG-2 decoding; programmable VSPs tend to support various video applications efficiently; and reconfigurable computing compromises flexibility and performance at the system level. Because the four approaches are targeted at different markets, each having both advantages and disadvantages, they will continue to coexist in the future. However, as standards become more complex, programmability will be important for even highly application-specific architectures. The past several years have seen limited programmability become a commonplace in the design of application-specific video processors. As just one example, all the major MPEG-2 encoders incorporate at least a RISC core on-chip.

Efficient transferring of data among many processing elements is another key point for VSPs, whether dedicated or programmable. To qualify for real-time video processing, VSPs must be able to accept a high-bandwidth incoming bit stream, process the huge amount of data, and produce an output stream. Parallel processing also requires communication between different modules. Therefore, all of the VSPs we have discussed use either a very wide bus (e.g., in Chromatic Research Mpact2, the internal data bus is 792-bit wide) or a crossbar (e.g., Texas Instruments' TMS320C8x), or some other extremely fast interconnect mechanisms to facilitate high-speed data transfer. External bandwidth is usually achieved by using Rambus RDRAM or synchronous DRAM.

A few VSPs (e.g., Sony CXD1930Q, MicroUnity Cronus, Philips TriMedia) are equipped with a real-time operating system or multitasking kernel to support different tasks in video applications. This brings video signal processing to an even more advanced stage. Usually in a multimedia system, many devices are involved. For example, in an MPEG-2 decoder, video and audio signals are separately handled by different processing units, and the coordination of different modules is very important. Multitasking will become an increasingly important capability as video processors are asked to handle a wider variety of tasks at multiple rates. However, the large amount of state in a video computation, whether it be in registers or main memory, creates a challenge for real-time operating systems. Ways must be found to efficiently switch contexts between tasks that use a large amount of data.

It is natural to ask which architecture will win in the long run: multimedia instruction set extensions, application-specific processors, programmable VSPs, or reconfigurable? It is safe to say that multimedia instruction set extensions for general-purpose CPUs are here to stay. These extensions cost very little silicon area to support, and now that they have been designed into architectures, they are unlikely to disappear. These extensions can significantly speed up video algorithms on general-purpose processors, but, so far, they do not provide the horsepower required to support the highest-end video applications; for example, although a workstation may be able to run MPEG-1 at this point in time, the same fabrication technology requires specialized processors for MPEG-2. We believe that the greatest impediment to video performance in general-purpose processors is the memory system. Innovation will be required to design a hierarchical memory system, which competes with VSPs yet is cost-effective and does not impede performance for traditional applications.

Application-specific processors are unlikely to disappear. There will continue to be high-volume applications in which it is worth the effort to design a specialized processor. However, as we have already mentioned, even many application-specific processors will be programmable to some extent because standards continue to become more complex.

Reconfigurable logic technology is rapidly improving, resulting in both higher clock rates and increased logic density. Reconfigurable logic should continue to improve for quite some time because it can be applied to many different applications, providing a large user base. As it improves, we can expect to see it used more frequently in video applications.

The wild card is the programmable VSP. It provides a higher performance than multimedia extensions for CPUs but is more flexible than application-specific processors. However, it is not clear what the "killer application" will be that drives VSPs into the marketplace. Given the cost of the VSP itself and of integrating it into a complex system like a PC, VSPs may not make it into wide use until some new application arrives which demands the performance and flexibility of VSPs. Home video editing, for example, might be one such application, if it catches on in a form that is sufficiently complex that the PC's main CPU cannot handle the workload. The next several years will see an interesting and, most likely intense battle between video architectures for their place in the market.

REFERENCES

1. JL Mitchell, WB Pennebaker, CE Fogg, DJ LeGall. MPEG Video Compression Standard. New York: Chapman & Hall, 1997.
2. Texas Instruments, TMS34010 graphics system processor data sheet, http://www-s.ti.com/sc/psheets/spvs002c/spvs002c.pdf
3. Philips Semiconductors. Data sheet—SAA9051 digital multi-standard color decoder.
4. Philips Semiconductors. Data sheet—SAA7151B digital multi-standard color decoder with SCART interface, http://www-us.semiconductors.philips.com/acrobat/2301.pdf
5. K Aono, M Toyokura, T Araki. A 30ns (600 MOPS) image processor with a reconfigurable pipeline architecture. Proceedings, IEEE 1989 Custom Integrated Circuits Conference, IEEE, 1989, pp 24.4.1–24.4.4.
6. T Fujii, T Sawabe, N Ohta, S Ono. Super high definition image processing on a parallel signal processing system. Visual Communications and Image Processing '91: Visual Communication, SPIE, 1991, pp 339–350.
7. KA Vissers, G Essink, P van Gerwen. Programming and tools for a general-purpose video signal processor. Proceedings, International Workshop on High-Level Synthesis, 1992.
8. T Inoue, J Goto, M Yamashina, K Suzuki, M Nomura, Y Koseki, T Kimura, T Atsumo, M Motomura, BS Shih, T Horiuchi, N Hamatake, K Kumagi, T Enomoto, H Yamada, M Takada. A 300 MHz 16b BiCMOS video signal processor. Proceedings, 1993 IEEE Int'l Solid State Circuits Conference, 1993, pp 36–37.
9. Intel Corp. i860 64-Bit Microprocessor, Data Sheet. Santa Clara, CA: Intel Corporation, 1989.
10. Superscalar techniques: superSparc vs. 88110, Microprocessor Rep 5(22), 1991.
11. R Lee, J Huck. 64-Bit and multimedia extensions in the PA-RISC 2.0 architecture. Proc. IEEE Compcon 25–28, February 1996.

12. R Lee. Subword parallelism with MAX2. IEEE Micro 16(4):51–59, 1996.
13. R Lee, L McMahan. Mapping of application software to the multimedia instructions of general-purpose microprocessors. Proc. SPIE Multimedia Hardware Architect 122–133, February 1997.
14. L Gwennap. Intel's MMX speeds multimedia. Microprocessor Rep 10(3), 1996.
15. L Gwennap. UltraSparc adds multimedia instructions. Microprocessor Rep 8(16): 16–18, 1994.
16. Sun Microsystems, Inc. The visual instruction set, Technology white paper 95-022, http://www.sun.com/microelectronics/whitepapers/wp95-022/index.html
17. P Rubinfeld, R Rose, M McCallig. Motion Video Instruction Extensions for Alpha, White Paper. Hudson, MA: Digital Equipment Corporation, 1996.
18. MIPS Technologies, Inc. MIPS extension for digital media with 3D, at http://www.mips.com/Documentation/isa5_tech_brf.pdf, 1997.
19. T Komarek, P Pirsch. Array architectures for block-matching algorithms. IEEE Trans. Circuits Syst 36(10):1301–1308, 1989.
20. M Yamashina et al. A microprogrammable real-time video signal processor (VSP) for motion compensation. IEEE J Solid-State Circuits 23(4):907–914, 1988.
21. H Fujiwara et al. An all-ASIC implementation of a low bit-rate video codec. IEEE Trans. Circuits Sys Video Technol 2(2):123–133, 1992.
22. http://www.8x8.com/docs/chips/lvp.html
23. http://products.analog.com/products/info.asp?product=ADV601
24. http://www.c-cube.com/products/products.html
25. T Sikora. MPEG Digital Video Coding Standards. In: R Jurgens. Digital Electronics Consumer Handbook. New York: McGraw-Hill, 1997.
26. ESS Technology, Inc. ES3308 MPEG2 audio/video decoder product brief, http://www.esstech.com/product/Video/pb3308b.pdf
27. http://www.chips.ibm.com/products/mpeg/briefs.html
28. W. Bruls, et al. A single-chip MPEG2 encoder for consumer video storage applications. Proc. IEEE Int. Conf. on Consumer Electronics, 1997, pp 262–263.
29. Philips Semiconductors. Data sheet—SAA7201 Integrated MPEG2 AVG decoder, http://www-us.semiconductors.philips.com/acrobat/2019.pdf
30. http://www-us.semiconductors.philips.com/news/archive.stm
31. P. Lippens, et al. Phideo: A silicon compiler for high speed algorithms. European Design Automation Conference, 1991.
32. Sony Semiconductor Company of America. CXD1922Q MPEG-2 technology white paper, http://www.sel.sony.com/semi/CXD1922Qwp.html
33. Sony Semiconductor Company of America. Press releases—virtuoso IC family, http://www.sel.sony.com/semi/nrVirtuoso.html
34. http://www.dvimpact.com/products/single-chipn.html
35. http://www.lsilogic.com/products/ff0013.html
36. http://eweb.mei.co.jp/product/mvd-lsi/me-e.html
37. T Araki et al. Video DSP architecture for MPEG2 codec. Proc. IEEE ICASSP 2: 417–420, April 1994.
38. http://www.mitsubishi.com/ghp_japan/TechShowcase/Text/tsText08.html
39. http://www.visiontech-dml.com/product/index.htm
40. http://www.mpact.com/

41. S Purcell. Mpact2 media processor, balanced 2X performance. Proc. SPIE Multimedia Hardware Architectures, 1997, pp 102–108.

42. C Hansen. MicroUnity's media processor architecture. IEEE Micro 16(4):34–41, 1996.

43. http://www.microunity.com/www/mediaprc.htm

44. R Hayes, et al. MicroUnity Software Development Environment. Proc. IEEE Compcon 341–348, February 1996.

45. E Holmann, et al. A media processor for multimedia signal processing applications. IEEE Workshop on Signal Processing Systems, 1997, pp 86–96.

46. T Yoshida, et al. A 2V 250MHz multimedia processor. Proc. ISSCC, 266–267:471, February 1997.

47. K Suzuki, T Arai, K Nadehara, I Kuroda. V830R/AV: Embedded multimedia superscalar RISC processor. IEEE Micro 18(2):36–47, 1998.

48. http://www.trimedia.philips.com/

49. JTJ van Eijndhoven, FW Sijstermans, KA Vissters, EJD Pol, MJA Tromp, P Struik, RHJ Bloks, P van der Wolf, AD Pimentel, HPE Vranken. TriMedia CPU64 architecture. In: Proceedings, ICCD '99. Los Alamitos, CA: IEEE Computer Society Press, 1999, pp 586–592.

50. L Nguyen, et al. Establish MSP as the standard for media processing. Proc. Hot Chips 8: A Symposium on High Performance Chips, 1996.

51. http://www.ti.com/sc/docs/dsps/products/c8x/index.htm

52. http://www.ti.com/sc/docs/dsps/products/c6x/index.htm

53. Z Wu, W Wolf. Parallelism analysis of memory system in single-chip VLIW video signal processors. Proc. SPIE Multimedia Hardware Architectures, 1998, pp 58–66.

54. P Faraboschi, G Desoli, JA Fisher. The latest word in digital and media processing. IEEE Signal Process Mag 15(2):59–85, 1998.

55. Xilinx Corporation, http://www.xilinx.com/

56. S Hauck. The roles of FPGAs in reprogrammable systems. Proc IEEE 615–638, April 1998.

57. K Compton, S Hauck. Configurable computing: A survey of systems and software. Technical Report. Northwestern University, 1999.

58. J Arnold, D Buell, E Davis. Splash II. Proc. 4th ACM Symposium of Parallel Algorithms and Architectures, 1992, pp 316–322.

59. PM Athanas, AL Abbott. Real-time image processing on a custom computing platform. IEEE Computer 28(2), 1995.

60. N Ratha, A Jain, D Rover. Convolution on Splash 2. Proc. IEEE Symposium on FPGAs for Custom Computing Machines, 1995, pp 204–213.

61. M Rencher, BL Hutchings. Automated target recognition on Splash II. Proc. 5th IEEE Symposium on FPGAs for Custom Computing Machines, 1997, pp 192–200.

62. J Vuillemin, P Bertin, D Roncin, M Shand, H Touati, P Boucard. Programmable active memories: Reconfigurable systems come of age. IEEE Trans. VLSI Syst 4(1): 56–69, 1996.

63. MJ Wirthlin, BL Hutchings. A dynamic instruction set computer. IEEE Workshop on FPGAs for Custom Computing Machines, 1995, pp 99–107.

64. JR Hauser, J Wawrzynek. Garp: A MIPS processor with a reconfigurable coproces-

sor. IEEE Workshop on FPGAs for Custom Computing Machines, 1997, pp 24–33.

65. T Miyamori, K Olukotun. A quantitative analysis of reconfigurable coprocessors for multimedia applications. Proc. IEEE International Symposium on FPGAs for Custom Computing Machines, 1998, pp 2–11.

66. SC Goldstein, H Schmit, M Moe, M Budiu, S Cadambi. PipeRench: A coprocessor for streaming multimedia acceleration. Proc. International Symposium on Computer Architecture, 1999, pp 28–39.

67. VM Bove Jr, JA Watlington. Cheops: A reconfigurable data-flow system for video processing. IEEE Trans. Circuits Syst Video Technol 5:140–149, April 1995.

6
OASIS: An Optimized Code Generation Approach for Complex Instruction Set PDSPs

Jim K. H. Yu* and Yu Hen Hu
University of Wisconsin–Madison, Madison, Wisconsin

1 INTRODUCTION

A programmable digital signal processor (PDSP) is a special-purpose micropro-cessor with specialized architecture and instruction set for implementing DSP algorithms. Typical architectural features include multiple memory partitions (on-chip, off-chip, data memory, program memory, etc.), multiple (generally pipe-lined) arithmetic and logic units (ALUs), nonuniform register sets, and extensive hardware numeric support [1,2]. Single-chip PDSPs have become increasingly popular for real-time DSP applications [3,4]. The newest introductions from sev-eral manufacturers offer peak computing power approaching that of supercom-puter systems of only a few years ago. However, to fully utilize the available computing power, the software designer must face the difficult task of program-ming in low-level assembly language.

High-level languages (HLLs) are attractive to software designers because they simplify the task of programming. Unlike assembly language, HLL pro-grams are readable, maintainable, and portable to other processors. All of these features contribute to increase productivity and reduce development cost.

An HLL compiler translates the instructions present in an HLL program into assembly instructions, more easily understood by the processor. Commer-cially available HLL compilers for PDSPs have existed for many years [5]. Un-fortunately, the performance of those compilers is still acceptable only to a few noncritical applications. The reason is twofold. First, the majority of those com-

---

* *Current affiliation*: Tivoli Systems, Austin, Texas

pilers are based on technologies developed with general-purpose applications in mind. For instance, compilation speed is an important issue for general-purpose programs, which can reach millions of lines of code, but completely immaterial to many embedded DSP applications. Second, PDSPs present peculiar architectural features that make good code generation difficult. Genin et al. [6] estimated that assembly codes written by human DSP experts perform 5–50 times faster than those obtained with conventional C compilers. Although the performance of current optimizing compilers has improved significantly, careful manual coding, typically with several iterations of ad hoc fine-tuning, is still the only effective approach for DSP applications with stringent constraints on execution time and/or code size.

In an attempt to diminish the speed and size penalties incurred by programming in HLL, some compilers resort to "external help." Examples of such external help are the support for in-line assembly language code in the source HLL program and the support for calls to subroutines from application libraries written in assembly language [7,8].

The problem with in-lining is obvious. It contradicts the main purpose of using HLL, which is to spare the software designer the burden of programming in assembly language. Chief among the problems associated with using assembly libraries is their lack of portability. Because assembly libraries are optimized for a given PDSP architecture and instruction set, even a simple PDSP upgrade from the same manufacturer would require the development of new libraries, if they were to fully utilize the features of the new architecture and instruction set. Another problem with using assembly language routines is the inherent compromise between generality and efficiency. For example, let us consider a routine for matrix multiplication. A general approach would pass the size of the matrix as a parameter. However, such a routine is obviously not optimal due to the inevitable overhead caused by loop control instructions. It is not uncommon for optimized application libraries to contain different routines to multiply matrices of size 2 × 2, 3 × 3, 4 × 4, etc. Furthermore, subroutines require an exact input/output parameter format, which the caller program must satisfy; this may translate into additional stack control overhead.

1.1 High-Level Languages and PDSP

Progammable digital signal processor designers never had HLL in mind; conversely, general-purpose HLL designers never had PDSP in mind. Despite successful attempts in controlled case studies [9], conventional HLLs are not suited for describing the semantics of DSP. They also cannot give the programmer direct access to unique features of PDSPs, such as modulo-addressing register arithmetic, special fixed-point scaling modes, dual data memory access, and fractional data types.

Undoubtedly, the best option is to develop a dedicated HLL for DSP. Although this approach certainly has its advantages [10–12], it would require the DSP programmer to learn yet another programming language. Compounding the disadvantages of adopting a dedicated HLL are the enormous effort in developing the dedicated HLL, its learning curve, and the need to rewrite common DSP routines, which are widely available in one of several general-purpose HLLs [13,14].

The alternative approach is to use an existing general-purpose HLL such as Ada [15], C [16], FORTRAN [17], or Pascal [18]. Among those languages, C is, arguably, the most flexible [12]. It has a sufficiently high-level syntax and yet it supports low-level bitwise operations. Due to its increasing popularity, a vast library of C routines is readily available. Furthermore, many commercial C compilers for PDSPs exist, which allows us to evaluate relative performances. Our HLL choice is further justified by the availability of high-level synthesis front ends [19], which can translate graphical object-oriented signal flow block designs directly into C programs.

By using a conventional HLL to implement DSP algorithms, we are handicapped in more than one way. First, as discussed in the beginning of this section, we are limited by the semantic representational power of the HLL. Second, a language can bias the implementation of an algorithm. Even if we could generate optimal code for a given HLL program, the latter may not be the most efficient implementation of a desired DSP algorithm. We will address the later issue using algorithm transformation techniques, which we will describe in more detail in Section 5.

1.2 Structure of a Compiler

The typical structure of a compiler is a front end (FE) feeding a back end (BE), as depicted in Figure 1. The FE processes the HLL source program, performing such tasks as lexical, syntactic, and semantic analysis, and builds an intermediate representation (IR) of that program. The BE then translates the IR into the desired assembly code, also known as code generation. The greatest advantage of such a structure is that if the IR is standardized, $N \times M$ compilers can be built with N FEs and M BEs.

Three important subtasks of code generation are *scheduling,** *instruction selection*, and *resource allocation*. Determining the order in which IR operators should be processed is known as scheduling. Determining a valid sequence of assembly instructions to translate a given IR operator is known as instruction

---

* Not to be confused with instruction scheduling, which is a fine-level optimization technique, discussed in Section 2.

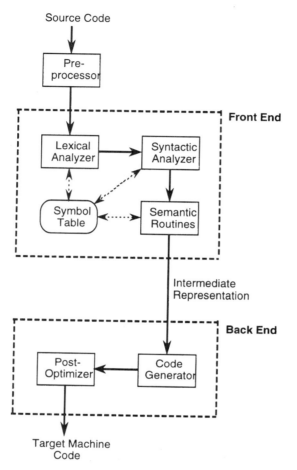

Figure 1 Typical structure of a compiler.

selection. Binding variables to physical resources (registers and memory addresses) is known as resource allocation.

Traditional code generation methods for processors with uniform register architectures usually perform scheduling, instruction selection, and register allocation independently, in separate compilation phases. However, for DSP architectures with multipartitioned memory and nonuniform register sets, in which certain registers and memory blocks are specialized for specific usage, those three phases affect each other [20]. Most importantly, the best translation for a given IR operator is run-time and context dependent. For instance, many modern PDSPs offer complex instructions that can accomplish several operations. In certain situations, by taking advantage of instruction side effects, a multiplication can be implemented with shifts and additions, at no cost. Hence, efficient code generation for PDSPs must be context sensitive and must combine scheduling, instruction selection, and register allocation into a single step.

2 BACKGROUND AND RELATED WORKS

The design of an FE is a well-understood problem and many tools, such as LEX, SCANGEN, LLGEN [21], and YACC [22], exist to automate that task. BE design is much more complex because the code generator must address the following issues:

1. Some architectures (especially PDSPs) are extremely nonuniform, making it difficult to design good tools to automate BE construction.
2. CISC (complex instruction set computing) machines (most PDSPs are CISC) have complex instructions with overlapping functionality. In case more than one instruction can accomplish a desired task, which one to use is architecture and context dependent.
3. Architecture-specific constraints may complicate optimizations that are considered machine independent in most general-purpose architectures. For example, common subexpression elimination stores the result of an expression in an attempt to avoid unnecessary recomputations of the same expression. However, in some PDSP architectures, it may be less expensive to recompute an expression than to save its results.*

Attempts to formalize [23,24] and automate [25–29] the process of BE construction have been only partially successful. As a whole, one could say that BE construction still remains an ad hoc problem, especially if the quality of the generated machine code is the primary factor to be considered.

2.1 Code Generation Methods

There are various approaches to code generation. A more detailed discussion of such techniques can be found in classic compiler textbooks such as Refs. 21 and 30. In this section, we will briefly introduce a few relevant methods to build the necessary background for the discussions in the following subsection. The interested reader is referred to the cited literature for further details.

---

* To a lesser extent, this is also true for other architectures.

2.1.1   Tree Traversal

In this method, the source code is represented by a parse (or syntax) tree. The code generator traverses the tree, usually bottom-up, and invokes small code-generating routines (or macros) for each operator in the tree. The simplicity of this approach has made it a popular code generation method. The main disadvantages of this approach are as follows:

1. Retargeting the code generator to a different architecture requires rewriting all the macros, a time-consuming task.
2. Code quality is directly related to the skill of the macro writer; hence, there is no guarantee that all available instructions on the target machine will be efficiently used.
3. The best sequence of assembly instructions for a given operator is both architecture and context dependent and cannot be captured by predetermined macros. For instance, depending on the operands, a multiplication could be implemented more efficiently as a series of shifts and additions, rather than by a simpler, but often costlier, MPY instruction. Moreover, some PDSPs, such as the TMS320 series, allow the shift to be bundled in many other instructions such as ADD and SUB, effectively providing a zero cost multiplication.

One approach to overcome disadvantage 3 is to enumerate all possible implementations of an IR operator and choose the best one for the current context through an exhaustive search. This method has been investigated by Krumme and Ackley [31]. Their compiler is driven by a set of code tables, each containing all possible (useful) ways to translate the corresponding operator. An interpreter traverses the program tree, one expression at a time, invokes the appropriate code table, and interprets its contents. Krumme's compiler returns the least expensive among the successful translations as the best implementation for the corresponding operator. In case an operand of an operator is itself an expression, the process recurs.

Despite the exhaustive search formulation, their compiler does not generate optimal codes. Its efficiency is reported to be only "as good as that of other optimizing compilers." The reason is simple: To overcome the combinatorial explosion of an exhaustive search, the original problem is broken down into subproblems, each corresponding to one statement in the input program. However, the combination of optimal solutions to subgoals does not necessarily yield the global optimal solution [32,33].

2.1.2   Template Pattern Matching

Another method of code generation is that of template pattern matching. In this approach, a set of basic elements (templates) is defined and must completely

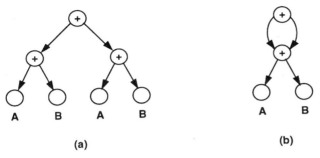

Figure 2   Two representations for expression $(A + B) + (A + B)$: (a) tree and (b) DAG.

describe the IR in question. A sequence of machine instructions is associated to each of these templates. The code generation problem now translates to a covering problem (i.e., finding a set of templates that can completely cover the IR). If a cost is also associated with each template, then we can define optimal code generation as finding the cheapest cover for the IR. Once a set of matching templates is selected, their associated machine instructions form the desired output code. Incidentally, if the IR is a tree, it has been shown that, for a certain class of machines,* a covering of least cost can be found in time linearly proportional to the size of the tree [23]. Hoffmann and O'Donnell [34] and Chase [35] give very efficient algorithms for tree pattern matching.

   A directed acyclic graph (DAG) [21] is a generalization of the tree, in which nodes may have more than one parent. Support for multiple parents allows sharing of common subexpressions and, consequently, yields a more compact representation. Figure 2 illustrates the tree and DAG representations for the expression $(A + B) + (A + B)$. The term *acyclic* implies that no cycles are allowed, as the latter may lead to infinite loops. Unfortunately, finding a least cost cover for a DAG is an NP-complete problem [36].

   The earliest works on this approach are those of Wasilew [37] and Weingart [38]. Ripken [39] and Johnson [40] also proposed code generation schemes based on this approach. Glanville [41] also used templates as a machine description, but automatically derived a transition table from them, resulting in a faster code generator.

   The dynamic programming algorithm used in Twig [28,42] is among the most efficient for template pattern matching in trees. It implements the algorithm proposed by Aho and Johnson [23]; consequently, optimal coverage can be accomplished in time linear with the size of the input tree. However, Twig is a

---

* The processor architecture assumed by Aho and Johnson [23] has n general-purpose registers and a finite number of memory locations. All registers are interchangeable, as are all memory locations.

compiler building tool and not a compiler itself. This means that the quality of the generated code is directly dependent on how faithfully the target machine's instruction set is represented in the Twig rules. Such rules must account for all possible side effects associated with individual instructions, so typical of PDSPs. The optimality of Twig's solution is heavily constrained on the following: (1) It is assumed that the target machine has a uniform-register architecture with r interchangeable registers and (2) evaluation is contiguous and over one expression tree. Many PDSP architectures are not uniform register and may not allow optimal contiguous evaluations. Twig does not provide means for high-level optimizations such as common subexpression elimination, although algebraic simplifications can be accomplished by tree rewrites (see Sec. 2.1.3); peephole optimization [43] must be used to handle boundaries of two code sequences. Register assignment and allocation must be done separately by a user-provided routine.

The works of Newcomer [44] and Cattell [26] are of particular interest to us. Although our system is structurally different from theirs, many similarities inevitably arise due to the application of the same conceptual tools, namely heuristic search and means-ends analysis (MEA).* Hence, it is appropriate to conduct a more detailed discussion of these works.

In 1975, Newcomer pioneered the idea of applying MEA [44] in the code generation domain. Probably because of the (lack of) performance of early computer systems, Newcomer (and Cattell) declared MEA "too expensive" and restricted its application in a template generation phase. His work laid the basis on which Cattell later extended and improved. For the purposes of this work and because both Newcomer and Cattell used the same underlying paradigm, it suffices to discuss Cattell's work. The interested reader is referred to Ref. 45 for an informal but detailed summary of Ref. 44.

Cattell divided the code generation process into a machine-dependent and a machine-independent phase. In the machine-dependent phase, at compiler compile time, the effects of each instruction are described using a register-transfer notation. Then, a template generator creates a finite set of patterns to be used in the code generation phase. For each template, the best sequence of machine instructions is obtained through MEA and heuristic search. This process is performed once for each target machine. In the machine-independent phase, at compile time, a tree pattern-matching algorithm called *maximal munching method* (MMM) is used. Briefly, MMM tries to cover the input tree recursively, in a top-down fashion, from the root to the leaves. At any node, MMM selects from the available templates the one that can cover (bite off) the largest possible subtree.

---

* Means-ends analysis is a well-established artificial intelligence technique that takes into consideration the available resources (means) and the goals of the problem (ends) in specifying a plan of action.

Once a template matches a portion of the tree, the sequence of machine instructions associated with that template is emitted as part of the generated code.

Generation of templates is based on a set of axioms, which expresses classical arithmetic and Boolean equivalencies. Rules are used to specify the equivalence of programs (e.g., fetch/store decompositions and sequencing semantics) and idiosyncrasies of the target architecture (e.g., side-effect compensation). These axioms are divided into three classes:

1. *Transformations*. These are the axioms concerned with arithmetic and Boolean equivalence. Transformations are used in conjunction with MEA.
2. *Decompositions*. These axioms are normally concerned with control constructs; they decompose constructs into sequences of other constructs, allowing the search to proceed recursively on subgoals. Decompositions are used in conjunction with a general heuristic search.
3. *Compensations*. These are the axioms concerned with side effects. No search is associated with these axioms.

Cattell's method has the advantage that the quality of the assembly instruction sequences associated with individual templates does not depend on the skills of a human writer.* Albeit suboptimal, those sequences are generated in a systematic, consistent, and automatic manner. Hence, portability is greatly enhanced. The main drawback is that it does not address the issue of the best sequence of machine instructions being compile-time context dependent. Furthermore, special subtrees not selected to be included in the set of patterns are not analyzed in the search process, resulting in suboptimal codes. Separate pass optimization techniques such as peephole must be applied. In addition, the original MEA (as used by Cattell), being table driven, poses some limitations to code generation. It is simply impossible to enumerate and rank all possible solutions in a table.

In Section 4, we will present a modified MEA that is context sensitive and able to evaluate the best candidates at compile time.

2.1.3   Tree Rewrite Systems

A tree rewrite system consists of a set of rewrite rules of the form $A \to B$, where A and B are tree patterns. Similar to the template pattern-matching method, each rule is associated with a sequence of machine code to be generated and, optionally, a cost. In such a framework, (optimal) code generation translates to solving

---

* In the tree/DAG traversal method, the code-generating routines (or macros) must be handwritten. In Twig, the assembly instructions associated with a given pattern are also handwritten.

the (C) reachability problem: given an input tree T and a goal tree G, determine if there exists a sequence of rewrite rules to rewrite T into G, and if so, obtain one such (the cheapest) sequence.

Pelegri-Llopart and Graham [46] implemented an optimal code generator for expression trees based on rewrite systems. In their paper, they describe a table-driven algorithm to solve reachability based on bottom-up rewrite systems (BURS) theory and submit that BURS-based code generation is faster than one based on Twig.

Wendt [29] developed a code generator, CHOP, based on rewrite rules for DAGs. CHOP reads nonprocedural descriptions of a target machine's instruction set and of a naive code generator for that machine. It then constructs an integrated code generator and peephole optimizer for that machine. A learning version of the compiler infers the optimization rules as it compiles a training suite and records them for translation into hard code and inclusion into the production version. CHOP is reported to run 30–50% faster than Johnson's landmark Portable C Compiler [47] and generates comparable code.

2.1.4 Knowledge-Based Systems

Many researchers have proposed providing the code generator with some form of expert knowledge about the processor's architecture and/or the application domain. One of the first attempts to incorporate human knowledge into code generation is described in Fraser's dissertation [25]. Fraser designed an expert system specialized to code generation. In his system, the IR is matched against a set of machine-independent rules; the action parts of those rules specify the assembly code to be generated. Retargetability is feasible only among similar processor architectures, because rules are not automatically generated and one must pay particular attention not to write new rules that may clash with existing rules.

Genin et al. [6] reported a system in which, by providing the compiler with knowledge about DSP and the target processor's instruction set, the generated code performs "5 to 50 times faster than the one produced with conventional C compilers and is comparable to the code generated by DSP experts." An internal signal flow graph is generated in which each node represents a DSP operation of a specific class. A pattern recognizer identifies and merges low-level nodes into semantically higher-level constructs. Each node is responsible for its own code generation and node activation is determined by a scheduler. The scheduler has knowledge of the global context through a set of *if–then* rules. The action parts of the rules contain the code to be generated. One of the key reasons for such an impressive performance is that this system requires the use of a specialized DSP HLL called Silage [10].

In DISSIPLE [48], the code generator is composed of two rule-based expert systems; one generates the control flow code and the other the computation code. The use of separate code generators is claimed to be advantageous for portability reasons and to allow the compiler to generate code for both single-processor and multiprocessor systems. Code optimization, however, is deferred until the last step of the compilation process, through conventional peephole techniques.

Kuroda et al. [20] developed a knowledge-based code generator for PDSPs with nonuniform registers and partitioned memory architectures. Such architectures are designed to allow multiple memory accesses in a single instruction. Hence, efficient memory assignment of DSP specific data types is critical to the quality of the generated code. A profiler analyzes the HLL program to obtain statistics about the execution times for instructions with multiple memory access. An access conflict graph, in which nodes represent data values and edges represent the number of conflicts, is constructed from the collected statistics. A heuristic algorithm then assigns variables to the several memory partitions, minimizing the number of conflicts. The code generator first translates the IR into a register-transfer-level representation by binding processor resources to each variable. Then, a recursive pattern-matching process is carried out to cover the entire IR. Patterns are *if–then* rules in which the *then* part is associated with instructions to be generated (including register allocation) and the *if* part is associated with a (sequence of) register-transfer operation(s) that realize the *then* part. By checking the number of additional steps required to generate a code, the code generator can select the best code among the alternatives stored in the rule base. Unfortunately, no quantitative results are presented, except the authors claim that their compiler "can generate almost the same code as programmers would have done by hand."

Azaria and Dvir [49] described an expert-system-based compiler for a special-purpose array processor. Because their proposed system (OC) can handle only the four elementary mathematical operations $(+, -, *, /)$, it was dedicated to solving linear constant systems and other problems that can be totally defined by a series of matrix operations. In such a restricted environment, expert knowledge was shown to be very effective. OC's high optimization performance is accomplished mainly through algorithm transformation. By consulting the rule base, OC identifies and eliminates redundancies in the source algorithm.

2.2 Optimizations

An optimizing compiler is one that applies optimization techniques such that the resultant machine code is faster and/or smaller than that produced by conventional compilers. Perhaps the most important tenet about optimizations is that they aim at improving code, rather than actually attaining optimal code.

Optimizations can be classified into two categories, based on the type of information that is used: *static* and *dynamic* optimizations.

2.2.1 Static Optimizations

Static optimizations rely on compile-time analysis information to alter the program. Static analyses are inherently limited and imprecise because many aspects of run-time behavior are undecidable at compile time. Nonetheless, static optimizations are relatively easy to perform, with little overhead in compile time and good potential for gains at run time.

Among the myriad of ingenious static optimization techniques implemented in current optimizing compilers, the most interesting include the following:

1. *Algebraic simplification and rearrangement*: Replacing expressions with simpler, equivalent expressions, or rearranging expressions to facilitate constant folding.
2. *Branch-tail merging*: Combining instructions common to branches that jump to the same location.
3. *Code motion (hoisting)*: Moving instructions common to the beginning of two branches of a conditional jump to a location before the jump.
4. *Common subexpression elimination*: The compiler computes the value of expressions that always yield the same result, saves the result as a temporary value, and uses that value instead of recomputing the subexpression each time it encounters them.
5. *Constant folding*: Replacement of arithmetic and logical operations by the constant to which they evaluate at compile time.
6. *Constant propagation*: Changing variable references to constants, if the variable has a constant value.
7. *Dead (unreachable) code removal*: Removal of code with no path. Other optimizations generally create dead code, not the programmer.
8. *In-lining*: In-line expansion of a function to save the overhead of the call, parameter passing, register saving, stack adjustment and value return.
9. *Instruction scheduling*: Rearranging instructions sequence to avoid stalling the instruction pipeline; includes overlapping CPU and FPU instructions and inserting useful instructions between data-dependent instructions that stall the pipeline.
10. *Loop rotation*: Rotation of the controlling expression of FOR and WHILE loops to the bottom of the loop. If the loop executes more than once, this optimization eliminates the initial test.

11. *Loop unrolling*: Using multiple copies of a loop's body statement instead of incrementing and checking the loop induction variable.

12. *Redundant code elimination*: Elimination of code sequences producing no real side effects; usually result from other optimizations.

13. *Register coloring*: Optimizes register usage throughout a function or routine, such as keeping local variables in registers at all times. Similar to coloring a map where no two adjacent areas can have the same color, the algorithm uses data flow (lifetime) analysis for each variable, mapping variables without overlapping lifetimes to the same register.

14. *Short-circuit (logical expression) evaluation*: Converting expressions containing AND, OR, and NOT operators into a series of conditional jumps.

15. *Strength reduction*: Replacing an operation with equivalent but less expensive (in size and/or speed) operations; for example, replacing multiplication with combinations of shifts and adds.

16. *Switch statement optimization*: Choosing among different code generation methods for switch statements based on code size or speed. Methods include linear test and branch, jump table, binary search, and parallel tables.

Few commercially available optimizing compilers for PDSPs have been introduced [7,8,50]. To varying degrees, they all perform static optimization techniques such as those listed above. The biggest challenge for an optimizing compiler is to apply optimizations in such an order that each optimization can uncover all possibilities for additional optimization opportunities and effectively exploit them. More importantly, the compiler must ensure that one optimization does not undo the results of previous optimizations. An interesting approach is that used in Benitez' Very Portable Optimizer (VPO) [51]. VPO uses an iterative approach: Each optimization is applied in separate phases; previously executed phases are reinvoked whenever new possibilities arise.

Static optimizations can be performed in three stages of the compilation process, as illustrated in Figure 3. Typically, optimizations performed in the FE or on the IR are machine independent. Their main purpose is to ease the code generator's job rather than to improve code quality. The BE has access to target-processor information, hence, the code generator can produce both machine-dependent and machine-independent optimizations. Certain optimizations, such as instruction scheduling and register coloring, are intended for postgeneration. These are inherently machine dependent.

Most commercially available optimizing compilers also perform peephole optimizations [24,43,52,53]. Peephole optimizers work on the assembly code by looking at a few (typically two or three) instructions at a time and attempt to

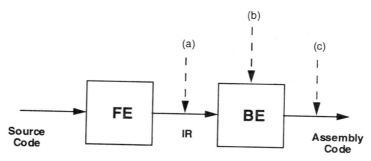

Figure 3 Three different stages of compilation where static optimizations can be applied: (a) before, (b) during, and (c) after code generation.

replace that sequence of instructions by another sequence, semantically equivalent but more efficient. The fundamental drawback of this technique is that it is localized in nature. Increasing the window size (looking at more instructions) increases computational complexity exponentially without much improvement in performance.

2.2.2 Dynamic Optimizations

Dynamic optimizations encompass all those optimizations that use run-time analysis to alter the program. They can be further subdivided into *context-dependent* and *data-dependent* optimizations. These two categories are not exclusive and can complement each other.

   Context-Dependent Optimizations.   The major shortcoming of static optimizations is that the compiler cannot decide which, among a large set of possible optimizations, will be beneficial and which will hurt code quality. A simple approach is to subdivide the set into smaller sets and classify them according to their generality (or innocuousness) and let the user decide when and which sets to apply. There are several problems with this approach. First, determining the generality of an optimization is a nontrivial task. As we will show in Section 5, an optimization such as common-subexpression elimination, which is typically considered general and therefore safe to perform whenever possible, is actually architecture and context dependent in the presence of run-time information. Second, the user must go through a tedious analysis of the generated code to determine which optimizations were the culprits for the code's poor quality and recompile with those optimizations disabled. What is worse, an optimization may be beneficial in one part of the program and harmful in others. Third, the order in which optimizations are applied is important and there is no easy way to determine the best order. More importantly, the best order may depend on the input program.

Context-dependent optimizations or CDAT, as we will refer to them in Section 5, use run-time analysis to determine the status of all resources, such as registers, various memory partitions, and status flags to determine if a particular optimization is advantageous to implement.

As we will see in Section 5, many interesting issues arise in interpretive code generation. We will discuss each one of them in detail and show that our heuristic search methodology for code generation can solve them quite naturally.

Data-Dependent Optimizations. Alternatively, one could consider optimizing a program based on the data that it is supposed to process. Because the actual data are unknown until run time, data-dependent optimizations and, consequently, code generation must be deferred until run time. The generated code, thus, contains instructions to emit code at run time. Many approaches have been proposed to reduce the overhead imposed by run-time code generation (RTCG), including the use of templates [54–56] and the use of a more general intermediate representation [57].

Interesting works on RTCG include that of the SELF group [92,93,58] and FABIUS from Carnegie-Mellon [94]. A good summary of other work on RTCG can be found in the technical report by Keppel et al. [56].

Run-time code generation has a high degree of flexibility in the sense that it can optimize the code for a particular set of data, possibly even for different iterations of the same loop. On the other hand, it is difficult to implement and the overhead added to the run time is recovered only for large input sizes. Many DSP-embedded applications value code size over execution speed,* which is substantial detriment to using RTCG methods.

An alternative approach to RTCG is to utilize run-time profiling information to optimize the program in subsequent compilations. University of Washington's SELF compiler has had some success using dynamic profile information to guide compile-time optimization of object-oriented code [59,60,93]. This approach would eliminate the code generation overhead in the executable code, at the expense of longer compilation times and complexity in the profiling analysis. The major shortcoming, however, is that for meaningful results, the input data must be either very predictable or known exactly.

---

* The available memory size is limited. In addition, the assumption that a small code will also execute fast is often valid.

2.3 Summary and Conclusions

In this section, we have presented four code generation methods: tree/DAG traversal, template pattern matching, tree rewrite, and knowledge based. We also surveyed many systems and research works within each of these methods and discussed their intrinsic advantages and disadvantages. None of the reviewed methods alone can generate code whose quality is truly comparable to that of handwritten code by DSP experts.

We also surveyed static and dynamic optimization techniques. We discussed why static optimizations are necessarily imprecise. Dynamic optimizations can be data and context dependent. The former can potentially produce better code than the latter. However, because the data are unknown until run time, the optimizations (and code generation) must be deferred to run time. This poses a heavy load on the executable code, because it now has to generate its own code. More importantly, RTCG increases the size of the executable code, a penalty not affordable by many embedded applications. A more suitable approach is that of context-dependent optimizations, which can produce very efficient code at the expense only of compilation time.

In the next section, we present our proposed approach, which combines tree/DAG traversal, template pattern matching, and knowledge based with other artificial intelligence techniques and context-dependent optimizations to attain the desired handwritten code quality.

3 OVERVIEW OF OASIS

Our formulation of code generation is a hybrid of the tree/DAG traversal, template pattern-matching, and knowledge-based approaches (introduced in Sec. 2). From the highest-level view, the algorithm, which we call OASIS (Optimized* Allocation, Scheduling, and Instruction Selection), works as illustrated in Figure 4. First, a DAG is built from the input program. Next, the DAG is augmented to include flow-of-control information, followed by scheduling of each node in the augmented DAG. This initial scheduling, however, does not determine the actual order in which nodes are covered. For the latter, a more complex heuristic algorithm is required because the quality of the generated code is directly related to the order in which nodes are covered. At each iteration, all coverable nodes are collected; each of those coverable nodes is evaluated and the least expensive node is selected and covered. If more than one node share the least expensive cost, then the node with the lowest initial schedule is selected. Node coverage

---

* We note the distinction between the terms *optimal* and *optimized*. The first refers to the absolute best. The latter is a frequently used term by compiler vendors, implying that optimization techniques have been applied and the result may or may not be optimal.

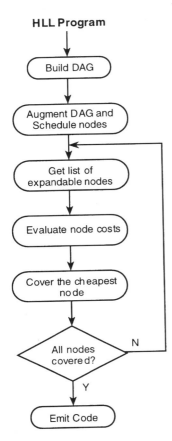

HLL Program

Build DAG

Augment DAG and Schedule nodes

Get list of expandable nodes

Evaluate node costs

Cover the cheapest node

All nodes covered?

N

Y

Emit Code

Figure 4 Simplified flowgraph of OASIS.

involves template pattern matching, whereas node evaluation involves heuristic search; both node coverage and node evaluation also involve means-ends analysis and hierarchical planning, which we will examine in more detail in the next section.

Unlike the approaches of Newcomer [44] and Cattell [26], there is no separate compiler compile phase. All analysis is done at compile time. The advantages of such an approach include the potential to achieve optimal code,* at the expense of longer compilation time and larger memory requirements.

---

* Due to the use of heuristics, an optimal solution cannot be guaranteed. However, for a sufficiently small DAG and given sufficient computational resources, if an optimal solution falls within the solution space defined by the heuristics, OASIS will find it.

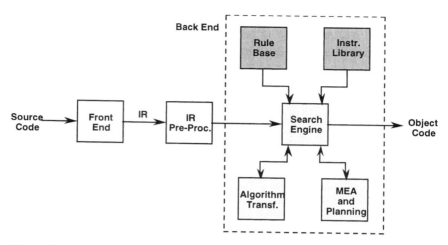

Figure 5   Functional block diagram of OASIS.

As we explained in Section 1, many embedded DSP applications can justify the costs of a more complex code generation algorithm and we will not concern ourselves with such issues as compilation time and memory requirements of the algorithm. Our main goal is to provide a means to obtain the best possible machine code. Nonetheless, the entire task must be realizable within practical limits of time. We will show in the next section how the application of artificial intelligence (AI) tools can reduce an essentially NP-complete problem into one that executes in polynomial time with the DAG size, albeit exponentially in the search depth K. Because K is user defined, our algorithm provides the flexibility to match computational resources with desired code quality. We will show in Section 6 that very good results can indeed be obtained with relatively small K and surprisingly short compilation times.

Figure 5 presents the structure of the entire system. The five gray blocks comprise the back end, where code generation occurs. The three lighter gray blocks are machine independent, whereas the two darker gray blocks are machine dependent. Porting this code generator to a new architecture then requires only a new rule base and a new instructions library.

3.1   The Front End

First, let us define a small example, which we will use throughout this section to illustrate specific issues. Consider the following fragment of C code:

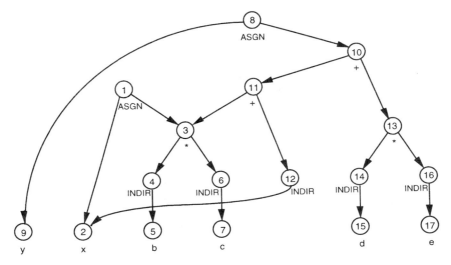

Figure 6 DAG representation for Example 1.

Example 1

```
S1:  x = b * c;
S2:  y = x + b * c + d * e;
```

In our current implementation, the front end (FE) is adapted from lcc compiler [61]. The FE takes an ANSI C program as input and generates a linearized DAG as the intermediate representation (IR). Because the linearized form is quite difficult to visualize, we will use a graphical representation of the FE's output for Example 1, as shown in Figure 6. Nodes of type INDIR represent the fetch of a variable value from a resource, and nodes of type ASGN write a new value to the variable stored in a certain resource. The translation from the linearized form to the graphical form loses some information, namely control and data dependencies. In Example 1, it implies that statements S1 must be executed prior to S2. However, in Figure 6, there is nothing that indicates that node 1 should be covered before node 12. In the next subsection, we show how that information can be recovered, as well as discuss other pertinent issues related to the IR.

3.2 The Intermediate Representation

Many static optimization techniques can be applied at the IR level, such as constant folding and relocation of invariant loop expressions [62]. Static algorithm transformation techniques can also be applied at the IR level to transform the input program into an equivalent representation that better matches the target processor's architecture (see Sec. 2.2). Our formulation of code generation calls for context-dependent transformations, which must be deferred until compile time, at the BE. Hence, the IR preprocessing block only augments the DAG with control and data dependency arcs, as well as scheduling information, which are needed by the BE.

3.2.1 Data Dependencies

A data dependency occurs between two statements when both reference the same memory location or access the same physical register. There are three kinds of data dependencies [63,64]:

1. True dependency (also called flow dependence or read-after-write hazard): characterized by

   ```
   S1:  x = . . .
   S2:  . . . = x
   ```

   The value of x at statement S2 depends on the value assigned to it at S1; hence, the assignment in S1 must be executed before the use of x in S2.

2. Antidependency (also called write-after-read hazard): characterized by

   ```
   S1:  . . . = x
   S2:  x = . . .
   ```

   A new value cannot be assigned to x until x's current value has been used; hence, the use of x in S1 must be executed before the assignment in S2.

3. Output dependency (or write-after-write hazard):characterized by

   ```
   S1:  x = . . .
   S2:  x = . . .
   ```

   The final value that x assumes depends directly on which statement is executed last. The order in which the statements appear in the source program must be preserved by the scheduling algorithm. In this case, the assignment in S1 must be executed before the assignment in S2.

   As the name implies, true dependencies are the only real dependencies; a variable required by the current statement must have its value evaluated by a previous statement. Both antidependence and output dependence can be eliminated by variable renaming [65,66]. Variable renaming usually increases the required memory for the program. Because our methodology supports all three types of data dependency, variable renaming is unnecessary.

The algorithm that follows provides a systematic method for introducing the necessary data dependency arcs to the DAG:

```
AddDataDependencyArcs(DAG) {
    for (every node i in DAG that is of type ASGN or INDIR) {
        for (every node j•i in DAG that is of type ASGN or INDIR) {
            if (no arc exists between nodes i and j) {
                Dep Type = DetermineDependencyType(DAG, i, j);
                case (DepType) {
                    TrueDep: add arc from node i to node j;
                    AntiDep: add arc from node j to node i;
                    OutputDep: if (i>j) add arc from node i to node j
                        else add arc from node j to node i;
                };
            };
        };
    };
};
```

Nodes of type INDIR represent the fetch of a variable value from a resource, and nodes of type ASGN write a new value to the variable stored in a certain resource. Hence, data dependency arcs occur only among INDIR and ASGN nodes. However, not all INDIR and ASGN nodes need a data dependency arc, only those that falls into a true, antidependency or output dependency category. Figure 7 illustrates one such case in which no data dependency arc is neces-

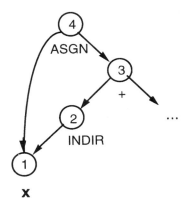

Figure 7 DAG representation of **x** = **x** +. . . .

sary between ASGN and INDIR operators. The dependency relationship in S1 is uniquely defined: The ASGN operator (node 4) must be scheduled *after* the + operator (node 3), which, in turn, must be scheduled after the INDIR operator (node 2).

3.2.2 Control Dependencies

Intuitively, a control dependence occurs from statement S1 to statement S2 when S2 is executed only if S1 produces a certain value [64,67]. This type of dependence occurs, for example, when S1 is a conditional statement and S2 is to be executed only if the condition evaluates to true. To illustrate when control dependencies might occur, let us consider Example 2:

Example 2

```
S1:  if (a) x = 1;
S2:  else x = 2;
S3:  y = x;
```

The linearized DAG generated by the FE is as follows:

```
. . .
node#3 ADDRLP count = 1 a
node#2 INDIRI count = 1 #3
node#4 CNSTI count = 1 0
node'1 EQI count = 0 #2 #4 2
node#6 ADDRLP count = 1 x
node#7 CNSTI count = 1 1
node'5 ASGNI count = 0 #6 #7
node#9 ADDRGP count = 1 3
node'8 JUMPV count = 0 #9
2:
node#12 ADDRLP count = 1 x
node#13 CNSTI count = 1 2
node'11 ASGNI count = 0 #12 #13
3:
node#16 ADDRLP count = 1 y
node#18 ADDRLP count = 1 x
node#17 INDIRI count = 1 #18
node'15 ASGNI count = 0 #16 #17
```

In this linearized DAG, node'5 ASGNI count = 0 #6 #7 reads as node 5 is the root of a subtree, with no arcs pointing to it (count = 0), its type is ASGNI

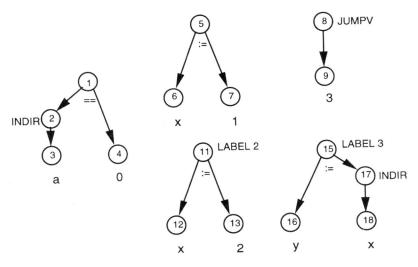

Figure 8 Graphical DAG representation for Example 2.

and it has arcs pointing to two other nodes, node 6 and node 7.* The same DAG, in a more legible graphical form is shown in Figure 8.

The lcc compiler transforms *if–then–else* structures into a one-dimensional description with GOTOs (JUMPs) to determine ordering. When a two-dimensional DAG is built from such description, the ordering information is lost. For example, in the linearized DAG, nodes 11 to 13 precede nodes 15 to 18. A JUMP in node 8 brings us to the subtree labeled **3**, but no JUMP is necessary at the end of the subtree labeled **2** because it is assumed that the subtree labeled **3** follows unconditionally. However in the graphical representation of the linearized DAG, shown in Figure 8, subtrees are spread around, with nothing to indicate the covering order. One way to recover the precedence information is to augment the DAG with control dependence edges. Such edges are processed by the IR Preprocessing block just like any other edge. Hence, if a node M must be covered before a node N, an edge must be added from N to M. This forces a partial scheduling order more in line with the one meant by lcc. The resultant augmented DAG is shown in Figure 9.

3.2.3 ALAP and ASAP Schedulings

Once the DAG has been augmented with data and control dependency arcs, the next step is to assign an initial schedule to each node in the DAG. We use an

---

* For a complete description of lcc's syntax, see Ref. 61.

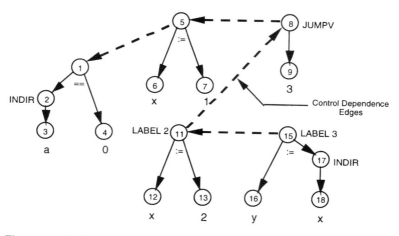

Figure 9 DAG for Example 2 augmented with control dependency arcs.

as-late-as-possible (ALAP) scheduling. In an ALAP schedule, if node M must be covered before node N, M is scheduled as close to N's schedule as possible. The rationale for using ALAP is as follows: If M must be covered before N, then there must be an arc from N to M. If it is a data dependency arc, then N depends on M's data and executing M immediately before N makes the least commitment of resources, yielding the positive effect of reducing spill instructions. If it is a control dependency arc, then any scheme that schedules M before N can be applied and ALAP provides a valid initial guess. Note that the actual coverage order will be determined by a more thorough node analysis, as we will see in the next sections.

The algorithm to perform as-soon-as-possible (ASAP) scheduling is as follows:

```
ASAP(DAG) {
    node[root].schedule = 0;
    N = set of all nodes of the DAG;
    p = 1; /* schedule # */
    while (N not empty) {
        for (every node j in N) {
            if (all parents of node j have been scheduled) {
                temp[j] = p;
                N = N-{j};
            }
            else temp[j] = -1;
        };
        for (i = 1; i< = number of nodes; i++) {
```

```
          if (temp[i] ! = -1) {
              node[i].schedule = temp[i];
          };
      };
      p++;
  };
};
```

ASAP is essentially the opposite of ALAP, as it tries to schedule nodes immediately after their parents have been scheduled. The ALAP scheduling is derived from the ASAP scheduling and its algorithm is as follows:

```
ALAP(DAG) {
    Reverse the directions of all edges in DAG;
    Assume the ''root'' is ''output'' and vice-versa;
    ASAP(DAG);
    n = largest schedule #;
    for (all nodes j) {
        node[j].schedule = n-(node[j].schedule) + 1;
    };
};
```

This ALAP initial scheduling has three purposes. First, it provides a basis for the actual node coverage ordering. At each iteration, the node evaluation functions (described in Section 4.1.3) evaluate the effective cost of all expandable nodes. The least expensive node is then selected for coverage. Second, it is common for more than one node to share the minimum effective cost. In that case, the decision is made based on the initial scheduling—the node with the lowest initial schedule is selected. It is still possible, however, that more than one node share both the minimum effective cost and initial schedule number. In that case, an arbitrary decision is made: The node with the lowest node number is selected.

As we will see, node evaluation is the single slowest component of the whole algorithm. OASIS allows the user to specify a limit, B, on the number of nodes to be evaluated. Again, the initial schedule plays an important role. Coverable nodes are sorted in increasing initial schedule and only the first B nodes are evaluated.

The scheduled DAG for Example 1 is shown in Figure 10. In Example 1, variable x in S2 depends on the value assigned to it in S1. This data dependency is indicated by the arc from node 12 to node 1, which causes node 12 to be scheduled after node 1.

3.3 The Back End

Code generation is performed in the BE, which comprises the last five blocks in Figure 5. There are three processor-independent modules—*search engine, algo-*

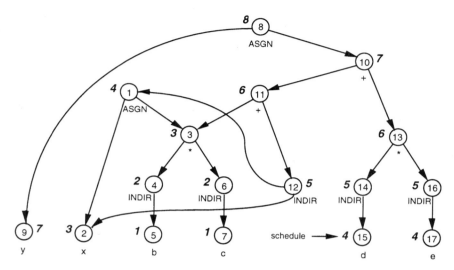

Figure 10 Augmented and scheduled DAG for Example 1, as received by the BE.

rithm transformation, and *means-ends analysis*—and two processor-dependent modules—*rule base* and the *instructions library.* This arrangement makes porting the code generator straightforward. All that is required is a new rule base, with heuristics, rules of thumb, or any expert information about the new architecture, and a new library of templates representing the new instruction set. The BE's main functional block is the *search engine,* which is assisted by the *rule base* and the *instructions library* modules and interacts with the *algorithm transformation* and *means-ends analysis* modules.

Each of the above functional blocks will be addressed in detail in Section 4.

3.4 Summary and Conclusions

In this section, we presented an overview of a proposed compiler. It is essentially an FE feeding a BE type of structure, as introduced in Section 1. FE generation is well defined using models such as context-free grammars; many tools exist to automate that task. Our FE is adapted from that of lcc [61]. Unlike FE design, good BE design remains more of an art than science and gives much room for ad hoc approaches.

An IR preprocessing module is introduced between the FE and the BE. That module transforms the IR generated by the FE into a form more suitable for the BE. This section discussed in great detail the tasks involved in the IR preprocessing module.

In this section, we briefly introduced our formulation of code generation as a heuristic search problem. Detailed discussion of our formulation of code generation will be conducted in the following sections.

4 CODE GENERATION IN OASIS

In the previous section, we discussed in detail the FE and the IR preprocessing modules. In this section, we will focus our discussion on the BE, where code generation is implemented. We will discuss in detail how artificial intelligence techniques such as means-ends analysis and hierarchical planning are combined in the heuristic search implementation. We will also conduct a time-complexity analysis to ensure that the entire algorithm is bounded.

4.1 The Search Engine

The search engine builds two types of heuristic search trees, illustrated in Figure 11. Each state (search node) in these search trees contains, among other information, the remaining DAG to be covered and the status and contents of all physical resources of the target processor. The latter information is encoded into a resource table (RT).

An instructions library containing all available instructions of the target processor assists the search engine. In the instructions library, one assembly instruction or a small sequence of assembly instructions can define a template pattern. Each template is described by a set of preconditions and a set of postconditions. Expanding a search node into a successor node is achieved by applying an instruction or matching a template* (see Section 4.1.1). A template matches if its preconditions can be fulfilled by the current node's RT. Postconditions represent the effects of matching the template and determine the contents of the successor's RT. In the following representation, we illustrate how the template **MPY_dma** can be represented. Note that we only indicate which resources hold operands and which are affected by the results of the instruction; we specify neither what kind of operation it performs nor the value of its results. Because templates are not linked to any particular operation, they could be considered for coverage of any IR operator, as long as the required rules are satisfied.

The search engine is also assisted by a *rule base*, which encapsulates expert knowledge about the code generation domain in the form of heuristics, context-dependent rules and architecture-dependent constraints. In the following simpli-

---

* Hereafter, we will·use both terms interchangeably.

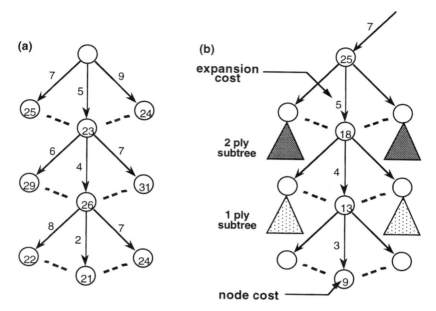

Figure 11 A heuristic search tree with depth $K = 3$. At each level p, heuristics are used to expand a node into its most promising successors. The search engine supports two kinds of trees: (a) Each successor is associated with a node cost and only the least expensive successor is selected for expansion at level $p + 1$; this kind of tree is used at the topmost level (see Fig. 4). (b) All promising successor nodes are expanded at level $p + 1$ and the node cost at level p is the sum of the expansion cost at level p and the minimum node cost at level $p + 1$; this kind of tree is used to evaluate the node costs of the tree in (a).

fied representation of template **MPY_dma**, text within brackets is treated as a comment:

```
{o = operand}
{l = left operand}
{r = right operand}
{s = result}
{n = not affected}
```

```
{Instruction}   MPY_dma            {PR ← TR * DM}
{Cost}          1
{Rules}         1 3 8
{Resources      PR  TR  AC  DM  PM  ARP  AR}
{PreCond}       n   ol  n   or  n   n    n
{PostCond}      s   o   n   o   n   n    n
```

As shown in the foregoing, each template has a field `Rules`, which contains all the rules from the rule base that are relevant to the current instruction. Because all rules listed in field `Rules` must be satisfied before a template can match, this field can also be used to differentiate two templates that affect the same set of resources identically (i.e., share the same preconditions and postconditions (more on rules in the next section).

Returning to Figure 11a, if a template matches, it expands the current search node into a new search node. The latter contains a new DAG, in which the covered portion is removed. The current search node is expanded into as many successors as there are matching templates. Each successor node is evaluated (see Section 4.1.3) and the least expensive one is selected for the next iteration.

Each search node also has a pointer to its parent and a field `Instructions` to store all the templates that matched during its expansion process. Once a goal node (a node in which the entire DAG has been covered) is reached, the pointers to parents are traced backward from the goal node. The contents of the `Instructions` fields along that path are collected to form the generated code.

4.1.1 Modified Means-Ends Analysis

Procedure **ExpandNode** encapsulates a technique known as means-ends analysis (MEA) [68] to expand a search node and return the cost of that expansion. Briefly, MEA works as follows. First, the difference between the preconditions of the candidate template T and the RT of the current node is established. This difference is passed to procedure **Apply**. If the difference is nil, T is immediately applicable and **Apply** expands the current node with T. Otherwise, procedure **Reduce** is invoked to reduce that difference. **Reduce** calls **Apply** with the new, hopefully smaller, difference. In the following algorithm for procedure **Apply**, **Apply** applies template T on node X to expand it into the successor node Y:

```
Apply(SearchNode X, Template T, SearchNode Y) {
    D = Difference(X, T);   /* evaluate difference between
                        X → RT and T → Preconditions */
    if (D is nil)
        Create Y(X, T, Y);   /* create output Y by modifying X → RT
                        and X → DAG to reflect the effects
                        of applying T to X*/
    else
        Reduce(X, T, D, Y);   /* reduce difference between X and T,
                        and generate output Y*/
};
```

To reduce a difference, the original MEA technique uses a connection table, which lists all possible candidate operators, ranked by order of efficacy, that can reduce each type of difference. The first problem is that such a table would be impractically large for our application. Any kind of context-insensitive, predetermined action would be a detriment to the goal of optimized code generation. To preserve the context-sensitive requirement, each difference, under all possible contexts, would have to be listed as a different entry. As for the second problem, ranking operators is a nontrivial and error-prone task, given the number of templates to be considered. Our version of MEA evaluates, at compile time, all relevant templates in the instructions library as to their potential to reduce the current difference. A profiling of the OASIS execution reveled that this on-line evaluation represents less than 1% of the total computation time.

In a straightforward implementation, MEA can lead to a dead end or to circular loop situations. **Reduce** may return a difference viewed by its evaluation function as smaller than the original difference and, nonetheless, when that difference is passed to **Apply**, it may not be further reducible, or when further reduced, it is identical to the original difference. In an analogy to game-playing programs, it is necessary to look ahead several moves to discover that a seemingly bad move is in fact the best one.

4.1.2   Hierarchical Planning

To address the circular cycle problem, we modify MEA by supplementing it with hierarchical planning (HP). HP is a technique introduced by the ABSTRIPS system [69], which has been shown to generate very little or no backtracking [70]. Korf [33] showed that by using an abstraction hierarchy, it is possible to reduce an exponential-time problem to a linear-time complexity. We have found that two abstraction levels suffice in the instruction selection domain. Once a difference is established, it becomes the goal for procedure **Reduce**. A plan (sequence of instructions that accomplishes a goal) is sought at the higher abstraction level. At that level, the candidate instruction's preconditions are ignored. Hence, the best instruction is the one that can bring the current RT closer to the desired goal, independent of its applicability. Once a plan is obtained, its implementation is attempted at the lower abstraction level, in which all preconditions are considered and must be satisfied. In the following algorithm for procedure **Reduce**, **Reduce** generates a plan of action to reduce the difference between template T and node X and applies each component of the plan until the difference is completely reduced:

```
Reduce(SearchNode X, Template T, Difference D, SearchNode Y) {
    Plan = GeneratePlan(X, T);   /* generate a list of templates,
                            which represents a plan of action */
    while (D is not nil and Plan is not empty) {
        P = First template in Plan;
```

```
    Apply(X, P, Y);
    D = Difference(Y, T);
    Plan = Plan-P;
    };
};
```

When a component of a plan is not applicable, the process recurs; subplans are sought to fill the gaps between components in the original plan. When developing subgoals, two important issues must be considered:

1. If more than one instruction is necessary to completely reduce a difference, in which order should those instructions be applied?
2. It is possible that the difference is completely reduced before all components of the plan are applied.

To illustrate both issues, let us consider statement S2 in Example 3:

Example 3

```
S2:  y = x + b * c + d * e;
```

Suppose instruction **ADD** is being considered for covering the subtree $(x + b \times c)$. **ADD**'s precondition requires that x be in the accumulator and $(b \times c)$ in data memory. Let us assume that the subtree $(b \times c)$ was covered by **MPY** and that the current RT indicates that x is in data memory and $(b \times c)$ is in register P. Consequently, **ADD** is not immediately applicable and **Reduce** is called with two subgoals:

Subgoal 1: move x from data memory to accumulator
Subgoal 2: move $(b \times c)$ from register P to data memory

The first issue is to determine which subgoal to pursue first. To address it, a new heuristic search is created. In other words, **ExpandNode** is recursed with all possible orders of the components of the plan and the least expensive order is returned as the best plan. The first choice is to move x and then move $(b \times c)$. However, no single instruction in the instructions library can move a value from register P directly to data memory. The best plan for subgoal 2 is to move $(b \times c)$ from register P to the accumulator and then from the latter to data memory. The resulting code would be as follows:

Subgoal 1:
```
    LAC      x      ; load accumulator with x
```
Subgoal 2:
```
    SACL     temp   ; store x in location temp in memory
    PAC             ; move (b × c) to accumulator
```

```
SACL    y       ;move (b × c) to location y in memory
LAC     temp    ;move x back to accumulator
```

The alternative is to move $(b \times c)$ first and then x. The resulting code would be as follows:

Subgoal 2:
```
PAC             ;move (b × c) to accumulator
SACL    y       ;move (b × c) to location y in memory
```
Subgoal 1:
```
LAC     x       ;load accumulator with x
```

The latter ordering is clearly less expensive and is returned as the better plan.

The second issue comes to play when the plan is executed. Note that after instruction **PAC** is applied, one operand to **ADD**, $(b \times c)$, is in the accumulator, whereas the other operand, x, is in data memory. **ADD**'s prerequisite requires exactly the opposite, but because + is a commutative operator, **ADD** is applicable at this point. Hence, only the first component of the plan is used and the rest is discarded.

4.1.3 Minimizing Search Space

Full search involves processing $O(B^K)$ nodes, where K is the depth of the search and B is the average branching factor. Hence, there are two obvious ways to reduce the combinatorial explosion: reduce K and reduce B.

In our approach, K determines the look-ahead factor and is a user-defined variable. The larger the K, the more accurate is the result returned by the evaluation functions, at the expense of exponentially increasing computation time.

To reduce the effective branching factor B, we apply domain knowledge and heuristics in three different ways:

1. Decide which node(s) can be pruned from the search tree. Our approach is to sort the list of expandable nodes according to increasing initial schedules and expand only the first B nodes (in OASIS, B is a user-defined variable).
2. Decide which node to expand next. As we discussed earlier, ALAP scheduling provides a starting point, but it is not sufficiently accurate for our purposes. We use look-ahead to evaluate all B candidates (detailed next) to determine their actual cost. Only the least expensive node is selected to be expanded (see Fig. 11a).
3. Decide which successor(s) to generate during node expansion. Here, domain knowledge is encoded into rules to eliminate unreasonable can-

didate templates, thus expanding a node into only the most promising
successors.

Evaluation Functions. As discussed in Section 4.1, the search engine implements two kinds of search tree. OASIS uses two evaluation functions to determine the cost of a node, one for each kind of tree.

In Figure 11a, the basic idea is to expand only the most promising node.
The term *most promising* connotes minimum cost, as defined by a cost function.
Because only one node is expanded, the evaluation function must be very accurate. To accomplish the desired accuracy, we perform look-ahead. Hence, to evaluate the cost of a node N, OASIS defines a heuristic function that is a weighted
sum of various attributes:

$$NC_1(N) = w_1 I(N) + w_2 G(N) + w_3 F(N) + w_4 NC_2(N, K) \tag{1}$$

where $NC_1(N)$ is node N's total node cost, $I(N)$ is the number of templates that
matched along the minimal path from the start node up to and including the
current node N, $G(N)$ is the total cost associated with those templates, $F(N)$ is
an estimated cost from N to a goal, and $NC_2(N, K)$ is a K-step look-ahead node
cost, computed by the second evaluation function. Currently in OASIS, $w_1 =
w_2 = w_3 = w_4 = 1$ and $F(N)$ is a function of the number of operators remaining
to be covered in the DAG.

The second evaluation function is computed by a K-step recursive call to
procedure **ExpandNode**. K is a user-defined value that tells **ExpandNode** how
many levels to look ahead in evaluating a node's cost. Each recursion decrements
K by 1, as one can observe in Figure 11b. When called with a node N as parameter, **ExpandNode** expands N into its successors S_i and returns N's node cost.
The node cost of a node N, with a K-step look-ahead, is defined by

$$NC_2(N, K) = EC_2(N) + \min_{S_i, 1 \leq i \leq BC} \{NC_2(S_i, K - 1)\} \tag{2}$$

$$NC_2(N, 1) = EC_2(N) \tag{3}$$

where $EC_2(N)$ is the expansion cost of node N, given by

$$EC_2(N) = w_1 T(N) + w_2 C(N) \tag{4}$$

where $T(N)$ is the number of templates that matched in the process of expanding
N and $C(N)$ is the total cost associated with those templates.

The Rule Base. The rule base is subdivided into three sets of rules. The
first set is consulted before the application of an instruction in order to determine
if the candidate instruction is applicable in the current context. In Section 4, we
noted that such rules are necessary to distinguish among instructions with the
same preconditions and postconditions. Rules are also useful to help reduce the
effective branching factor. For instance, both instructions **APAC** (add contents

of register P to the contents of the accumulator) and **ADD** (add contents of a memory location to the contents of the accumulator) could be used to cover $(x + b \times c)$ in Example 3. **APAC**'s precondition requires that one operand be in register P and the other in the accumulator. However, the architecture of TMS320 does not provide an easy way for loading register P with a desired value d. In the best case, d is in register T and one could use instruction **MPYK_1** to multiply d by 1, with the result stored in register P. Hence, one rule specifies that **APAC** be considered only when one of the operands is already in either registers P or T. Otherwise, **ADD** is certain to be less expensive and **APAC** is pruned from the search.

The second set of rules is consulted *after* the application of an instruction, to determine if additional actions should be taken. For example, consider instruction **MAC_pma_dma**. This instruction first adds the contents of register P with the contents of the accumulator, leaving the sum in the accumulator; then, it replaces the contents of register P with the product of two operands, one addressed by PMA and the other by DMA. Now consider the two statements in Example 1. After covering statement S1 with **MPY**, the product $(b \times c)$ is left in register P and the new value of x is left in the accumulator. In statement S2, after covering $(d \times e)$ with **MAC**, one of the postaction rules matches, because the portion $(x + b \times c)$ can be readily covered by **MAC**. Hence, the action part of that rule will also mark node 11 covered, as a postaction to marking node 13 covered.

The third set of rules is used *during* algorithm transformations and will be discussed in the next section.

4.2   Time-Complexity Analysis

Although we stated in Section 1 that compilation speed should not be a major design constraint, we must not overlook that the entire algorithm must be executable within practical time limits. Therefore, we must analyze the time complexity of the proposed algorithm.

THEOREM 1   *The evaluation function defined by expression (1) executes in time linearly proportional to the DAG size S and exponentially proportional to the look-ahead factor K.*

Proof.   Expression 4 involves a count of the number of templates and their associated costs. It is independent of the DAG size or the look-ahead factor. Assuming that a node expands, in the worst case, into B successor nodes, expression (2) involves $O(B^K)$ calls to expression (4) and hence it will execute in $O(B^K)$

steps.* In expression (1), F(N) involves a DAG traversal to count the number of operands and hence will execute in O(S); $NC_2(N)$ is computed by expressions (2) and (3), and thus will execute in $O(B^K)$. Analogous to (4), the computation of I(N) and G(N) is independent of S and K. Consequently, expression (1) executes in $O(S + B^K)$ steps.

THEOREM 2 *The entire search algorithm can be accomplished in polynomial time with the DAG size S and exponentially proportional with the look-ahead factor K.*

Proof.   The worst case is when each template covers only one DAG node. From the search tree in Figure 11a, if a node expands (in the worst case) into B successors and the time necessary to evaluate the cost of a successor node is T, one can readily derive that the (worst case) total search time is given by

$$\tau = \text{total search time} = S \text{ (time to expand one node)} = S(BT) \qquad (5)$$

Because T is proportional to $O(S + B^K)$ (from Theorem 1), the total search time is proportional to $O(BS^2 + B^{K+1}S)$.

Hence, for reasonably small (fixed) values of K, the proposed algorithm executes in polynomial time with the DAG size.

4.3   Summary and Conclusions

In this section, we have presented in detail a methodology for code generation for PDSPs. Our approach combines traditional code generation methods, namely tree/DAG traversal and template pattern matching, with artificial intelligence techniques, namely MEA, HP, expert system, and heuristic search, to mimic the approach used by human PDSP assembly programmers. A time–cost analysis ensured that the proposed algorithm is bounded and feasible within practical limits of time. OASIS executes in time proportional to a second-order polynomial on the program size.

So far, we have introduced an efficient approach for code generation. By combining the subproblems of scheduling, instruction selection, and resource allocation into a single heuristic search framework, our approach can potentially generate optimal code for a given DAG. The need for post-code-generation optimization is obviated by the heuristic search formulation. Better yet, important

---

* A K-level tree rooted at one node, with each node expanding into B successors, has exactly $(B^K + B^{K-1} + \ldots + B + 1)$ nodes. For simplicity, it is common practice to approximate the total number of nodes to $O(B^K)$. This approximation becomes more accurate as K increases.

optimization issues that conventional optimizing compilers must deal with, such as how to ensure that an optimization technique will not undo the results of a previous optimization, are absent.

However, handwritten quality code generation still hinges on one assumption—that the input HLL program is a good representation of the intended DSP algorithm. In other words, the generated code can only be as good as the input code allows it to be.

To alleviate this constraint, we propose the use of algorithm transformations. In the next section, we go one step further and introduce the idea of context-sensitive algorithm transformations, examine some complicating issues involved, and show how it can be easily incorporated into our methodology.

5 CONTEXT-DEPENDENT ALGORITHM TRANSFORMATIONS

Even if a code generator could generate optimal code for a given HLL program, that code may still be inferior to handwritten assembly code. The reason is simple. Although the IR can uniquely represent a given sequence of HLL instructions (program), the latter is not a unique implementation of a desired algorithm. In fact, there are infinite programs that evaluate a given expression. For example, one could implement $(a \times b)$ as $[(a + b)^2 - a^2 - b^2]/2$. Incidentally, Massalin's Superoptimizer [71] can generate optimal instruction sequences given a function to be performed. Superoptimizer, however, is not a code generator. It takes as input a machine language program and returns another (smaller) program, in the same machine language, which computes the same function as the input. It does not understand the intended function. It simply performs an exhaustive search over the target machine's instruction set for all possible equivalent programs, first of length 1, then of length 2, and so on. For each function, the user must define a set of test vectors to be used to verify program equivalence and manually inspect the programs that pass such a test for equivalence with the original program. Due to its exponential nature, its application is limited to very small programs. Massalin wrote: ''The current version of Superoptimizer has generated programs 12 instructions long in several hours running time. . . . Therefore, the Superoptimizer has limited usefulness as a code generator for a compiler.''

The use of HLLs emphasizes the need for *algorithm transformation*. Unlike assembly languages, HLLs are, in principle, processor independent. Without target architecture information, HLL programmers cannot bias the program toward certain constructs, as experienced assembly programmers often do. Hence, it is the responsibility of the compiler to perform that task.

The term *algorithm transformation* has a broad meaning. Research on transformations can be traced in many related areas, such as numerical analysis and algebra [72,73], high-level synthesis [74,75], CAD design [76,77], and software compilers [64,78–80]. We will obviously focus our attention on transformations in software compilers, also known by another equally broad term—*optimizations*. In this section, we will discuss only context-dependent transformations. See Section 2 for a survey of optimizations and their classifications in a more general sense.

In static transformations, the compiler uses static analysis to determine the merits of a transformation. For instance, the apparently more complex implementation $[(a + b)^2 - a^2 - b^2]/2$ for $(a \times b)$ may actually make sense if the target processor is an analog computer. In such a machine, there may not be direct hardware support for multiplication. On the other hand, addition and division can be easily implemented with resistors and operational amplifiers, and the exponential characteristics of certain analog elements can be exploited to implement the squaring operation. In such an environment, the transformation $(a \times b) \rightarrow [(a + b)^2 - a^2 - b^2]/2$ is context independent, and as such, the compiler could simply replace, at parse time, all instances of $(a \times b)$ with $[(a + b)^2 - a^2 - b^2]/2$.

In contrast, context-dependent transformations are those which depend on either the values of the operands or the contents and status of the available resources such as registers and memory addresses at run time. Such transformations are usually undecidable at parse time. For example, consider a processor in which subtraction and right shift take one machine cycle each, whereas multiplication takes four machine cycles. In most circumstances, implementing $(a \times b)$ as $[(a + b)^2 - a^2 - b^2]/2$ would not make sense in such a machine. However, if the partial results $(a + b)^2$, a^2, and b^2 have been evaluated in previous statements and are still available in appropriate registers, the code for $[(a + b)^2 - a^2 - b^2]/2$ could execute in three machine cycles as opposed to four machine cycles for $(a \times b)$.

In the framework for code generation presented in Sections 3 and 4, the HLL program is first parsed into a DAG. Our system makes use of artificial intelligence techniques (means-ends analysis, hierarchical planning, expert system, and heuristic search) to mimic the reasoning and planning processes used by human assembly programmers. Similar to template pattern matching (TPM) (see Sect. 2.1.2), all evaluations are performed at compile time. Unlike TPM, however, template costs are dynamically obtained. This means that a template may have different costs for different program contexts. Nevertheless, the ability of human assembly programmers to understand the algorithm being coded and modify certain parts of it to suit the target architecture was not supported. We now present an extension to our methodology for supporting algorithm transformations and describe how it can be easily integrated into our system.

5.1 Algorithm Transformations

Algorithm transformation in software compilation is the process of rewriting a given implementation of an algorithm (or parts of it) into another implementation that allows the generation of more efficient code. For our purposes, efficient means compact and fast. Twaddell [81] writes ''the long-term trend in optimizations is for the compiler to effectively rewrite the code however it pleases with the user's code representing just and expression of intent.''

Algorithm transformations can be classified into four major groups, as shown in Table 1. In this subsection, we present a framework that can support algebraic and Boolean transformations. In our approach for code generation, register allocation, instruction selection, and scheduling (of IR operations) are handled concurrently. The context-dependent search formulation assures that the sequence of instructions generated is optimized in its ordering. Hence, the value of a separate execution reordering phase is much reduced. However, that is not to say that separate instruction reordering is unnecessary. It is possible that executing certain instructions out of their original order, as long as the new order does not disturb the content of particular resources, can reduce pipeline stalls in some architectures. The need for instruction scheduling is further reduced by modern out-of-order-execution design technology, in which the processor is capable of executing any given sequence of instructions in the order that it considers best, without affecting the correctness of the result [82,83].

5.1.1 Context-Dependent Transformations

Some of the transformations in Table 1 are performed by conventional optimizing compilers. In these compilers, however, transformations are static and context

Table 1 Major Classes of Algorithm Transformations

Classes	Transformations
Algebraic transformations	• Associativity, commutativity
	• Identity element, neutral element
	• Strength reduction, constant folding
	Example: $a + (b - c) = b - (c - a)$
Boolean transformations	• Associativity, commutativity
	• DeMorgan's Law
	Example: $A + A \cdot B = A + B$
Execution reordering	• Instruction scheduling
Loop restructuring	• Loop fission, loop fusion,
	• Unimodulo transformations
	• Loop unfolding, retiming

insensitive. For example, if multiplication executes much slower than addition in the target processor, subtree $(a \times 2)$ could be replaced by subtree $(a + a)$. This transformation can be done at parse time and is context independent. However, if both operations perform in the same number of cycles, the advantage of such transformation is unrecognizable at parse time.

A common approach in these compilers is to parse the IR before code is generated and perform transformations wherever possible, assuming that such transformations will always benefit code generation. If that assumption proves to be incorrect, the user must disable specific transformations and recompile. Such an approach has two shortcomings. First, determining which transformations are responsible for the resulting inefficiency of the code is not trivial. Second, it cannot address the issue of a given transformation being advantageous in certain parts of the program but undesirable in others. We call the latter context-dependent transformations.

Another difficulty with context-dependent transformations is that the code generator must decide not only if a certain transformation is worth performing but also, in case several alternative transformations exist, which one to perform.

Our heuristic search framework can be easily extended to support this issue. Currently, each search node contains a copy of the remaining DAG to be covered. A pointer called `CurrentNode` points to the node in that DAG currently being covered. If N alternative patterns exist for the node pointed to by `CurrentNode`, the search node is replicated N times. Each replica contains a copy of the DAG, with the sub-DAG pointed to by `CurrentNode` replaced by one of the alternative patterns. A look-ahead heuristic search is then conducted for each replica and the least expensive among all replicas is preserved and others are discarded.

5.1.2 Reducing Search

Figure 12 lists the transformations currently implemented in OASIS. The column Condition in Figure 12 conveys heuristics to minimize the search space. A transformation is considered only if the listed condition is satisfied. If the condition is not satisfied, the transformation most likely yields either less compact or a slower code, and hence is ignored.

For example, in transformation 6, there are two alternatives: $Op1 - (Op2 + Op3) \rightarrow (Op1 - Op2) - Op3$ and $Op1 - (Op2 + Op3) \rightarrow (Op1 - Op3) - Op2$. First, before these transformations can even be considered, the required condition that the accumulator contains $Op1$ must be satisfied. Second, if the condition is satisfied, which one of the two alternative transformations will yield better code depends on the run-time context. Both alternatives are evaluated (with K-step look-ahead) together with the original expression and the least expensive among the three implementations is retained and the others are discarded.

In another example, transformation 7 also has two possible alternatives: $Op1 - (Op2 - Op3) \rightarrow (Op1 - Op2) + Op3$ and $Op1 - (Op2 - Op3) \rightarrow$

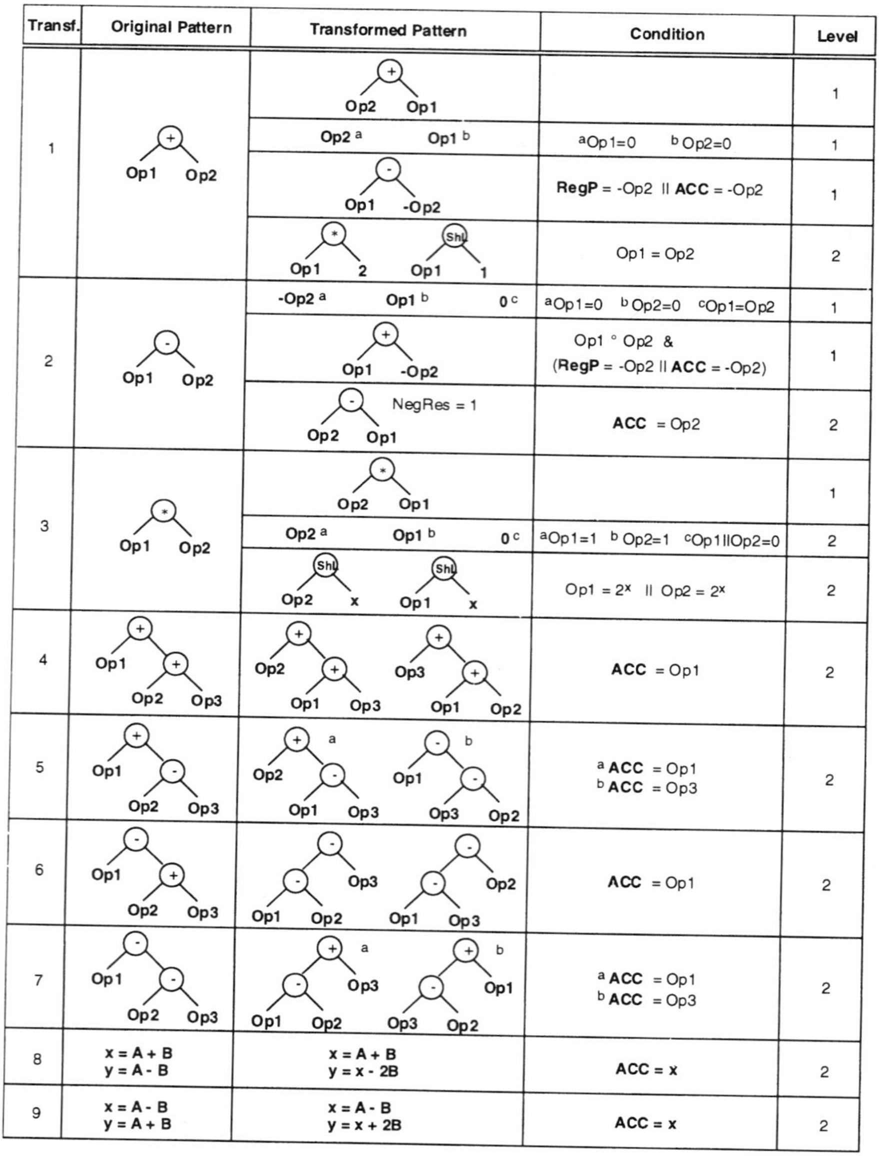

Transf.	Original Pattern	Transformed Pattern	Condition	Level
1	(+) → Op1, Op2	(+) → Op2, Op1		1
		Op2 [a] Op1 [b]	[a]Op1=0 [b]Op2=0	1
		(-) → Op1, -Op2	RegP = -Op2 ‖ ACC = -Op2	1
		(*) → Op1, 2 / (Shl) → Op1, 1	Op1 = Op2	2
2	(-) → Op1, Op2	-Op2 [a] Op1 [b] 0 [c]	[a]Op1=0 [b]Op2=0 [c]Op1=Op2	1
		(+) → Op1, -Op2	Op1 ° Op2 & (RegP = -Op2 ‖ ACC = -Op2)	1
		(-) → Op2, Op1 NegRes = 1	ACC = Op2	2
3	(*) → Op1, Op2	(*) → Op2, Op1		1
		Op2 [a] Op1 [b] 0 [c]	[a]Op1=1 [b]Op2=1 [c]Op1‖Op2=0	2
		(Shl) → Op2, x / (Shl) → Op1, x	Op1 = 2^x ‖ Op2 = 2^x	2
4	(+) → Op1, (+) → Op2, Op3	(+) → Op2, (+) → Op1, Op3 ; (+) → Op3, (+) → Op1, Op2	ACC = Op1	2
5	(+) → Op1, (-) → Op2, Op3	(+) [a] → Op2, (-) → Op1, Op3 ; (-) [b] → Op1, (-) → Op3, Op2	[a] ACC = Op1 [b] ACC = Op3	2
6	(-) → Op1, (+) → Op2, Op3	(-) → Op3, (-) → Op1, Op2 ; (-) → Op2, (-) → Op1, Op3	ACC = Op1	2
7	(-) → Op1, (-) → Op2, Op3	(+) [a] → Op3, (-) → Op1, Op2 ; (+) [b] → Op1, (-) → Op3, Op2	[a] ACC = Op1 [b] ACC = Op3	2
8	x = A + B y = A - B	x = A + B y = x - 2B	ACC = x	2
9	x = A - B y = A + B	x = A - B y = x + 2B	ACC = x	2

Figure 12   Algorithm transformations performed by OASIS during instruction selection.

(Op3 − Op2) + Op1. The content of the accumulator dictates which alternative (if any) should be used. The search process then determines if the original or the transformed expression will yield better code.

Transformations are classified into two levels. A level-1 transformation is one that can be implemented without modifying the DAG, whereas a level-2 transformation involves DAG modification. For example, transformation 2, Op1 − Op2 → Op1 + (−Op2), can be accomplished by adding such templates as **ADD_dma** and **APAC** into the connection table associated with the subtraction operator, under the condition that Op2's **Negative** flag is set (by a previous operation). No modification in the DAG is necessary. However, the alternative Op1 − Op2 → − (Op2 − Op1) involves modification of the original DAG and is therefore classified as level 2. The main reason for classifying transformations according to changes in the DAG is to give the user more flexibility in matching computational resources. Level-1 transformations are computationally much less demanding than level-2 transformations. The former adds a new branch in the search tree, whereas the latter requires replicating the entire search tree.

5.1.3 Delayed Common Subexpression Elimination

Common subexpression elimination (CSE) is a popular technique applied by many optimizing compilers. In CSE, the compiler first determines the existence of subexpressions that are common to several parts of the program. If such subexpressions exist, the representation of the program is altered such that only the first occurrence of that subexpression is evaluated. The result is saved as a temporary value to be used by all other occurrences, eliminating multiple recomputations of a same expression.

In performing context-dependent optimizations, we must deal with three important issues. First, applying transformations during code generation can lead to what we call delayed CSE. For example, consider the code fragment in Example 4.

Example 4

```
S1:   f = a + (b − c);
S2:   g = a − (b + c);
```

Static CSE would not detect any common subexpressions. However, if we apply transformations 5 and 6 from Figure 12, the above code would result in

```
S1':   f = b + (a − c);
S2':   g = (a − c) − b;
```

and the common subexpression $(a - c)$ is created. In our code generation formulation, performing delayed CSE is a simple matter of traversing the DAG after each transformation is applied and merging common subexpressions as they are encountered. This leads to the second issue.

Depending on the subexpression, the resource status and the target architecture, recomputing the expression could be less expensive than storing and reloading a value from memory. The correct decision, to perform CSE or not, can only be made after both alternatives are evaluated. Our code generation formulation supports such evaluations through heuristic search.

The third issue is that of look-ahead, which must be used to further refine the evaluation. Let us consider Example 5.

Example 5

```
S1:   f = b + (a − c);
S2:   g = a − (b + c);
```

Suppose we are processing statement S1 and that transformation 5 is applicable (i.e., the accumulator holds the value of variable b). Indeed, transforming S1 into

```
S1′:   f = a + (b − c);
```

is advantageous because it eliminates the need to store b into a temporary location and reloading the accumulator with the value of a to compute $(a - c)$. A naive evaluation function would opt to transform S1. However, only by looking ahead can one see that if transformation 5 is not applied and transformation 6 is applied on S2 instead, the result would be

```
S1:   f = b + (a − c);
S2′:  g = (a − c) − b;
```

and performing the delayed CSE on $(a - c)$ could be potentially more advantageous, especially if a and c are complicated subexpressions.

OASIS supports heuristic search with user specified look-ahead levels.

5.2. Time-Complexity Analysis

With the implementation of algorithm transformations as replication of search trees, it would seem that context-sensitive transformations could be too expensive in practice. In fact, one could wonder if the algorithm now is still bounded. In other words, is the new algorithm still realizable within feasible amounts of time? To ensure the feasibility of the algorithm, we conduct the following time–cost analysis

THEOREM 3 *With CDAT, the proposed algorithm still executes in polynomial time with the DAG size S and time exponentially proportional with the look-ahead factor K.*

Proof. In the worst case, every search node has Z alternative transformations and each transformation requires replication of the entire branch rooted at that search node. Following the rationale in the proof of Theorem 2, the time necessary to expand one node is multiplied by Z, because each node now can have Z replicas. Similarly, the time necessary to evaluate the cost of a successor node, T is also multiplied by Z. Hence,

$$\tau_x = \text{total search time with transformations}$$
$$= SZ[B(ZT)] \propto O(Z^2[BS^2 + B^{K+1}S]) \tag{6}$$

5.3 Summary and Conclusions

In this subsection, we presented an extension to the heuristic search framework for code generation presented in Sections 3 and 4. This extension allows the code generator to consider context-dependent algebraic and Boolean algorithm transformations. The implementation is straightforward, based on recursive calls to the same search engine introduced in Section 4.

The only drawback is that more look-ahead levels are required when transformations are involved. Although our heuristic search formulation executes in (second order) polynomial time with the program size, it is exponential with the look-ahead factor. For time-critical applications, transformations are well worth the increased compilation time.

In the next section, we present some empirical results obtained with OASIS, a prototype code generator based on the ideas described in this and the previous two sections. The results will demonstrate quantitatively the efficacy of our methodology.

6 EXPERIMENTAL RESULTS

A prototype code generator, OASIS, following the ideas discussed in Sections 3–5, has been implemented. We targeted it to a simplified architecture, based on the Texas Instruments TMS3202x/5x [84]. The primary purpose of OASIS is to evaluate the feasibility of the proposed approach. It is not intended to be a fully working compiler.

6.1 Quality of the Generated Code

Two sets of benchmarks involving common DSP algorithms have been designed to evaluate the performance of our code generation approach. Whenever possible, we selected published DSP routines that have been coded in assembly by DSP experts. In the absence of published material, we coded the routines in assembly, with a single purpose in mind: to produce the most efficient code possible. For that purpose, we utilized all possible transformations that we could apply. This set of handwritten assembly codes serves as our benchmarks.

The DSP algorithms implemented by those routines (not necessarily the assembly codes) were coded in C and supplied to both OASIS and the Texas Instruments TMS3202x/5x Optimizing C Compiler [85].

Table 2 contains the results of the first set of benchmarks. The codes gener-

Table 2 Quality (Without Algorithm Transformations) of the Assembly Code Generated by OASIS Compared to That Generated by a Conventional Optimizing Compiler and Handwritten Assembly Code

| | OASIS | | | TI COMP.[a] | DSP experts | |
| | Code size (# instr.) | Exec. time[b] (μsec) | Comp. time[c] (sec) | Code size (# instr.) | Code size (# instr.) | Exec. time[b] (μsec) |
Benchmark						
FFT2	23	4.6	23.2	63	22[d]	4.4[d]
IIR	16	3.2	12.5	43	17[e]	3.4[e]
FIR	14	2.8	14.7	37	15[e]	3.0[e]
BIQUAD[f]	13	2.6	1.7	49	13	2.6
ROTATION	36	10.8	93.5	128	31[g]	14.4[g]
ILATTICE[h]	24	4.8	150.6	56	26	5.2
FLATTICE[h]	26	5.2	42.7	58	25	5.0

Note: The number of instructions excludes assembler directives and routine initialization instructions.
[a] TI TMS3202x/C5x Optimizing C Compiler, with global optimizations enabled (level 2).
[b] Estimate for TMS32020 at 200 nsec per instruction cycle.
[c] Average (three runs) user CPU time on a SUN SPARCStation IPX.
[d] Results obtained from Ref. 86.
[e] Results obtained from Ref. 87 (straight code implementation).
[f] Example adapted from Ref. 88.
[g] Results obtained from Ref. 89.
[h] Example adapted from Ref. 90, pp. 12–79.

ated by OASIS were obtained without algorithm transformations. The TI compiler ran with all optimizations enabled.

The code generated by OASIS is up to 3.8 times smaller (in benchmark BIQUAD) than that obtained from the TI compiler. Furthermore, OASIS is capable of generating code whose quality is comparable, in some cases even superior, to that generated by human DSP experts.

Note that in the benchmark ROTATION, although OASIS generated a few more instructions, the program executes faster than the handwritten version. This is, in part, due to our heuristic function, which minimizes the sum of code size and execution time.

Except for FFT2, the other six benchmarks used in Table 2 do not lend themselves well to testing CDAT (discussed in Sec. 5). Hence, a new set of benchmarks was devised, specifically to test the efficacy of the proposed context-dependent transformations and the issues involved when delayed CSE is present. Table 3 summarizes the results. Along with the size of the generated code, it also lists the smallest look-ahead factor K necessary to obtain those results.

Without transformations, the code generated by OASIS is already much smaller than that obtained from the TI compiler, but not quite as small as the handwritten version. The reason is that when coding in assembly, the programmer has access and, indeed, takes advantage of all kinds of "tricks" (algorithm trans-

Table 3 Performance of OASIS Compared with TI Compiler and Handwritten Assembly Code

HLL program	TI compiler[a] (# instr.)	OASIS				Handwritten assembly (# instr.)
		Without transf.		With transf.		
		(# instr.)	K	(# instr.)	K	
xform1	18	9	2	7	4	7
xform2	18	9	3	7	3	7
xform3	27	14	6	11	6	12
xform4	24	9	2	8	5	8
xform5	23	10	4	8	6	8
xform6	27	14	6	12	6	12
FFT2	63	23	1	21	2	22[b]
FFT4	190	107	3	89	3	77[c]

Note: The number of instructions excludes assembler directives and routine initialization instructions. Results obtained with all optimizations enabled (level 2).
[a] The TI compiler is capable of static arithmetic transformations.
[b] Results obtained from Ref. 86.
[c] Results obtained from Ref. 91.

formations) that one can envision. However, with context-dependent transformations enabled, the prototype generates codes that are truly comparable to their handwritten counterparts. FFT4 is an exception. Due to its size (over 100 nodes in its DAG), we exceeded the available computational resources before we reached the limitations of our code generator. Analysis of the generated code shows that not all possible optimization opportunities have been performed by OASIS. On the other hand, OASIS generated smaller codes for xform3 and FFT2 than the assembly programmers.

Careful examination of the generated assembly code for xform3 reveals that OASIS took advantage of delayed common subexpression elimination and the programmer missed that optimization opportunity in the handwritten version. This example illustrates well the problems faced by programmers. Hand coding in assembly language gives the programmer tremendous flexibility in performing algorithm transformations. Although human ingenuity is far superior than what current technology can give to a computer program, the latter can perform certain tasks much faster, more reliably, and more thoroughly than we can. An automated and consistent approach can uncover optimizations that are hidden deep within the program.

6.2 Summary and Conclusions

The results presented in this section show that our code generator is able to generate code that is as good as that generated by human DSP experts. Our approach allows a great degree of flexibility in matching available computational resources to the desired output code quality. By varying the weights of individual features in the heuristic function, one can generate code that executes fastest (e.g., for real-time applications) or which has the smallest size (e.g., to fit into a limited size PROM), or both. By increasing the depth of the search, our approach is capable of yielding globally optimized code, albeit in time exponentially proportional to the search depth. From the previous results, without algorithm transformations, a search depth of 3 is adequate for most cases and compiles in time polynomial in the input program size.

The experimental results also confirm the efficacy of context-dependent algorithm transformations. The only drawback is that more look-ahead levels may be required when transformations are involved. Although our heuristic search formulation executes in (second-order) polynomial time with the program size, it is exponential with the look-ahead factor. For time-critical applications, transformations are well worth the increased computational time.

Previous approaches are restricted in time and scope. An exhaustive code generation approach may require more computational resources than currently available [31,71]. A fast approach developed for today's processing power cannot generate the desired quality [7,8]. Worse yet, it may be very quickly outdated given the pace at which computational power is advancing.

Although no specific limits exist on input program size, this model is especially suited for tasks characterized by small routines with very stringent constraints on execution time and/or code size.

Advantages of the proposed approach include the following:

1. No need for postgeneration optimizations. Unlike conventional compilers, instruction selection and resource allocation are combined into one single pass, assisting and influencing each other. Many optimization techniques implemented in conventional optimizing compilers are trivialized by the search formulation. A few optimizations are implemented in the form of CDAT. Unlike static optimizations, CDAT can guarantee to improve code quality. The problem of applying optimizations in the correct order so that one optimization does not undo the results of previous optimizations is also trivialized.

2. Potential for obtaining optimal code. The search strategy is flexible, in the sense that it allows a trade-off between code quality and compilation time.

3. Enhanced portability. Retargeting the code generator to different PDSPs requires modifying only two modules: the instruction library and the rule base.

REFERENCES

1. EA Lee. Programmable DSP architectures: Part I. IEEE ASSP Mag 5(4):4–19, 1988.
2. EA Lee. Programmable DSP architectures: Part II. IEEE ASSP Mag 6(1):4–14, 1988.
3. H Ahmed. Recent advances in DSP systems, IEEE Commun Mag 29(5):32–45, 1991.
4. D Bursky. DSPs expand role as costs drops and speed increases, Electron Des 39(19):53–65, 1991.
5. D Shear. HLL compilers and DSP run-time libraries make DSP-system programming easy. EDN. 33(13):69–74, 1988.
6. D Genin, EV deVelde, JD Moortel, D Desmet. System design, optimization and intelligent code generation for standard digital signal processors. Proceedings of the IEEE International Symposium on Circuits and Systems. New York: IEEE, 1989, pp 565–569.
7. J Hartung, SL Gay, SG Haigh. A practical C language compiler/optimizer for real time implementations on a family of floating point DSPs. Proceedings of the International Conference on Acoustics, Speech and Signal Processing. New York: IEEE, 1988, pp 1674–1677.

8. R Simar Jr, A Davis. The application of high-level languages to single-chip digital signal processors. Proceedings of the International Conference on Acoustics, Speech and Signal Processing. New York: IEEE, 1988, pp 1678–1681.

9. P Lawlis, TW Elam. Ada outperforms assembly: A case study. Proceedings of the Tri-Ada. Orlando, FL: ACM, 1992, pp 334–337.

10. PN Hilfinger, J Rabaey, D Genin, C Sheets, HD Mar. DSP specification using the Silage language. Proceedings of the International Conference on Acoustics, Speech and Signal Processing, Albuquerque, NM, 1990, pp 1057–1060.

11. CD Covington, GE Carter, DW Summers. Graphic oriented signal processing language—GOSPL. Proceedings of the International Conference on Acoustics, Speech and Signal Processing. Dallas, TX, 1987, pp 1879–1882.

12. KW Leary, W Waddington. DSP/C: A standard high level language for DSP and numeric processing. Proceedings of the International Conference on Acoustics, Speech and Signal Processing, San Francisco, CA, 1990, pp 1065–1068.

13. Digital Signal Processing Committee IEEE Acoustics, Speech, and Signal Processing Society. Programs for Digital Signal Processing. Wiley, New York, 1979.

14. WH Press, SA Teukolsky, WT Vetterling, BP Flannery. Numerical Recipes in C: the Art of Scientific Computing. 2nd ed. Cambridge: Cambridge University Press, 1992.

15. JGP Barnes, Programming in Ada. 2nd ed. Reading, MA: Addison-Wesley, 1984.

16. BW Kernighan, DM Ritchie. The C Programming Language. 2nd ed. New Delhi: Prentice-Hall of India, 1989.

17. EB Koffman, FL Friedman. FORTRAN. Reading, MA: Addison-Wesley, 1993.

18. EB Koffman. Pascal. 3rd ed. Reading, MA: Addison-Wesley, 1989.

19. L Nachtergaele, I Bolsens, HD Man. Specification and simulation front-end for hardware synthesis of digital signal processing applications. Int J Comp Simulat 2(2): 213–229, 1992.

20. I Kuroda, A Hirano, T Nishitani. A knowledge-based compiler enhancing DSP internal parallelism. Proceedings of the 1991 IEEE International Symposium on Circuits and Systems, 1991, pp 236–239.

21. CN Fischer, RJ LeBlanc Jr. Crafting a Compiler. Menlo Park, CA: Benjamin/Cummings, 1988.

22. SC Johnson, YACC—Yet Another Compiler Compiler. Murray Hill, NJ: Bell Telephone Laboratories, Technical report 32, 1975.

23. AV Aho, SC Johnson. Optimal code generation for expression trees. J ACM 23(3): 488–501, 1976.

24. R Giegerich. A formal framework for the derivation of machine specific optimizers. ACM TOPLAS 5(3):191–202, 1983.

25. CW Fraser. PhD dissertation. Automatic generation of code generators, Yale University, 1977.

26. RG Cattell. PhD dissertation. Formalization and automatic derivation of code generators. Carnegie-Mellon University, 1978.

27. AS Tanenbaum, HV Staveren, EG Keizer, JW Stevenson. A practical tool kit for making portable compilers. Commun ACM 26(9):654–660, 1983.

28. SWK Tjiang. Twig Reference Manual. Murray Hill, NJ: Bell Telephone Laboratories, Technical report 120, 1986.

29. AL Wendt. PhD dissertation. An Optimizing Code Generator Generator. The University of Arizona, 1989.

30. AV Aho, R Sethi, JD Ullman. Compilers: Principles, Techniques, and Tools. Reading, MA: Addison-Wesley, 1986.

31. DW Krumme, DH Ackley. A practical method for code generation based on exhaustive search. Proceedings of the ACM SIGPLAN '82 Symposium on Compiler Construction. White Plains, NY: ACM SIGPLAN, 1982, pp 185–196.

32. PP Chakrabarti, S Ghose, SC DeSarkar. Heuristic search through islands. Artif Intell 29:339–347, 1986.

33. RE Korf. Planning as search: A quantitative approach. Artif Intell 33:65–88, 1987.

34. CM Hoffmann, MJ O'Donnell. Pattern matching in trees. J ACM 29(1):68–95, 1982.

35. DR Chase. An Improvement to bottom-up tree pattern matching. Proceedings of the 14th Annual Symposium on Principles of Programming Languages, 1987, pp 168–177.

36. AV Aho, SC Johnson, JD Ullman. Code generation for expressions with common subexpressions. J ACM 24(1):146–160, 1977.

37. SG Wasilew. PhD dissertation. A compiler writing system with optimization capabilities for complex order structures, Northwestern University, 1971.

38. SW Weingart. PhD dissertation. An efficient and systematic method of compiler code generation, Yale University, 1973.

39. K Ripken. PhD dissertation. Formale Beschreibung von Maschinen, Implementierungen und Optimierender Maschinen-codeerzeugung aus Attributierten Programmgraphen, Technische Universitat Munchen, 1977 (in German).

40. SC Johnson. A portable compiler: Theory and practice. Proceedings of the 5th Conference on Principles of Programming Languages, 1978, pp 97–104.

41. R Glanville. PhD dissertation. A new method for compiler code generation, University of California, Berkeley, 1977.

42. AV Aho, M Ganapathi, SWK Tjiang. Code generation using tree matching and dynamic programming, AT&T Bell Laboratories, 1986, pp 1–22.

43. WM McKeeman. Peephole Optimization. Commun ACM 8(7):443–444, 1965.

44. JM Newcomer. PhD dissertation. Machine independent generation of optimal local code, Carnegie-Mellon University, 1975.

45. RG Cattell. A survey and critique of some models of code generation, Carnegie-Mellon University, Technical report November 1977.

46. E Pelegri-Llopart, SL Graham. Optimal code generation for expression trees: An application of BURS theory. Proceedings of the 15th Annual Symposium on Principles of Programming Languages, 1988, pp 294–308.

47. SC Johnson. A Tour through the portable C compiler, Bell Telephone Laboratories, 1977.

48. JE Peters, SM Dunn. A compiler that easily retargets high level language programs for different signal processing architectures. Proceedings of the IEEE International Symposium on Circuits and Systems. Portland, OR, 1989, pp 1103–1106.

49. H Azaria, A Dvir. An optimizing compiler for an SPAP architecture using AI tools. Computer 25(6):39–48, 1992.

50. Tartan, Inc. Technical Summary, Ada C40-Targeted Development System. Monroeville, PA: Tartan, Inc., Document TL-EXT-92-9, 1992.

51. ME Benitez. MS dissertation. A global object code optimizer, University of Virginia, School of Engineering and Applied Science, 1989.

52. JW Davidson, CW Fraser. The design and application of a retargetable peephole optimizer. ACM Trans Program Lang Systems 2(2):191–202, 1980.

53. DA Lamb. Construction of a peephole optimizer. Software Pract Exper 11:638–647, 1981.

54. H Massalin. PhD dissertation. Synthesis: An efficient implementation of fundamental operating system services, Department of Computer Science, Columbia University, 1992.

55. D Keppel, SJ Eggers, RR Henry. A case for runtime code generation, Department of Computer Science, University of Washington, Technical report 91-11-04, 1991.

56. D Keppel, SJ Eggers, RR Henry. Evaluating runtime-compiled value-specific optimizations, Department of Computer Science, University of Washington, Technical report 93-11-02, 1993.

57. DR Engler, TA Proebsting. DCG: An efficient, retargetable dynamic code generation system, Proceedings of the ASPLOS-VI, 1994, pp 263–272.

58. C Chambers, D Ungar. Making pure object-oriented languages practical. OOPSLA 91, 26(11):1–15, 1991.

59. CD Garrett, J Dean, D Drove, C Chambers. Measurement and application of dynamic receiver class distributions, Department of Computer Science and Engineering, University of Washington, Technical report 94-03-05, 1994.

60. J Dean, G Chambers, D Grove. Identifying profitable specialization in object-oriented languages, Department of Computer Science and Engineering, University of Washington, Technical report 94-02-05, 1994.

61. CW Fraser, DR Hanson. A code generation interface for ANSI C, Department of Computer Science, Princeton University, Research report CS-TR-270-90, 1990.

62. AS Tanenbaum, H Staveren, JW Stevenson. Using peephole optimization on intermediate code. ACM Trans Prog Lang Syst 4(1):21–36, 1982.

63. D Padua, M Wolfe. Advanced compiler optimizations for supercomputers. Commun ACM 29(12):1184–1201, 1986.

64. MJ Wolfe. Optimizing Supercompilers for Supercomputers. Cambridge, MA: MIT Press, 1989.

65. RM Tomasulo. An efficient algorithm for exploiting multiple arithmetic units. IBM J Res Devel 11(1):25–33, 1967.

66. M Lam. Software Pipelining: An effective scheduling technique for VLIW machines. Proceedings of the SIGPLAN 88 Conference on Programming Language Design and Implementation. New York: ACM, 1988, pp 318–328.

67. U Banerjee. Dependence Analysis for Supercomputing. Norwell, MA: Kluwer Academic, 1988.

68. G Ernst, A Newell. GPS: A Case Study in Generality and Problem Solving. ACM Monograph Series. New York: Academic Press, 1969.

69. A Barr, EA Feigenbaum, eds. The Handbook of Artificial Intelligence. Los Altos, CA: Morgan Kaufmann, 1981, Vol. 1.

70. ED Sacerdoti. Planning in a hierarchy of abstract spaces. Artif Intell 5:115–135, 1974.

71. H Massalin. Superoptimizer: A look at the smallest program, Proceedings of the 2nd International Conference on Architecture Support for Programming Language and Operating Systems, Palo Alto, 1987, pp 122–126.

72. GH Golub, CF von Loan. Matrix Computations. 2nd ed. Baltimore, MD: Johns Hopkins University Press, 1989.

73. D Goldberg. What every computer scientist should know about floating-point arithmetic. ACM Computing Surveys 23(1):5–48, 1991.

74. R Hartley, A Casavant. Tree-height minimization in pipelined architectures. IEEE Trans Computer-Aided Design 8:112–115, 1989.

75. J Bhasker, H Lee. An Optimizer for Hardware Synthesis. IEEE Design and Test, October 1990.

76. WH Joyner, Jr, LH Trevillyan, D Brand, TA Nix, SC Gunderson. Technology adaptation in logic synthesis. Proceedings of the 23rd ACM/IEEE Design Automation Conference, 1986, pp 94–100.

77. D Gregory, K Bartlett, G Hachtel. SOCRATES: A system for automatic synthesis and optimizing combinational logic. Proceedings of the 23rd ACM/IEEE Design Automation Conference, 1986, pp 79–85.

78. ME Wolf, and MS Lam. A loop transformation theory and an algorithm to maximize parallelism. IEEE Trans Parallel Distrib Syst 2:452–471, 1991.

79. MJ Wolfe. The TINY loop restructuring research tool. Proceedings of the Parallel Processing, 2:47–53, 1991.

80. SWK Tjiang, JL Hennessy. Sharlit—A tool for building optimizers. ACM SIG-PLAN' 92 Conference on Programming Language Design and Implementation, San Francisco, CA, 1992.

81. WB Twaddell. Optimizations ignite new battle in the compiler wars. Personal Engineering and Instrumentation News. 10(2):25–32, 1993.

82. TR Halfhill. AMD vs. Superman. Byte 19(11):95–104, 1994.

83. TR Halfhill. T5: Brute Force. Byte 19(11):123–128, 1994.

84. Texas Instruments. TMS32020 User's Guide. Houston, TX: Texas Instruments, 1986.

85. Texas Instruments. TMS320C2x/C5x ANSI C Compiler. Houston, TX: Texas Instruments, 1992.

86. P Papamichalis, J So. Implementation of fast Fourier transform algorithms with the TMS32020. In: K-S Lin, ed. Digital Signal Processing Applications with the TMS320 Family. Englewood Cliffs, NJ: Prentice-Hall, 1987, pp 69–168.

87. A Lovrich, R Simar Jr. Implementation of FIR/IIR Filters with the TMS32010/TMS32020. In: K-S Lin, ed. Digital Signal Processing Applications with the TMS320 Family. Englewood Cliffs, NJ: Prentice-Hall, 1987, pp 27–67.

88. DB Powell, EA Lee, WC Newman. Direct synthesis of optimized DSP assembly code from signal flow block diagrams. Proceedings of the International Conference on Acoustics Speech and Signal Processing. New York: IEEE, 1992, pp V553–V556.

89. C Crowell. Matrix multiplication with the TMS32010 and TMS32020. in: K-S Lin,

ed. Digital Signal Processing Applications with the TMS320 Family, Englewood Cliffs, NJ: Prentice-Hall, 1987, pp 291–306.

90. Texas Instruments. Third-Generation TMS320 User's Guide. Houston, TX: Texas Instruments, 1988.

91. CS Burrus, TW Parks. DFT/FFT and Convolution Algorithms Theory and Implementation. New York: Wiley, 1985.

92. U Hölzle, D Ungar. Optimizing dynamically-dispatched calls with runtime type feedback. Proceedings of the ACM SIGPLAN '94 Conference on Programming, 1994, pp 326–336.

93. J Dean, G Chambers. Toward better inlining decisions using inlining trials. Proceedings of the 1994 Conference on LISP and Functional Programming, 1994, pp 273–282.

94. M Leone, P Lee. Lightweight runtime code generation. Proceedings of the ACM SIGPLAN Workshop on Partial Evaluation and Semantics-Based Program, 1994, pp 97–106.

7
Digital Signal Processing on MMX Technology

Yen-Kuang Chen, Nicholas Yu, and Birju Shah
Intel Corporation, Santa Clara, California

Algorithm-level optimization and instruction-level optimization are tightly coupled with each other. Many programmers can optimize the implementation of a specific algorithm using MMX™ technology. However, without algorithm optimization, the speedup of the optimization will be limited. On the other hand, many algorithm developers can optimize the digital signal processing (DSP) algorithm in terms of the number of operations (multiplications or additions), but without implementation details, the number of operations cannot be directly translated into the number of clock cycles spent in CPU. There are also many algorithms that can accomplish the same task. For the best performance of DSP/multimedia applications on personal computers we should consider algorithm-MMX™ technology co-optimization.

One way to increase the performance of digital signal processing algorithms is to execute several computations in parallel. MMX technology is one of the techniques that speed up software performance by performing the same operation on multiple data elements in parallel using a single instruction. However, MMX programming and designing DSP algorithms for MMX technology are full of twists and turns. Implementation of digital signal processing algorithms using MMX technology is a mix of art and science. Matching the algorithms to MMX instruction capabilities is the key to extracting the best performance. This chapter covers algorithm design and algorithmic optimization for MMX technology. In this chapter, besides showing how to optimize your code and algorithm from a scientific view, we will also show how we go about optimizing ours from an artistic perspective.

1 INTRODUCTION

1.1 Developing History

As digital signal processing finds broader areas of application, more processors are adapting to the need for DSP operations [1]. One way to boost the performance is to execute several computations in parallel using single instruction multiple data (SIMD) techniques. Formerly, a computationally intensive, real-time, multimedia application required application specific integrated circuits (ASICs) or digital signal processors. With the introduction of SIMD extensions to general-purpose processors, things have changed.

In 1996, the Intel Corporation introduced MMX technology into Intel Pentium® processors [2]. A set of 57 instructions was added to treat data in a SIMD fashion. These instructions exploit the data parallelism that is often available in DSP and multimedia applications. These instructions can multiply, shift, add/subtract, load/store, and do logical operations on 64 bits of data at a time. These 64 bits can be packed as two 32-bit (doublewords), four 16-bit (words), or eight 8-bit (bytes) groups of integers. The parallel operations can yield a speedup of up to eight times over existing integer implementation [3–6].

Following a similar underlying concept, in 1999, the Intel Corporation introduced a new generation of the IA-32 microprocessors called the Pentium® III processor. This processor introduced 70 new instructions, which include MMX technology enhancements, SIMD floating-point instructions, and cacheability instructions [7,8]. The introduction of Streaming SIMD Extensions by the Intel Corporation has further improved floating-point performance by executing up to four single-precision operations per cycle. Since 2000, Intel Pentium 4 processors have used 128-bit XMM registers for multiple integer, single-precision floating-point, and double-precision floating-point operations [9–11]. Readers are encouraged to take a look at the ''MMX™-Enabled Dolby Digital Decoder'' by Abel and Julier [41]. Although an implementation using integer MMX instructions and an implementation using floating-point Streaming SIMD Extensions take about the same execution time, using floating-point Streaming SIMD Extension produces slightly better results. (This is because additional instructions are required to maintain sufficient accuracy in the integer MMX implementation). Because this technology is constantly under development, this chapter will use basic MMX technology as the example. Readers can extend their knowledge into advanced MMX technologies (even onto different microprocessors) on their own.

1.2 Motivation

Although MMX technology can offer much higher performance, programming with MMX technology is easier said than done. First, different algorithms have different characteristics for parallel processing. Algorithms with repetitive, regular data movement and computation found in graphics, filtering, compression, and decompression typically can be improved tremendously by using MMX technology. On the other hand, code such as decision trees does not obtain great performance increases because it lacks that repetitive (looping) behavior [12].

Algorithm, software, and hardware must be designed together for a good implementation. A multimedia application can be implemented using many different algorithms (some regular, some irregular). Conventional software and hardware codesign may not be suitable for today's multimedia applications. For optimal implementation, we should consider architectural style when we design or choose the algorithm for a specific task [13].

Choosing the right algorithm depends on users' needs. For example, let algorithm $i = 1, ..., N$ be the solutions for a multimedia signal processing task and the execution time $T_i = T_{is} + T_{ip}/P$, where T_{is} is the nonparallelizable execution time, T_{ip} is the parallelizable execution time, and P is the number of parallel execution units. In a uniprocessor system where $P = 1$, algorithm i is better if $T_{is} + T_{ip}$ is minimal. For a multiprocessor system, algorithm j is better if $T_{js} + T_{jp}/P$ is minimal. As P become larger and larger in the near future, the design/choice of algorithms should be changed. Using motion estimation as a design example, we found that the full-search block-matching algorithm is more efficiently implemented in a systolic array [14] than the hierarchical-search block-matching algorithm [15,16], although the full-search block-matching algorithm needs more operations [13].

Consequently, when implementing digital signal processing algorithms using MMX technology, one should also consider the underlying MMX media-enhancement technology architecture and possibly redesign the algorithms. In this chapter, we will show the following:

1. Basic MMX instructions and programming methodology
2. Adaptation of algorithm design for MMX technology

1.3 Chapter Organization

Considering the underlying MMX media-enhancement technology architecture for designing the best algorithm, we would like to show four basic principles of implementing DSP algorithms on MMX technology:

1. Put multiple identical operations in one instruction (i.e., vectorize discrete scalar operations into SIMD operations) and use the smallest possible data type to enable more parallelism.
2. Convert conditional executions into logic operations to reduce hard-to-predict conditional branches and increase parallelism.
3. Put data into the right format for parallel execution and arrange the data structure for better memory access.

4. Reduce shuffling and maximize grouping of operations into one instruction*; do not overoptimize the algorithms in terms of the number of scalar operation.

We will show some of our work in MPEG video processing as examples of these principles. MPEG video decompression consists of (1) high memory bound elements, like motion compensation, (2) highly computationaly intensive parts, like IDCT (inverse discrete cosine transform), and (3) highly irregular program flows in the Huffman decoder. This mix is a good representative for many multimedia and digital processing algorithms.

This chapter is organized as the following. Section 2 first describes some basics about the MMX programming environment (MMX registers, data types, and instruction sets) and guidelines for instruction-level optimization. We will also demonstrate how to convert scalar operations into SIMD operations in Section 2. In Section 3, we will convert conditional executions into logic operations using the MPEG-4 pixel padding procedure as an example. Section 3 will also show that efficient memory access is important in algorithm designs because of the one-dimensional nature of the computer memory. In Section 4, we will then show an example of algorithm-level optimization of the MPEG-4 Shape-Adaptive-DCT for MMX technology. We could use an entire book to cover the details of MMX programming. In this chapter, we focus more on guiding the algorithmic development of DSP on MMX technology. In order to help readers understand more about these topics, we also list some examples. In Section 5, we will give pointers to additional resources for readers.

2 OVERVIEW OF THE MMX TECHNOLOGY AND BASIC OPTIMIZATION PRINCIPLES

The first rule of MMX technology optimization is to vectorize the operations (i.e., execute multiple identical operations in one instruction). Although the name MMX media-enhancement technology implies its use in multimedia applications, the new instruction set is a general-purpose implementation of the SIMD concept, as shown in Figure 1. With MMX instructions, we can execute eight 8-bit integer operations per clock cycle. MMX technology provides benefits for all applications that perform the same operation repetitively on contiguous blocks of data.

Because MMX instructions can operate on multiple operands in parallel,

---

* For example, one instruction that we use frequently in DCT/IDCT is PMADDWD, Packed Multiply and Add instruction. This instruction multiplies four operands by another four operands, adds the results of the first two multiplications, and adds the results of the other two multiplications at once. If the operands are in the right place, four multiplications take the same time as two or three multiplications. In this case, overoptimized algorithms with data shuffling may result in a slower performance.

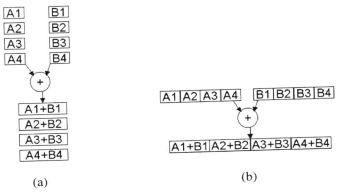

Figure 1 (a) Conventional scalar operation. To add two vectors together, we have to add each pair of the components sequentially. (b) In processors with SIMD capability, we can add two vectors in parallel using one instruction.

the fundamental principle of MMX technology optimization is to vectorize the operation: (1) arrange multiple operands to be execute in parallel; (2) use the smallest possible data type to enable more parallelism with the use of a longer vector; and (3) avoid the use of conditionals. In this section, we will accomplish the following:

1. Introduce the MMX programming environment—MMX™ registers, data types, and instruction sets
2. Show general code optimization on Intel Architecture
3. List the rules of MMX instruction-level optimization
4. Give a simple example of MMX technology to compute a reduced-resolution image

Our main goal is to give readers a high-level understanding of the MMX programming environment. Then, we demonstrate the first rule of MMX technology optimization by optimizing a procedure to compute reduced-resolution images.

2.1 Programming Environment

MMX technology provides the following new extensions to the Intel Architecture programming environment:

1. Eight MMX registers (MM0 through MM7).
2. Four MMX data types (packed bytes, packed words, packed doublewords, and quadwords).
3. A set of MMX instructions.

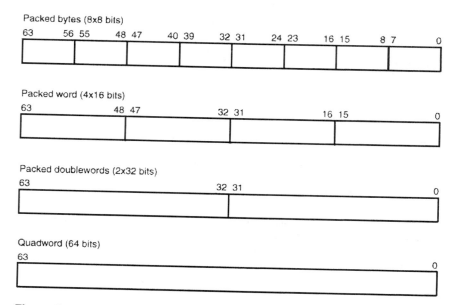

Figure 2 Data formats for MMX instructions.

MMX technology introduces eight new 64-bit MMX™ registers (MM0, MM1, . . . , MM7). MMX instructions operate in parallel on byte (8 bits), word (16 bits), doubleword (32 bits), and quadword (64 bits) data types packed in these 64-bit registers. As shown in Figure 2, you can interpret the 64-bit data format in an MMX register according to the instruction that you use. MMX technology defines some new instructions that allow manipulation of these quantities in parallel. When an operation is performed on 1 byte in a 64-bit register, the same operation may be performed on the other 7 bytes simultaneously.

The MMX instruction set consists of arithmetic, comparison, conversion, logical, shift, and data transfer instructions. MMX instructions follow the format shown in Figure 3. In Table 1, you can find a list of the MMX instructions with

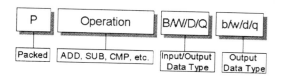

Figure 3 The MMX instruction format.

Table 1 Summary of MMX Instruction Set

Category		Wraparound	Signed saturation	Unsigned saturation
Arithmetic	Addition	PADDB, PADDW, PADDD	PADDSB, PADDSW	PADDUSB, PADDUSW
	Subtraction	PSUBB, PSUBW, PSUBD	PSUBSB, PSUBSW	PSUBUSB, PSUBUSW
	Multiplication	PMULL, PMULH		
	Multiply and Add	PMADD		
Comparison	Compare for Equal	PCMPEQB, PCMPEQW, PCMPEQD		
	Compare for Greater Than	PCMPGTPB, PCMPGTPW, PCMPGTPD		
Conversion	Pack		PACKSSWB, PACKSSDW	PACKUSWB
	Unpack High	PUNPCKHBW, PUNPCKHWD, PUNPCKHDQ		
	Unpack Low	PUNPCKLBW, PUNPCKLWD, PUNPCKLDQ		
		Packed		**Full quadword**
Logical	And			PAND
	And Not			PANDN
	Or			POR
	Exclusive OR			PXOR
Shift	Shift Left Logical	PSLLW, PSLLD		PSLLQ
	Shift Right Logical	PSRLW, PSRLD		PSRLQ
	Shift Right Arithmetic	PSRAW, PSRAD		
		Doubleword transfers		**Quadword transfers**
Data transfer	Register to Register	MOVD		MOVQ
	Load from Memory	MOVD		MOVQ
	Store to Memory	MOVD		MOVQ
Empty MMX™ State		EMMS		

a brief description of each. For a detailed description, please refer to Refs. 3 and 4. The following list shows the complete set of new instructions available in MMX technology for manipulating 64-bit data.

- ADDITION/SUBTRACTION: Add or subtract 8 bytes, 4 words, or 2 doublewords in parallel; also includes saturation hardware to prevent overflow or underflow wraparound.
- COMPARE: Compare bytes, words, or doublewords to build a Boolean mask, which can be used to selectively pick elements in subsequent operations.
- MULTIPLY: Multiply four 16-bit words in parallel, producing four 16-bit truncated products.
- MULTIPLY/ACCUMULATE: Multiply four pairs of 16-bit operands and sum the first two products and the last two products to give two 32-bit sums.
- SHIFT: Arithmetic and logical shifts and rotates by word, doubleword, or quadword.
- PACK/UNPACK: Converting data between 8-, 16-, and 32-bit format.
- LOGICAL: AND, OR, XOR; up to 64 bits.
- MOVE: Move 32 or 64 bits between MMX registers and memory or other MMX registers, or move 32 bits between MMX registers and integer registers.

MMX instructions PCMPEQ, PCMPGT, and so on produce a redundant binary mask per data element that depends on the comparison's result. This approach allows programs to use the mask with subsequent logic operations (AND, ANDN, OR, XOR) to perform conditional moves. These instructions simplify data-dependent branching. (We will see how to convert conditional executions into logic operations in Section 3.)

In addition, a fourwide-parallel multiply–accumulator allows four 16-bit quantities to be multiplied by four other 16-bit quantities and partially summed into two sums of two multiplies each in a single instruction. This powerful operation speeds up image processing functions such as convolution and morphology.

One of the best ways to understand the operation of these instructions and data types is by example. We will show some examples soon.

2.2 Guidelines for Code Optimization

2.2.1 Understand Where the Application Spends Most of Execution Time

Before changing a line of code, it is important to identify the possible areas for MMX technology optimization first. The following are three basic steps [5,17–19]:

1. Understand where the application spends most of its execution time
2. Understand which algorithm is best suited for MMX technology in the application
3. Understand where data values in the application can be converted to integer (fixed-point) or single-precision floating-point while maintaining the required range and precision*

The benefit of optimizing computationally intensive parts is larger than that of optimizing less intensive parts. We should optimize the most computationally intensive components first. There are some tools that can help in isolating the computationally intensive sections of code. In this work, we use the Intel VTune Performance Analyzer to identify system bottlenecks [12,20,21].

2.2.2 General Instruction-Level Code Optimization

After we identify possible code segments for MMX technology optimization, we can begin code optimization. First, we review some general code optimization techniques before we examine the MMX code optimization guidelines. General code optimization guidelines are as follows:

1. Use a current generation compiler that will produce an optimized application. This will help you generate good code from the start.
2. Maximize memory access performance:
   1. Minimize memory references.
   2. Maximize register usage.
   3. If possible, prefetch data that will not otherwise be brought into cache.†
   4. Arrange code to minimize instruction cache misses and optimize prefetch.
   5. Make sure all data are aligned.
   6. Align frequently executed branch targets on 16-byte boundaries.
   7. Avoid partial register stalls [5].
   8. Load and store data to the same area of memory using the same data sizes and address alignments.

---

* There is a trade-off between precision and parallelism. Although the MMX instructions on Intel Pentium II processors or earlier can operate on multiple integer operands, the Streaming SIMD Extensions on Intel Pentium III processors or later can operate on multiple single-precision floating-point operands. On Intel Pentium 4 processors, we can even operate on multiple double-precision floating-point operands. If we give more precision to the operands, then we can work on fewer operands at once.

† Intel Pentium III processors have prefetch instructions to help applications bring operands into caches before real operations need the data. An application example can be found in Ref. 22. Intel Pentium 4 processors have hardware prefetch unit so that applications do not have to prefetch data that will be brought into caches.

3. Minimize branching penalties:
    1. Minimize branch instructions; for instance, unroll small loops.
    2. Arrange code to minimize the misprediction in the branch prediction algorithm. For example, forward conditional branches are not taken and backward conditional branches are taken by default.
4. Use software pipelining to schedule latencies and the utilization of the different functional units. Unroll small loops to schedule more instructions.

2.2.3 Matching the Algorithms to MMX Instruction Capabilities

Matching algorithms to MMX instruction capabilities is the key to extracting the best performance. Once the computationally intensive sections of code are identified, an evaluation should be done to determine whether the current algorithm or a modified one would give the best performance. In some cases, it is possible to improve performance by changing the types of operations in the algorithm. There could be many algorithms for one DSP application. Usually, an algorithm with fewer multiplications is better because multiplications are more expensive in most implementations. However, in MMX technology, grouped multiplications can be done easily and so an algorithm that uses more multiplications can be better than others.

MMX instructions offer the best support for 8-bit and 16-bit integer data types. Although some DSP algorithms must be computed in the floating-point domain, some can be done in the integer domain. MMX technology can provide a significant speedup in certain DSP and multimedia applications that use integers. MMX technology is well suited for many image processing applications where the data are contiguous, 8-bit, and rarely require precision beyond 16 bits. On the other hand, some signal processing applications seem to cause problems due to their higher-precision requirement [23].*

Because MMX instructions can operate on multiple operands in parallel, the fundamental principle of MMX technology optimization is to vectorize the operation: (1) arrange multiple operands to be execute in parallel; (2) avoid the use of conditional operations; (3) use the smallest possible data type to enable more parallelism with the use of a longer vector. For example, use single precision instead of double precision where possible.

2.2.4 MMX Instructions-Level Code Optimization

The following is a list of rules for MMX instruction-level optimization:

1. It is not recommended to intermix MMX instructions and floating-point instructions. MMX instructions do not mix well with floating-point

---

* In Streaming SIMD Extensions, multiple floating-point operands can be calculated together.

instructions because MMX registers and state are aliased onto the floating-point registers and state. This was done so that no new registers or states were introduced by MMX technology.†

2. MMX instructions that reference memory or integer registers do not mix well with integer instructions referencing the same memory or registers.

3. It is important to arrange data in the best way for MMX technology processing (e.g., structure of array, array of structure, rowwise or columnwise arrangements). Columnwise processing in general is better than sequential rowwise processing.

The following are two processor-specific rules (on Intel Pentium III or earlier):

4. MMX shift/pack/unpack instructions do not mix well with each other. In general, two MMX instructions can be executed in the same clock cycle. However, only one MMX shift/pack/unpack instruction can be executed on one clock cycle because there is only one shifter unit.

5. MMX multiplication instructions "pmull/pmulh/pmadd" do not mix well with each other because there is only one multiplication unit.

2.3 A Simple MMX Programming Example

The first rule of MMX technology optimization is to execute multiple identical operations in one instruction. In this example we use MMX instructions to speed up the computation of the reduced-resolution image by 2.3 times. Computing a reduced-resolution image I' from a given image I is used in many applications (e.g., hierarchical motion estimation [16]). In [Refs. 22 and 24], we describe a spatial-domain watermarking detection scheme, which starts by computing a reduced resolution image. Based on that image, we compute a pseudo-random-noise pattern. Then, we correlate the low-resolution image and the image-dependent pseudo-random signal to extract the watermark message.

Our first implementation of the watermark detection module is purely in C. From the Intel VTune™ Analyzer profiling of the first C implementation, we found that the most computationally intensive component of the detection system

---

† MMX code sections should end with "emms" instructions if floating-point operations are to be used later in the program. In order to maintain operating system compatibility, MMX registers are aliases to the scalar floating-point registers. As we read or write to an MMX register, we read and write to one of the floating-point registers and vice versa. Thus, we cannot guarantee what the contents of the floating-point register will be after we execute an MMX piece of code, or vice versa. Mixing MMX instructions and floating-point code fragments in the same application is challenging. To guarantee that no floating-point errors will occur when we switch from MMX to floating point, we must use the new MMX instruction EMMS (Empty MMX Technology State), which marks all the floating-point registers as *Empty*. Using EMMS at the end of every MMX function may not be the most efficient; we just need to use EMMS before floating-point operations.

is the subroutine that calculates the reduced-resolution image. The subroutine takes 47% of the time of the watermark detection scheme, as shown in Figure 4a. We exclude the MPEG decoding in the CPU time breakdown because our MPEG decoder is already optimized [25]. Although this subroutine takes most of the CPU cycles, the operations of this subroutine are extremely simple as shown in the following:

```
void get_dc_image_c(void)
{
  int y, x, j, k, temp;
  for (y = 0; y < height_in_blocks; y++){
    for (x = 0; x < width_in_blocks; x++)
    {
      temp = 0;
      for (j = 0; j < block_size; j++){
        for (k = 0; k < block_size; k++){
          temp += image_data[(y * block_size + j) * image_width
              + x * block_size + k];
        }
      }
      dc_image_data[y * width_in_blocks + x] = (temp/(block_size
      *block_size));
    }
  }
}
```

The inner loops (loop j and loop k) are simple additions of 8-bit integers. Because of the extreme regularity, the inner loops of this subroutine can be implemented efficiently in MMX technology. We unrolled the inner loop so that the operation can be executed in parallel. The following is a high-level, conceptual code after unrolling. (We define image_data [y, x, j, k] as image_data[(y * block_size + j) * image_width + x * block_size + k].)

```
for (k = 0; k < block_size; k++){
  temp[k] = 0;
}
for (j = 0; j < block_size; j++)
{  /* execute the following additions in parallel */
  temp[0] += image_data [y, x, j, 0];
  temp[1] += image_data [y, x, j, 1];
  temp[2] += image_data [y, x, j, 2];
  . . .
  temp[block_size-1] += image_data [y, x, j, block_size-1];
}
temp = 0;
for (k = 0; k < block_size; k++){
  temp += temp [k];
}
```

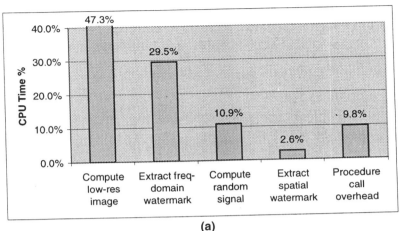

(a)

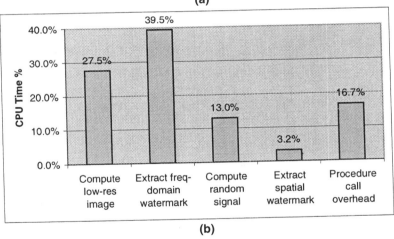

(b)

Figure 4 Breakdowns of CPU time in our watermark detection scheme: (a) C implementation and (b) MMX technology optimized implementation. We divide our watermark detection scheme into five major parts: (1) compute reduced-resolution images, (2) compute image-dependent pseudo-random noises, (3) extract spatial-domain watermark, (4) extract frequency-domain watermark, and (5) function call overheads. In the C implementation, about half of the CPU time is spent in the module that calculates low-resolution images. After we optimize this module with MMX technology, this module is no longer the major bottleneck. We achieve 2.3× speedup in this module, which makes the speedup of our whole watermark detection 1.4×.

If we choose the block size to be 8×8 pixels, then, in theory, we can use the MMX instruction, PADDB (_mm_add_pi8), to add two sets of 8 pixels together. Instead of having eight independent and sequential additions, we just need one MMX instruction. The inner loop conceptually will be the following:

```
for (j = 0; j < block_size; j++)
{
  temp [*] =_mm_add_pi8 (temp[*], image _data [y, x, j, *]);
}
```

We can potentially have an 8× speedup. However, the precision require-ments reduce the speedup. Adding multiple 8-bit data together requires 9-bit or higher precision. Instead of packed addition of eight 8-bit data, one method of adding 8-bit data together is (1) to convert those 8-bit data into 16-bit words and (2) to add the data in the packed-word format. The sample code is shown as follows:

Input: MM0: unsigned source1 value (8-bit)
MM1: unsigned source2 value (8-bit)
MM7: a zero register (PXOR MM7, MM7)
Output: MM0: four 16-bit words from adding the four LOW end bytes
MM1: four 16-bit words from adding the four HIGH end bytes

```
MOVQ         MM2, MM0
PUNPCKLBW    MM0, MM7
PUNPCKHBW    MM2, MM7   ; [mm2 mm0] ← mm0
MOVQ         MM3, MM1
PUNPCKHBW    MM1, MM7
PUNPCKLBW    MM3, MM7   ; [mm1 mm3] ← mm1
PADDW        MM0, MM3
PADDW        MM1, MM2   ; [mm1 mm0] ← mm1 + mm0
```

After this MMX technology optimization, the subroutine that calculates the reduced-resolution image takes only 27.5% of the CPU time, as shown in Figure 4b. Also, after we optimize this module with MMX technology, the execution time is distributed more evenly to the other modules so this module is no longer the major bottleneck. We achieve a 2.3× speedup in this module.

2.4   What We Have Learned

1.   Because MMX instructions can operate on multiple operands in paral-lel, the first rule of MMX technology optimization is to arrange multi-ple operands to be executed in parallel.

2. After vectorizing the operations, we can speed up the processing by up to 8×. In the above example, we find that the actual speed up is around 2.3×.

3. The overhead of packing and unpacking slows down the maximal performance. In the above example, we have eight 8-bit data as input. However, we cannot use 8-bit data as the internal representation because of the potential for arithmetic overflow. Thus, we have to convert those 8-bit data into 16-bit data. A number of unpacking instructions are used for this conversion and slow down the processing.

4. Choosing the right vector size and data arrangement is important. We choose a block size of 8×8 pixels instead of others (e.g., 9×9 pixels). MMX technology can process eight operands in one instruction. It is very natural to process 8×8 blocks (or 4×4 blocks). It would be much slower if we need to process 3×3 or 9×9 blocks.

5. Sometimes, we have to make trade-offs between the speed and the quality of the algorithm. In our watermarking scheme [24], the robustness and picture quality of the watermark depend on the block size. Our algorithm using an 8×8-pixel block is faster than using 2×2-pixel blocks and 4×4-pixel blocks, but using 8×8-pixel blocks makes the watermarks more visible. Similarly, there are many other engineering trade-offs depending on the application requirements.

3 BASIC MMX PROGRAMMING: VECTOR AND LOGIC OPERATIONS

In this section, we will show two more principles of MMX technology optimization by using the MPEG-4 pixel padding procedure as an example. The *second rule* of MMX technology optimization is to transform conditional executions into logic operations. The first rule of MMX technology optimization is to execute multiple identical operations in one instruction, as shown in Section 2.3. However, there are conditional operations like the following:

```
for (i = 0; i < block_size-1; i++)
  for (j = 0; j < block_size; j++)
  {
    if (mask[i + 1][i] == 0)
      pixel[i + 1][j] = pixel[i][j];
  }
```

or after we unroll the inner loop as the following:

```
for (i = 0; i < block_size-1; i++)
{
  if (mask [i + 1] [0] == 0)
    pixel[i + 1] [0] = pixel[i] [0];
  if (mask [i + 1] [1] == 0)
    pixel[i + 1] [1] = pixel[i] [1];
  ...
  if (mask [i + 1] [block_size-1] == 0)
    pixel [i + 1] [block_size-1] = pixel[i] [block_size-1];
}
```

Because mask[i][j] and mask[i][j + 1] could be different, we cannot execute pixel[i + 1][j] = pixel[i][j] and pixel[i + 1][j + 1] = pixel[i][j + 1] together. Currently, MMX technology does not allow conditional execution in one instruction. In Section 3.2, we will demonstrate that the conditional operations can be expressed as logic operations so that they can be executed in parallel.

The *third rule* of MMX technology optimization is to put the data into the right format for parallel execution. The example in Section 3.2 assumes that the data can be loaded into the MMX register in one instruction. However, there is a challenge when the data array is in row major and we want to process the data in column major order like the following:

```
for (i = 0; i < block_size; i++)
  for (j = 0; j < block_size-1; j++)
  {
    if (mask [i][j + 1] == 0)
      pixel[i][j + 1] = pixel[i][j];
  }
```

MMX instructions can load or store multiple operands if operands are placed in a row. However, multiple operands in a column are harder to access. For faster execution, we should arrange data to be processed in a row-major order or change the algorithm. On the other hand, for some two-dimensional image/video processing operations, an algorithm needs to process data in both directions. In Section 3.4, a matrix-transpose procedure will be demonstrated to provide the flexibility of choosing row-major or column-major processing.

3.1   MPEG-4 Pixel Padding Procedure

In this subsection, we use the MPEG-4 pixel padding procedure as an example for MMX technology optimization. In addition to the conventional frame-based functionality in MPEG-1 and MPEG-2, the new MPEG-4 video coding standard supports arbitrary shaped video objects [26]. In MPEG-4, the video input is no longer considered as a rectangular region. On the other hand, similar to the MPEG-1 and MPEG-2, the MPEG-4 video coding scheme processes the successive images of a VOP (video–object–plane) sequence in a block-based manner (e.g., motion estimation, motion compensation, DTC, IDCT). Therefore, before motion estimation, motion compensation, and DCT, nonobject pixels of contour macroblocks (which contain the shape edge of an object) are filled using the padding technique. The padding operation turns out to be a computationally complex and irregular operation [27]. In the following example, we have created a new procedure using MMX technology to speed up the MPEG-4 padding process by 1.5× to 2.0×.

First, for each arbitrary shaped object, a minimum-sized rectangular bounding box is defined. The box is divided and expanded to an array of macroblocks with the natural number of macroblocks in horizontal and vertical directions. Because of the arbitrary shape of the object, not all pixels inside this bounding box contain valid object pixel values. There are macroblocks that lie completely inside the object, macroblocks that lie completely outside, and macroblocks that cover the border of the video object, as shown in Figure 5. Macroblocks that lie inside the object remain untouched. Macroblocks that cover the object boundary are filled using the repetitive padding algorithm.

Padding is accomplished by copying the pixels that lay on the edge of the mask outward. First, the pixels are padded in the horizontal direction, with boundary pixels propagated both leftward and rightward. On the second pass, pixels are padded in the vertical direction. In both cases, if a pixel that lies outside of the mask is bounded by two masked pixels on opposite sides, the unmasked pixel should be assigned the average of both bounding pixels.

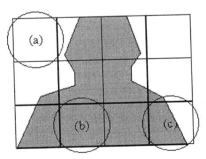

Figure 5 Bounding box and macroblocks of an arbitrary shaped video object: (a) outside the object, (b) inside the object, and (c) on the boundary.

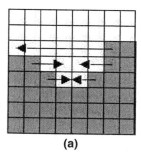
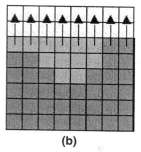

(a)   **(b)**

Figure 6 Repetitive padding: (a) horizontal step and (b) vertical step.

In the horizontal padding step, the macroblock is processed row by row, as shown in Figure 6a. All nonobject pixels inside a row of the macroblock are filled with the next border value of the object. As shown in the following code, if there are shape edges on both sides of a nonobject pixel, the average of the border values is taken (e.g., Table 2B). Afterward, these filled pixel positions are marked as object pixel for the next padding step.

```
for (y = 0; y < block_size; y++)
{
    if (mask[y][x] == 1)
        hv_pad[y][x] = hor_pad[y][x];
    else
    {
        if ( mask[y'][x] == 1 && mask[y"][x] == 1)
          hv_pad[y][x] = (hor_pad[y'][x] + hor_pad[y"][x])//2;
        else if (mask[y'][x] == 1)
          hv_pad[y][x] = hor_pad[y'][x];
        else if (mask[y"][x] == 1)
          hv_pad[y][x] = hor_pad[y"][x];
    }
}
```

Table 2 An Example of Pixel Padding

A				B				C			
*	*	*	40	40	**40**	40	**40**	40	40	40	40
*	*	*	*	*	*	*	*	**29**	**29**	**30**	**36**
*	17	19	32	**17**	17	19	32	17	17	19	32
20	*	*	14	20	**16**	**16**	14	20	16	16	14

A shows the original 4 × 4 matrix. Pixels labeled with an asterisk are outside of the pixel mask. B shows the matrix after the horizontal padding stage. Pixels in bold are the changed values. C. The final matrix after the vertical padding stage.

In the previous reference pseudocode of the vertical pixel padding algorithm, y' is the location of the nearest valid sample ($\text{mask}[y'][x] == 1$) above the current location y at the boundary of hv_pad, and y'' is the location of the nearest boundary sample below y. The input is a horizontally padded block.

In the vertical padding step, the macroblock is processed column by column, as shown in Figure 6b. All nonobject pixels inside the column are filled with the next border value of the object or with the next pixel value filled in the horizontal padding step. If there are shape edges or previously filled pixel on both sides of a nonobject pixel, the average of these values is taken. If all columns of the macroblock are processed, the padding algorithm is done (e.g., Table 2C).

3.2 They Are All Logic Operations

Currently, MMX technology does not allow conditional execution in one instruction. In this section, we demonstrate how to use logic operations to perform conditional operations. By assuming that the object is convex in shape, we do not need to average pixel values for padding. In Figure 7, by assuming we are performing vertical padding, we simplified the pixel-padding procedure as the following:

```
for (i = 0; i < block_size-1; i++) // vertical direction
  for (j = 0; j < block_size; j++)    // horizontal direction
  {
    if (mask[i + 1][j] == 0)
      pixel[i + 1][j] = pixel[i][j];
  }
```

Furthermore, we assume that it is always true that $\text{mask}[i][j] = 0$ or $\text{mask}[i][j] = 1$. In this case,

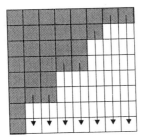

Figure 7 A simplified case of the pixel padding procedure. We assume that we are performing vertical padding without average pixel values.

```
if (mask[i + 1][j] == 0)
    pixel[i + 1][j] = pixel[i][j];
```

is equivalent to

$$\text{pixel}[i + 1][j] = \begin{cases} \text{pixel}[i][j] & \text{if mask}[i + 1][j] = 0 \\ \text{pixel}[i + 1][j] & \text{if mask}[i + 1][j] = 1 \end{cases}$$

Then, it is clear that

```
pixel[i + 1][j] = mask[i + 1][j] * pixel[i + 1][j] + (1 − mask[i +
1][j]) * pixel[i][j];
```

carries the same functionality as

```
if (mask[i + 1][j] == 0)
    pixel[i + 1][j] = pixel[i][j];
```

Thus, the above conditional procedure can be expressed as follows:

```
for (i = 0; i < block_size-1; i++)
  for (j = 0; j < block_size; j++)
  {
    pixel[i + 1][j] = mask[i + 1][j] * pixel[i + 1][j] + (1 − mask[i +
1][j]) * pixel[i][j];
  }
```

This can be done easily in MMX technology because the above algorithm can be executed without any knowledge of the pixel or mask values. Note that there are no branch statements. This algorithm can be sped up using MMX instructions by computing all eight pixels in a row concurrently. The following is our code for this simplified pixel padding procedure:

```
              mov        edi, pixelsptr
              mov        esi, maskptr
              mov        ecx, 7              ; loop counter
Simple_loop:
              movq       mm1, [esi + 8]      ; mm1 ← M2
              movq       mm0, [edi + 8]      ; mm0 ← P2
              pandn      mm1, [edi]          ; mm1 ← P1 AND (NOT M2)
                                             ; do bitwise not of the mask of
                                             ; the current line, and it with
                                             ; the pixels of the last line
```

```
        por           mm0, mm1              ; mm0 ← P2 + [P1 AND (NOT M2)]
                                            ; and add them together. This
                                            ; ''smears'' the pixels in the
                                            ; down direction.

        movq          [edi + 8], mm0
        add           esi, 8
        add           edi, 8
        loop          Simple_loop
```

3.3 Data Arrangement Optimization for MMX Technology

Efficient processing of a continuous stream of data requires both a faster CPU and fast memory throughput. In addition to algorithmic and code optimization, data optimization is another important issue to consider when using MMX technology. Multimedia applications are memory-intensive in nature and exert a huge demand on the memory subsystem. Although a scalar multiplication takes as long as 10 cycles, an MMX instruction with 4 multiplies takes as little as 1 cycle.* Fast calculation makes memory access a bottleneck.

Depending on the nature of the data, certain access patterns are more efficient than others [20]. The performance can be improved by changing the data organization (if possible) or by changing the way data is processed. It is important to arrange data so that MMX instructions can perform as many operations as possible in parallel with the least amount of load/store operations.

First, in the previous example, we assumed that the pixel values and the mask values are both stored in 8-bit formats. We made this assumption so that we could perform eight pixel-padding operations on each iteration. In theory, this would give us an eight times speedup using MMX instructions. If the data are stored in 16-bit values, then we can perform only four operations on each iteration (only $4\times$ speedup in theory).

It is clear that there are trade-offs between precision and speed in some applications. In some applications, precision is more important than the speed. Even for 9-bit values, it is better to store the data in a 16-bit format. In this case, we can only process four 16-bit values in MMX instructions. On the other hand, when speed is more critical than precision, it may be better to truncate the 9-bit values and store them into the 8-bit format for greater parallelism.

Second, the data array for pixel values and the data array for mask values are stored separately instead of interleaving the pixel data with the mask data. Hence, one load instruction can move 8-pixel data into one register and another load instruction can move 8-mask data into another register. We can easily perform logic operations on them. If the data are interleaved, then we need additional

---

* Execution cycles are processor-specific.

operations to perform the logic operations. Clearly, it is important to design the data arrangement to match the capability of the MMX instructions.

Data alignment is important for efficient data processing. Misaligned data incurs a penalty on quadword boundaries or an even longer penalty on cache-line boundaries.* The best performance is achieved when data for MMX operations are aligned on 8-byte boundaries.

3.4 Matrix Transpose: Transforming Column-Major to Row-Major Processing

MMX technology is very powerful for padding eight pixels in a row in parallel. Previously, we assumed that we were performing vertical padding. In this case, multiple data in a row can be processed in parallel (columnwise processing) by reading the data row by row. Nonetheless, we need to pad the pixels horizontally as well; that is, multiple data in a column must be processed in parallel. In row-major memory organization, pixel[i][j] and pixel[i][j + 1] are adjacent to each other, but pixel[i][j] and pixel[i + 1][j] are not. In today's memory subsystem, loading pixel[i][j], pixel[i + 1][j], pixel[i + 2][j], . . . into one MMX register often takes a great amount of effort.

The solution to our difficulties in horizontal padding is to transpose the data matrix. After transposition, vertical padding, which processes multiple pixels in a row, is equivalent to the original horizontal padding. Figure 8 shows the flowchart of our implementation of the pixel-padding procedure. Figure 9 shows an example. First, we transpose the input macroblocks and perform vertical padding processing. Then, we transpose the intermediate macroblocks back and perform the vertical padding process again. Although a matrix transposition takes a significant amount of time, two matrix transpositions and one MMX vertical padding is still faster than a horizontal padding using scalar operations.

Because MMX registers can store only a one-dimensional array, the challenge now is to transpose the matrix efficiently using MMX technology. In Section 2.1, we only briefly explained the unpacking instructions PUNPCKL and PUNPCKH. The unpack instructions shuffle half of the contents of one register with half of the contents of another register (or 64-bit memory location). The PUNPCKH instructions interleave the high-order data elements of the two operands and ignore the low-order data elements. Similarly, PUNPCKL instructions interleave the low-order data elements of both operands. The PUNPCKHBW instruction interleaves the four high-order bytes from both operands. PUNPCKHWD interleaves the two high-order words from each operand as shown in Figure 10. If the second operand contains all zeros, the result is a zero

---

* Penalties are processor-specific.

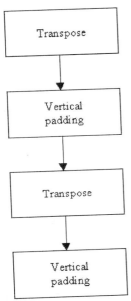

Figure 8 Our pixel padding procedure contains four parts: (1) transpose of the block, (2) vertical padding, (3) transpose of the block, and (4) vertical padding.

extension of the high-order elements of the destination operand. In this case, the instructions can convert data into a higher-precision representation.

In addition to gathering data from different memory locations, interleaving planar and duplicating data, the unpacking/packing instructions can transpose rows and columns of data. Figure 11 illustrates a method for performing 4×4 transpositions on 16-bit packed values [28]. The basic idea behind this method

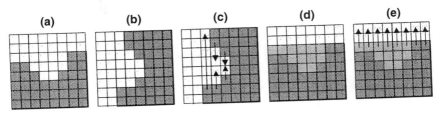

Figure 9 An example of our pixel padding procedure: (a) original block, (b) transpose of the original block, (c) vertical padding, (d) transpose of the vertically padded transposed block (which is equivalent to the horizontally padded block), and (e) vertical padding.

PUNPCKHWD mm, mm/m64

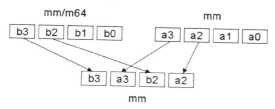

mm

Figure 10 Shuffling the most significant two words from two 64-bit operands into one 64-bit operand with PUNPCKHWD.

follows. First, we collect the higher-order data into a set of registers. Then, collecting the higher-order data from the registers, which contains higher-order data, is equivalent to collecting the highest-order data from each row; that is, we have the data originally in a column now in a MMX register. Transposing 8×8 transpositions on 8-bit packed values is left as an exercise for the readers.

We measure the performance of the MMX technology-optimized implementation. Our simulation results show that the new pixel padding routines run between 1.5 and 2 times faster than the original scalar-instruction-only version.

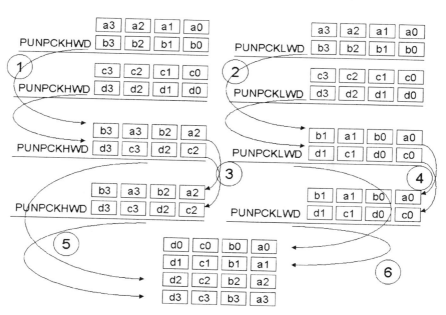

Figure 11 Performing a 4×4 transposition using PUNPCKH and PUNPCKL.

3.5 Key Points

To illustrate MMX technology optimization, we optimized the MPEG-4 pixel padding algorithm in this section. During the optimization, we learned the following implications:

1. Algorithms should be designed to take advantage of the SIMD operations (i.e., to execute identical operations in parallel). It is important to avoid conditional operations if possible. If it is not, use logic operations.

2. A deterministic algorithm is more preferable than a nondeterministic algorithm. In a MMX technology implementation, although logic operations can replace conditional operations, logic operations are overheads.

3. Algorithms should be designed to take advantage of the natural data arrangement. (For example, arrange the data into row-major processing order.) In order to achieve MMX technology data optimization, we should determine the optimal data packing format for SIMD processing (array of structures, structure of arrays, or data layout).

4. We observed that in many implementations, too many of the instructions in the code are data format conversions (e.g., the implementations spend too much time packing and unpacking data). The format of input data to be processed by an algorithm may need to be rethought to make it optimal for MMX technology. In some cases, the optimal format may be quite strange and not obvious.

5. We should design algorithms in the row-major processing order (i.e., process multiple data in a row in parallel). Although MMX technology is good at processing row-major data, MMX technology is less efficient at processing column-major data.

6. If column-major processing is necessary, we can use matrix transposition before we process the columns of data.

7. Two-dimensional or high-dimensional DSP algorithms are better designed into separable one-dimensional operations. Because of the one-dimensional nature of MMX instructions, inseparable high-dimensional operations are inherently difficult to be implemented efficiently in MMX technology.

4 ALGORITHM DESIGN FOR MMX TECHNOLOGY IN SA-DCT AND SA-IDCT

The *fourth rule* of MMX technology optimization is to reduce shuffling and maximize grouping of operations into one instruction. In this section, we demonstrate this rule by optimizing the SA-DCT (shape-adaptive inverse discrete cosine transform) and the SA-IDCT (shape-adaptive inverse cosine transform). The 8×8

DCTs and 8×8 IDCTs are widely used in image and video compression/decompression. Optimizing 8×8 DCT/IDCT or eight-point DCT/IDCT in single-processor/VLSI implementations has been studied extensively [29,30]. The MMX technology optimization of 8-point DCT/IDCT has been implemented as well [31–34].

Most of the fast DCT implementations reduce computational complexity by factoring out the discrete cosine transformation matrix into butterfly and shuffle matrices. The butterfly and shuffle matrices can be computed with fast integer addition. For example, Chen's algorithm [29] is one of the most popular algorithms of this kind. Conventionally, optimization usually focuses on reducing the number of DCT arithmetic operations, especially the number of multiplications.

However, algorithmic optimization for MMX technology is different than the traditional optimization. MMX instructions can perform multiple arithmetic operations in one instruction. MMX instructions work best when the operations are regular and the data elements are adjacent. Furthermore, individual bytes and words within an MMX register are difficult to manipulate. Data shuffling is more expensive than actual arithmetic. In this case, "overreducing" the number of arithmetic operations may not reduce the total computational time. Instead of re-examining the 8×8 DCT/IDCT, this section uses the SA-DCT/IDCT as examples for applying the fourth rule.

4.1 Shape-Adaptive DCT and Shape-Adaptive IDCT

One of the building blocks for Version 2 of the MPEG-4 Visual Coding Standard is the SA-DCT for arbitrary-shaped objects [16,26,35]. The new MPEG-4 video coding standard supports arbitrary-shaped video objects in addition to the conventional frame-based functionalities in MPEG-1 and MPEG-2. In a MPEG-4 image, there are some blocks called contour macroblocks, which contain the shape edge of an object (e.g., Figure 5c). Instead of performing an 8×8 DCT after filling the nonobject pixels, the new standard adaptively performs an N-point DCT based on the shape. Compared to the 8×8 DCT, the SA-DCT provides a significantly better rate-distortion trade-off, especially at high bit rates [36].

Similar to the standard DCT, forward and inverse SA-DCT convert $f(x, y)$ to $F(u, v)$ and vice versa. In contrast to the standard 8×8 DCT, only the opaque pixels within the boundary blocks are really transformed and coded. The pixels transformed and coded are controlled by shape parameters. As a consequence, the SA-DCT does not require the padding technique and the number of SA-DCT coefficients is identical to the number of opaque pixels in the given boundary block.

Figure 12 outlines the concept of the SA-DCT baseline algorithm for coding an arbitrarily shaped image segment that is contained within an 8×8 block.

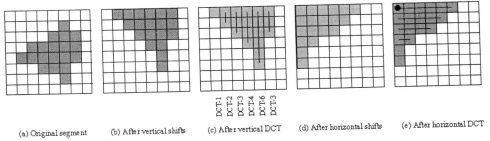

(a) Original segment (b) After vertical shifts (c) After vertical DCT (d) After horizontal shifts (e) After horizontal DCT

Figure 12 Successive steps involved for performing a SA-DCT forward transform on a block of arbitrary shape.

The SA-DCT algorithm is based on predefined orthonormal sets of DCT basis functions. The forward two-dimensional SA-DCT processes columns first and rows next. The inverse two-dimensional SA-DCT applies the one-dimensional transform to the rows first.

Figure 12a shows an example of an image block segmented into two regions. The foreground is gray and background is white. To perform the vertical transform of the foreground, the length (vector size N, $0 < N < 9$) of each column j ($0 < j < 9$) of the foreground segment is calculated, and the columns are shifted and aligned to the upper border of the 8×8 reference block, as shown in Figure 12b.

Dependent on the vector size N of each particular column of the segment, a one-dimensional N-point DCT (a transform kernel containing a set of N basis vectors) is selected for each particular column and applied to the first N pixels of the column. For example, in Figure 12b, the rightmost column is transformed using three-point DCT.*

---

* One-dimensional N-point DCT is given by the following equation:

$$y_n = c_n \sum_{k=0}^{N-1} \cos\left(\frac{n(2k + 1)}{2N}\pi\right)x_k$$

where

$c_0 = 1/\sqrt{N}$

and

$c_n = \sqrt{2/N}$

for $n = 1, \ldots, N - 1$.

As shown Figure 12d, before the horizontal DCT transformation, the rows are shifted to the left border of the 8 × 8 reference block. Figure 12e shows the final location of the resulting DCT coefficients within the 8 × 8 image block.

In this algorithm, the final number of DCT coefficients is identical to the number of pixels contained in the image segment. Also, the coefficients are located in comparable positions as in a standard 8×8 block. The DC coefficient is located in the upper left border of the reference block, and dependent on the actual shape of the segment, the remaining coefficients are concentrated around the DC coefficient.

Because the contour of the segment is transmitted to the receiver prior to transmitting the macroblock information, the decoder can perform the shape-adapted inverse DCT as the reverse operation in both the horizontal and vertical segment directions on the basis of the decoded shape data.

4.2 Two-Point DCT

The computational complexity of the two-point DCT can be simplified as follows:

$$\begin{bmatrix} y_0 \\ y_1 \end{bmatrix} = \begin{bmatrix} 1/\sqrt{2} & 1/\sqrt{2} \\ \cos(\pi/4) & \cos(3\pi/4) \end{bmatrix} \begin{bmatrix} x_0 \\ x_1 \end{bmatrix} = \begin{bmatrix} 1/\sqrt{2} & 1/\sqrt{2} \\ 1/\sqrt{2} & -1/\sqrt{2} \end{bmatrix} \begin{bmatrix} x_0 \\ x_1 \end{bmatrix}$$

$$= 1/\sqrt{2} \begin{bmatrix} 1 & 1 \\ 1 & -1 \end{bmatrix} \begin{bmatrix} x_0 \\ x_1 \end{bmatrix}$$

where x_i is the input data and y_i is the transformed data. In conventional algorithmic optimization, we minimize the number of additions and multiplications. Thus, we define

$$z_0 = x_0 + x_1$$
$$z_1 = x_0 - x_1$$

Then,

$$y_0 = \frac{1}{\sqrt{2}} z_0$$

$$y_1 = \frac{1}{\sqrt{2}} z_1$$

In this way, we need only two additions and two multiplications instead of two additions and four multiplications. The following is the C code for this algorithm.

```
void fsadct2_float (float in[2], float out[2])
{
    static float f0 = 0.707107;

    out[0] = (in[0] + in[1]) * f0;
    out[1] = (in[0] - in[1]) * f0;
}
```

In MMX technology, two additions and four multiplications can be done quickly with only one PMADDWD instruction. It only takes one instruction to execute: out[0] = in[0] * f0 + in[1] * f0 and out[1] = in [0] * f0 + in[1] * (f0). Although the following code uses four multiplications instead of two, its MMX technology implementation is faster than the preceding one:

```
void fsadct2_float (float in[2], float out[2])
{
    static float f0 = 0.707107;

    out[0] = in[0] * f0 + in[1] * f0;
    out[1] = in[0] * f0 + in[1] * (-f0);
}
```

The following is the MMX code for the two-point DCT:

```
void fsadct2_mmx (short in[2], short out[2])
{
    static _int64 xstatic1 = 0xA57E5A825A825A82; // - f0 f0 f0 f0
    static _int64 rounding = 0x0000400000004000;

    _asm {
      mov eax, in
      mov ecx, out
      movd mm0, [eax]            // mm0 = xx, xx i1, i0
      pshufw mm1, mm0, 01000100b // mm1 = i1, i0, i1, i0
      pmaddwd mm1, xstatic1      // mm1 = i0 * f0 -i1 * f0,
                                 // i0 * f0 + i1 * f0
      paddd mm1, rounding        // do proper rounding
      psrad mm1, 15
      packssdw mm1, mm7          // mm1 = x, x, o1, o0
      movd [ecx], mm1
    }
}
```

4.3 Four-Point DCT

The computational complexity of the four-point DCT can be simplified as follows:

$$
\begin{bmatrix} y_0 \\ y_1 \\ y_2 \\ y_3 \end{bmatrix} =
\begin{bmatrix}
\frac{1}{\sqrt{4}} & \frac{1}{\sqrt{4}} & \frac{1}{\sqrt{4}} & \frac{1}{\sqrt{4}} \\
\sqrt{\frac{2}{4}}\cos\left(\frac{\pi}{8}\right) & \sqrt{\frac{2}{4}}\cos\left(\frac{3\pi}{8}\right) & \sqrt{\frac{2}{4}}\cos\left(\frac{5\pi}{8}\right) & \sqrt{\frac{2}{4}}\cos\left(\frac{7\pi}{8}\right) \\
\sqrt{\frac{2}{4}}\cos\left(\frac{2\pi}{8}\right) & \sqrt{\frac{2}{4}}\cos\left(\frac{6\pi}{8}\right) & \sqrt{\frac{2}{4}}\cos\left(\frac{10\pi}{8}\right) & \sqrt{\frac{2}{4}}\cos\left(\frac{14\pi}{8}\right) \\
\sqrt{\frac{2}{4}}\cos\left(\frac{3\pi}{8}\right) & \sqrt{\frac{2}{4}}\cos\left(\frac{9\pi}{8}\right) & \sqrt{\frac{2}{4}}\cos\left(\frac{15\pi}{8}\right) & \sqrt{\frac{2}{4}}\cos\left(\frac{21\pi}{8}\right)
\end{bmatrix}
\begin{bmatrix} x_0 \\ x_1 \\ x_2 \\ x_3 \end{bmatrix}
$$

$$
= \sqrt{\frac{1}{2}}
\begin{bmatrix}
1 & 0 & 0 & 0 \\
0 & 0 & 1 & 0 \\
0 & 1 & 0 & 0 \\
0 & 0 & 0 & 1
\end{bmatrix}
\begin{bmatrix}
\frac{1}{\sqrt{2}} & \frac{1}{\sqrt{2}} & 0 & 0 \\
\frac{1}{\sqrt{2}} & -\frac{1}{\sqrt{2}} & 0 & 0 \\
0 & 0 & \cos\left(\frac{3\pi}{8}\right) & \cos\left(\frac{\pi}{8}\right) \\
0 & 0 & -\cos\left(\frac{\pi}{8}\right) & \cos\left(\frac{3\pi}{8}\right)
\end{bmatrix}
\begin{bmatrix}
1 & 0 & 0 & 1 \\
0 & 1 & 1 & 0 \\
0 & 1 & -1 & 0 \\
1 & 0 & 0 & -1
\end{bmatrix}
\begin{bmatrix} x_0 \\ x_1 \\ x_2 \\ x_3 \end{bmatrix}
\tag{1}
$$

The total number of operations is reduced from 16 multiplications to 12 multiplications in this case. The widely referenced Chen's algorithm introduced in 1977 is optimized for the number of multiplications [29]. Chen's algorithm is based on further factorizing the inner matrix from

$$
\begin{bmatrix}
\frac{1}{\sqrt{2}} & \frac{1}{\sqrt{2}} \\
\frac{1}{\sqrt{2}} & -\frac{1}{\sqrt{2}}
\end{bmatrix}
$$

to

$$
\frac{1}{\sqrt{2}}
\begin{bmatrix}
1 & 1 \\
1 & -1
\end{bmatrix}.
$$

Equation (1) can be expressed as follows:

$$
\begin{bmatrix} y_0 \\ y_1 \\ y_2 \\ y_3 \end{bmatrix} =
\sqrt{\frac{1}{2}}
\begin{bmatrix}
1 & 0 & 0 & 0 \\
0 & 0 & 1 & 0 \\
0 & 1 & 0 & 0 \\
0 & 0 & 0 & 1
\end{bmatrix}
\begin{bmatrix}
\frac{1}{\sqrt{2}}\begin{bmatrix} 1 & 1 \\ 1 & -1 \end{bmatrix} & 0 & 0 \\
& 0 & 0 \\
0 & 0 & \cos\left(\frac{3\pi}{8}\right) & \cos\left(\frac{\pi}{8}\right) \\
0 & 0 & -\cos\left(\frac{\pi}{8}\right) & \cos\left(\frac{3\pi}{8}\right)
\end{bmatrix}
\begin{bmatrix}
1 & 0 & 0 & 1 \\
0 & 1 & 1 & 0 \\
0 & 1 & -1 & 0 \\
1 & 0 & 0 & -1
\end{bmatrix}
\begin{bmatrix} x_0 \\ x_1 \\ x_2 \\ x_3 \end{bmatrix}
$$

The total number of multiplications is reduced from 12 to 10.

Nevertheless, the extra factorization above does not reduce the clock cycle count of the implementation in MMX technology although it has fewer multiplications. Because the PMADDWD instruction always multiplies and adds, it is impossible to get four discrete 32-bit values from four sets of 16-bit multiplies. Multiplying one set of values consumes the same amount of processor time as multiplying four sets of values. Even if we further factor the lower right cosine block in the original matrix, leaving three total multiplies, it would still need two PMADDWD operations, plus a substantial amount of additional instructions to shuffle and add the results. Because of this, it is important to reduce shuffling and maximize pairing.

We chose to implement Eq. (1) and have implemented in the following code:

```
void fsadct4_mmx (short in[4], short out[4])
{
    static _int64 xstatic1 = 0x4000C00040004000; // f0 - f0 f0 f0
    static _int64 xstatic2 = 0xDD5D539F22A3539F; // -f1 f2 f1 f2
    static _int64 rounding = 0x0000400000004000;

    _asm {
        mov       eax, in
        mov       ecx, out
        movq      mm0, [eax]          // i3 i2 i1 i0
        pshufw    mm1, mm0, 00011011b // i0 i1 i2 i3
        movq      mm2, mm1
        paddsw    mm2, mm0            // b0 b1 b1 b0
        psubsw    mm0, mm1            // -b3 -b2 b2 b3
        pmaddwd   mm2, xstatic1       // o1 << 15, o0, << 15
        pmaddwd   mm0, xstatic2       // o3 << 15, o2 << 15
        paddd     mm2, rounding       // proper rounding
        paddd     mm0, rounding       // proper rounding
        psrad     mm2, 15
        psrad     mm0, 15
        packssdw  mm2, mm0            // o3 o1 o2 o0
        pshufw    mm3, mm2, 11011000b // o3 o2 o1 o0
        movq      [ecx], mm3
    }
}
```

The rest of the N-point DCTs and IDCTs are left as exercise for the readers. Our final implementation boosts the SA-DCT/SA-IDCT process by 1.1–1.5 times in the MPEG-4 VOP-based coding scheme. The MMX technology versions of

Table 3 Performance Comparison of *N*-Point DCT and *N*-Point IDCT Implementations, Using Floating-Point, Integer, and MMX Instructions

		Time (μsec)			Speedup	
	N	Floating point	Integer	MMX instructions	From floating point	From integer
DCT	2	126	83	60	2.10	1.38
	3	110	104	71	1.55	1.46
	4	143	138	71	2.01	1.94
	5	181	170	105	1.72	1.61
	6	214	203	110	1.94	1.84
	7	407	302	120	3.39	2.51
	8	440	346	115	3.82	3.00
IDCT	2	83	77	60	1.38	1.28
	3	110	104	71	1.55	1.46
	4	154	143	71	2.16	2.01
	5	192	187	88	2.18	2.12
	6	231	214	121	1.90	1.76
	7	368	346	115	3.20	3.00
	8	374	296	126	2.96	2.34

the *N*-point DCTs performed from 1.3 to 3.0 times faster than the fixed-point versions, as shown in Table 3.

We also compare the performance of our MMX technology-optimized SA-DCT/SA-IDCT implementation and the performance of an MMX technology optimized 8×8 DCT/IDCT. SA-DCT is 1.5 times faster than the 8×8 DCT, even with the mask shifting overhead.* The eight-point DCT is slower than the lower-order DCTs.

4.4 Key Points Learned

1. It is important to maximize grouped operations. Sometimes, overreducing the computation in algorithms may result in poorer performance.
2. Our MMX technology-optimized four-point DCT runs as fast as the three-point DCT; our optimized eight-point DCT runs faster than the seven-point DCT. Several difficulties arise in the three-point DCT. The four-point DCT has groups of operations of two and four, each of which fits nicely into a single MMX instruction. The three-point DCT

---

* Assuming that all *N*-point routines are called with equal probability.

has a more irregular pattern that necessitates additional operand shuffling. Although the two PMADDWD instructions are effectively performing eight multiplies and four additions, only six multiplies and
three additions are actually needed. Thus, the three-point DCT in our
implementation is not faster than the four-point DCT. For all of the
N-point DCT and IDCT transforms, the more regular transforms
(8,4,6,2) will benefit more than the irregular transforms (3,5,7) from
MMX technology.

3. Vector lengths that are not multiples of four words or 8 bytes may be
problematic for MMX technology. Code can be simplified by padding
arrays of data with zero so that vector lengths become multiples of the
basic data.

4. There is a significant overhead associated with shuffling the operands
and the rounding necessary to minimize inaccuracies due to fixed-point
arithmetic.

5. The larger N-point transforms benefit more from MMX technology
optimization because using MMX instructions for any problem necessitates the additional overhead of packing, unpacking, and shuffling
data.

6. Mask shifting is more data dependent in the overall SA-DCT/
SA-IDCT process. It is harder to design this data-dependent module
using MMX instructions. Currently, instead of using MMX instructions, we use the scalar instructions to shift the mask and pixels.

5   CONCLUSIONS

MMX technology provides SIMD operations that operate on multiple operands
in one instruction and can produce up to an $8\times$ speedup (up to a $16\times$ speedup
in Intel Pentium 4 processors). One the other hand, there are some restrictions
in using MMX technology. Designing DSP algorithms for MMX technology is
an engineering challenge in a constrained space. MMX technology code generation and optimization is a complex and time-consuming process. The process
requires an understanding of the algorithms and the processor architecture. It is
an art and a science.

MMX instruction-level optimization steps can be summarized as follows:

1. Unroll loops, reorder/pair instructions, and use logic operations to improve parallelism.

2. Determine the minimal suitable precision for the processed values and
the corresponding packed data type for SIMD processing.

3. Arrange data in the best way for SIMD processing (rowwise arrangements).

4. Use an optimization tool (such as Intel VTune™ Performance Analyzer) to help fine-tune the code.

In order to accomplish the best performance, it is important to use algorithm-level optimization along with MMX instruction-level optimization. The key to MMX technology algorithmic optimization is to match the algorithms with the MMX instruction capabilities.

This chapter covers the basic principles of DSP–MMX technology cooptimization and provides some examples. For more MMX programming information and examples, readers are encouraged to take a look at *The Complete Guide to MMX Technology* by Bistry et al. [28]. In addition to detailed information on MMX instructions and programming, it has sample code for Viterbi sorting, finite impulse response (FIR) filter, a 3×3 linear separable image filter, the auto-correlation step of linear predictive coding (LPC) analysis for speech processing, and so forth. Additionally, DirectX®, RDX, RSX, and MMX Technology by Coelho and Hawash [20] also provides some valuable information on Intel processors and performance optimization. It also demonstrates the use of the Intel VTune Performance Enhancement Environment in detail.

Readers who are interested in software implementation of MPEG video and audio decoders are encouraged to take a look at Refs. 37, 38, and 42. In "Optimizing Software MPEG-2 Video Decoder" [37], Chen et al. provide a step-by-step approach to improve the performance of the reference decoder created by the MPEG Software Simulation Group. Their method exploits various ways to take advantage of the features of MMX technology. In "A Fast Integer-Based, CPU Scalable, MPEG-1 Layer-2 Audio Decoder" [38], Hans et al. developed a new IDCT algorithm which uses integer operations instead of floating-point operations for audio. Although maintaining similar accuracy, the integer implementation leads to an efficient implementation in MMX technology.

Finally, the Intel Corporation offers free implementation examples and optimized libraries to provide developers a jumpstart to building new applications using MMX technology. More implementation examples, including fast implementation of DCT/IDCT, can be found in the Intel Application Notes [7,31,32,39]. The free MMX technology-optimized Performance Library Suite covers a variety of image processing, mathematical, and DSP functions. It is available on-line [40].

ACKNOWLEDGMENT

The authors would like to thank James C. Abel, Intel Corporation, for his extensive and precious suggestions in the early stage of this work.

REFERENCES

1. RE Owen, D Martin. A uniform analysis method for DSP architectures and instruction sets with a comprehensive example. Proceedings of IEEE Workshop on Signal Processing Systems, 1998, pp 528–537.
2. A Peleg, U Weiser. The MMX technology extension to the Intel architecture. IEEE Micro, 16(4):42–50, 1996.
3. Intel Corp. Intel® Architecture MMX™ Technology Developer's Manual. IL: Intel Corporation, 1996. (Order No. 243006-001.)
4. Intel Corp. Intel® Architecture MMX™ Technology Programmer's Reference Manual. IL: Intel Corporation, 1996. (Order No. 243007-002.)
5. Intel Corp. Intel® Architecture Optimization Manual. IL: Intel Corporation, 1997.
6. Intel Corp. Intel® Architecture Software Developer's Manual. IL: Intel Corporation, 1997. (Order No. 243191.)
7. Intel Corp. Intel Streaming SIMD Extensions Application Notes. IL: Intel Corporation (available on-line: http://developer.intel.com/software/products/itc/strmsimd/ sseappnots.htm).
8. S Thakkar, T Huff. Internet Streaming SIMD Extensions. IEEE Computer 32(12): 26–34, 1999.
9. C Dichter, GJ Hinton. Optimizing for the Willamette processor. Intel Developer UPDATE Mag 6:3–5, March 2000 (http://developer.intel.com/update/departments/ initech/it03003.pdf).
10. Intel Corp. IA-32 Intel® Architecture Software Developer's Manual Volume 1: Basic Architecture. IL: Intel Corporation; 2000. (Order No. 245470.)
11. Intel Corp. IA-32 Intel® Architecture Software Developer's Manual Volume 2: Instruction Set Reference. IL: Intel Corporation, 2000. (Order No. 245471.)
12. M Atkins, R Subramanism. PC software performance tuning. IEEE Computer 29(9): 47–54, 1996.
13. Y-K Chen, SY Kung. Multimedia signal processors: An architectural platform with algorithmic compilation. J VLSI Signal Process Syst 20(1/2):183–206, 1998.
14. SY Kung. VLSI Array Processor, Englewood Cliffs, NJ: Prentice-Hall, 1988.
15. M Bierling. Displacement estimation by hierarchical block matching. Proceedings of SPIE Visual Communications and Image Processing, 1988, vol. 1001, pp 942–951.
16. J Ju, Y-K Chen, SY Kung. A fast algorithm for very low bit rate video coding. IEEE Trans Circuits Syst Video Technol 9(7):994–1002, 1999.
17. M Fomitchev. MMX technology code optimization. Dr. Dobb's J 303:38–48, September 1999.
18. J Khazam, B Bachmayer. Programming strategies for Intel's MMX. BYTE 21(8): 63–64, 1996.
19. JE Lecky. Using MMX technology to speed up machine vision algorithms. Imaging Technology Tutorials (available on-line: http://www.imaging.com/tutorials/ 00000009/tutorial.html).
20. R Coelho, M Hawash. DirectX®, RDX, RSX, and MMX™ Technology: A Jumpstart Guide to High Performance APIs. Reading, MA: Addison-Wesley, 1998.

21. Intel Corp. Intel® Vtune™ Performance Analyzer. IL: Intel Corporation (available on-line: http://developer.intel.com/software/products/vtune/).

22. Y-K Chen, M Holliman, W Macy, M Yeung. Real-time detection of video watermark on Intel architecture. Proceedings of SPIE Conference on Security and Watermarking on Multimedia Contents, 2000, Vol. 3971, pp 198–208.

23. R Bhargava, LK John, BL Evans, R Radhakrishnan. Evaluating MMX technology using DSP and multimedia applications. Proceedings of International Symposium on Microarchitecture, 1998, pp 37–46.

24. M Holliman, W Macy, MM Yeung. Robust frame-dependent video watermarking. Proceedings of SPIE Conference on Security and Watermarking of Multimedia Contents, 2000, Vol. 3917.

25. MM Yeung. MPL: MPEG processing library—Tools and advanced technology for video-centric applications. Intel Developer Forum, 1999.

26. International Standard Organization. Information Technology—Coding of Audio-Visual Objects, Part 2—Visual. ISO/IEC 14496-2.

27. C Heer, K Migge. VLSI hardware accelerator for the MPEG-4 padding algorithm. Proceedings of SPIE Conference on Media Processors, 1999, Vol. 3655, pp 113–119.

28. D Bistry, C Dulong, M Gutman, M Julier, M Keith, L Mennemeier, M Mittal, A Peleg, U Weiser. The Complete Guide to MMX Technology. New York: McGraw-Hill, 1997.

29. W-H Chen, CH Smith, SC Fralick. A fast computational algorithm for the discrete cosine transform. IEEE Trans Commun COM-25 (9):1004–1009, 1977.

30. P Pirsch, N Demassieux, W Gehrke. VLSI architectures for video compression—a survey. Proc IEEE 83(2):220–246, 1995.

31. Intel Corp. A fast precise implementation of 8×8 discrete cosine transform using the Streaming SIMD Extensions and MMX instructions. Intel Application Notes AP-922, 1999.

32. Intel Corp. Using MMX Instructions in a Fast iDCT Algorithm for MPEG Decoding. Intel Application Notes AP-528. IL: Intel Corporation, 1996.

33. E Murata, M Ikekawa, I Kuroda. Fast 2D IDCT implementation with multimedia instructions for a software MPEG2 decoder. Proceedings of ICASSP, 1998, Vol. 5, pp 3105–3108.

34. Y-S Tung, C-C Ho, J-L Wu. MMX-based DCT and MC algorithms for real-time pure software MPEG decoding. Proceedings IEEE International Conference on Multimedia Computing and Systems, 1999, Vol. 1, pp 357–362.

35. M Owzar, M Talmi, G Heising, P Kauff. Evaluation of SA-DCT hardware complexity. ISO/IEC JTC1/SC29/WG11 MPEG99/M4407, March 1999.

36. P Kauff, K Schuur, M Zhou. Experimental results on a fast SA-DCT implementation. ISO/IEC JTC1/SC29/WG11 MPEG99/M3569, July 1998.

37. H Chen, K Li, B Wei. Optimizing software MPEG-2 video decoder, unpublished, 2000.

38. M Hans, V Bhaskaran. A fast integer-based, CPU scalable, MPEG-1 layer-II audio decoder. Proceedings of 101st AES Convention, 1996.

39. Intel Corp. MMX Technology Application Notes. IL: Intel Corporation (available

on-line: http://developer.intel.com/software/idap/resources/technical_collateral/mmx).

40. Intel Corp. Intel® Performance Library Suite. IL: Intel Corporation (available on-line: http://developer.intel.com/software/products/perflib/index.htm).

41. JC Abel, MA Julier, MMX-enabled Dolby digital decoder. IEEE Signal Process 17(2):36–42, 2000.

42. F Casalino, GD Cagno, R Luca. MPEG-4 video decoder optimization. Proceedings of International Conference on Multimedia Computing and Systems, 1999, Vol. 1, pp 363–368.

8

Hardware/Software Cosynthesis of DSP Systems

Shuvra S. Bhattacharyya
University of Maryland at College Park, College Park, Maryland

This chapter focuses on the automated mapping of high-level specifications of digital signal processing (DSP) applications into implementation platforms that employ programmable DSPs. Since programmable DSPs are often used in conjunction with other types of programmable processor, such as microcontrollers and general-purpose microprocessors, and with various types of hardware module, such as field programmable gate arrays (FPGAs), and application-specific integrated circuit (ASIC) circuitry, this mapping task, in general, is one of *cosynthesis*—the joint synthesis of both hardware and software—for a heterogeneous multiprocessor.

Because a large variety of cosynthesis techniques have been developed to date, it is not possible here to provide comprehensive coverage of the field. Instead, we focus on a subset of topics that are central to DSP-oriented cosynthesis—application modeling, hardware/software partitioning, synchronization optimization, and block processing. Some important topics related to cosynthesis that are not covered here include memory management [1–5], which is discussed in Chapter 9; DSP code generation from procedural language specifications [6], which is the topic of Chapter 6; and performance analysis [7–9].

Additionally, we focus on synthesis from *coarse-grain data flow models* due to the increasing importance of such modeling in DSP design tools and the ability of such modeling to expose valuable, high-level structure of DSP applications that are difficult to deduce from within compilers for general-purpose programming models and other types of model. Thus, we do not explore techniques for fine-grain cosynthesis [10], including synthesis of application-specific instruction processors (ASIPs) [11], nor do we explore cosynthesis for control-dominant systems, such as those based on procedural language specifications [12], communicating sequential processes [13], and finite-state machine models [14]. All of these are important directions within cosynthesis research, but they do not fit centrally within the DSP-oriented scope of this chapter.

Motivation for coarse-grain data flow specification stems from the growing trend toward specifying, analyzing, and verifying embedded system designs in terms of domain-specific concurrency models [15], and the increasing use of data-flow-based concurrency models in high-level design environments for DSP system implementation. Such design environments, which enable DSP systems to be specified as hierarchies of block diagrams, offer several important advantages, including intuitive appeal, and natural support for desirable software engineering practices such as library-based design, modularity, and design reuse.

Potentially, the most useful benefit of data-flow-based graphical programming environments for DSP is that carefully specified graphical programs can expose coarse-grain structure of the underlying algorithm, and this structure can be exploited to facilitate synthesis and formal verification in a wide variety of ways. For example, the cosynthesis tasks of *partitioning* and *scheduling*—determining the resources on which the computations in an application will execute and the execution ordering of computations assigned to the same resource—typically have a large impact on all of the key implementation metrics of a DSP system. A data-flow-based system specification exposes high-level partitioning and scheduling flexibility that is often not possible to deduce manually or automatically from procedural language (e.g., assembly language or C) specifications. This flexibility can be exploited by cosynthesis tools to streamline an implementation based on the given set of performance and cost objectives. We will elaborate on partitioning and scheduling of data-flow-based specifications in Sections 3, 4, and 6.

The organization of the remainder of this chapter is as follows. We begin with a brief summary of our notation in working with fundamental, discrete math concepts. Then, we discuss the principles of coarse-grain data flow modeling that underlie many high-level DSP design tools. This discussion includes a detailed treatment of synchronous data flow and cyclo-static data flow, which are two of the most popular forms of data flow employed in DSP design. Next, we review three techniques—GCLP, COSYN, and the evolutionary algorithm approach of CodeSign—for automated partitioning of coarse-grain data flow specifications into hardware and software. In Section 5, we present an overview of techniques for efficiently synchronizing multiple processing elements in heterogeneous multiprocessor systems, such as those that result from hardware/software cosynthesis, and in Section 6, we discuss techniques for optimizing the application of *block processing*, which is a key opportunity for improving the throughput of cosynthesis solutions. Finally, we conclude in Section 7 with a summary of the main developments in the chapter. Throughout the chapter, we occasionally incorporate minor semantic modifications of the techniques that we discuss—without changing their essential behavior—to promote conciseness, clarity, and more uniform notation.

1 BACKGROUND

We denote the set of non-negative integers $\{0, 1, 2, \ldots\}$ by the symbol $\aleph$, the set of extended non-negative integers ($\aleph \cup \{\infty\}$) by $\overline{\aleph}$, the set of positive integers by Z^+, the set of extended integers ($\{-\infty, \infty\} \cup \{\ldots, -1, 0, 1, \ldots\}$) by $\overline{Z}$, and the cardinality of (number of elements in) a finite set S by $|S|$. By a *directed graph*, we mean an ordered pair (V, E), where V is a set of objects called *vertices* and E is a set of ordered pairs, called *edges*, of elements in V. We use the usual pictorial representation of directed graphs in which circles represent vertices and arrows represent edges. For example, Figure 1 represents a directed graph with vertex set $V = \{a, b, c, d, e, f, g, h\}$ and edge set

$$E = \{(c, a), (b, c), (a, b), (b, h), (d, f), (f, e), (e, f), (e, d)\} \tag{1}$$

If $e = (v_1, v_2)$ is an edge in a directed graph, we write $\text{src}(e) = v_1$, and $\text{snk}(e) = v_2$, and we say that $\text{src}(e)$ is the *source* vertex of e and $\text{snk}(e)$ is the *sink* vertex of e; e is *directed from* $\text{src}(e)$ *to* $\text{snk}(e)$; e is an *outgoing edge* of $\text{src}(e)$; and e is an *incoming edge* of $\text{snk}(e)$.

Given a directed graph $G = (V, E)$ and a vertex $v \in V$, we define the *incoming* and *outgoing edge sets* of v by

$$\text{in}(v) = \{e \in E | \text{snk}(e) = v\} \quad \text{and} \quad \text{out}(v) = \{e \in E | \text{src}(e) = v\} \tag{2}$$

respectively. Furthermore, given two vertices v_1 and v_2 in G, we say that v_1 is a *predecessor* of v_2 if there exists $e \in E$ such that $\text{src}(e) = v_1$ and $\text{snk}(e) = v_2$; we say that v_1 is a *successor* of v_2 if v_2 is a predecessor of v_1; and we say that

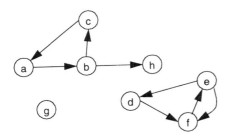

Figure 1 An example of a directed graph.

v_1 and v_2 are *adjacent* if v_1 is a successor or predecessor of v_2. A *path* in (V, E) is a finite sequence $(e_1, e_2, \ldots, e_n) \in E$ such that for $i = 1, 2, \ldots, (n - 1)$,

$$\text{snk}(e_i) = \text{src}(e_{i+1}) \tag{3}$$

Thus, $((a, b))$, $((d, f), (f, e), (e, f), (f, e))$, $((b, c), (c, a), (a, b))$, and $((a, b), (b, h))$ are examples of paths in Figure 1.

We say that a path $p = (e_1, e_2, \ldots, e_n)$ *originates at* the vertex $\text{src}(e_1)$ and *terminates at* $\text{snk}(e_n)$, and we write

$$\begin{aligned}\text{edges}(p) &= \{e_1, e_2, \ldots, e_n\} \\ \text{vertices}(p) &= \{\text{src}(e_1), \text{src}(e_2), \ldots, \text{src}(e_n), \text{snk}(e_n)\}\end{aligned} \tag{4}$$

A *cycle* is a path that originates and terminates at the same vertex. A cycle $(e_1, e_2, \ldots, e_n)$ is a *simple cycle* if $\text{src}(e_i) \neq \text{src}(e_j)$ for all $i \neq j$. In Figure 1, $((c, a), (a, b), (b, c))$, $((a, b), (b, c), (c, a))$, and $((f, e), (e, f))$ are examples of simple cycles. The path $((d, f), (f, e), (e, f), (f, e), (e, d))$ is a cycle that is not a simple cycle.

By a *subgraph* of a directed graph $G = (V, E)$, we mean the directed graph formed by any subset $V' \subseteq V$ together with the set of edges $\{e \in E | (\text{src}(e), \text{snk}(e) \in V')\}$. For example, the directed graph

$$(\{e, f\}, \{(e, f), (f, e)\}) \tag{5}$$

is a subgraph of the directed graph shown in Figure 1.

Given a directed graph $G = (V, E)$, a sequence of vertices $(v_1, v_2 \ldots, v_k)$ is a *chain* that joins v_1 and v_k if v_{i+1} is adjacent to v_i for $i = 1, 2, \ldots, (k - 1)$. We say that a directed graph is *connected* if for any pair of distinct members A and B of V, there is a chain that joins A and B. Thus, the directed graph in Figure 1 is not connected (e.g., because there is no chain that joins g and b), whereas the subgraph associated with the vertex subset $\{a, b, c, h\}$ is connected.

A *strongly connected* directed graph C has the property that between every distinct pair of vertices w and v in C, there is a directed path from w to v and a directed path from v to w. A *strongly connected component* (SCC) of a directed graph is a maximal strongly connected subgraph. The directed graph in Figure 1 contains four SCCs. Two of these SCCs, $(\{g\}, \varnothing)$ and $(\{h\}, \varnothing)$, are called *trivial* SCCs because each contains a single vertex and no edges. The other two SCCs in Figure 1 are the directed graphs (V_1, E_1) and (V_2, E_2), where $V_1 = \{a, b, c\}$, $E_1 = \{(a, b), (b, c), (c, a)\}$, $V_2 = \{d, e, f\}$, and $E_2 = \{(e, f), (f, e), (e, d), (d, f)\}$.

Many excellent textbooks, such as Refs. 16 and 17, provide elaboration on the graph-theoretic fundamentals summarized in this section.

2 COARSE-GRAIN DATA FLOW MODELING FOR DSP

2.1 Data Flow Modeling Principles

In the data flow paradigm, a computational specification is represented as a directed graph. Vertices in the graph (called *actors*) correspond to computational modules in the specification. In most data-flow-based DSP design environments, actors can be of arbitrary complexity. Typically, they range from elementary operations such as addition or multiplication to DSP subsystems such as fast Fourier transform (FFT) units or adaptive filters.

An edge (v_1, v_2) in a data flow graph represents the communication of data from v_1 to v_2. More specifically, an edge represents a FIFO (first-in first-out) queue that buffers data values (*tokens*) as they pass from the output of one actor to the input of another. When data flow graphs are used to represent signal processing applications, a data flow edge e has a non-negative integer delay $del(e)$ associated with it. The delay of an edge gives the number of initial data values that are queued on the edge. Each unit of data flow delay is functionally equivalent to the z^{-1} operator in DSP: the sequence of data values $\{y_n\}$ generated at the input of the actor $snk(e)$ is equal to the shifted sequence $\{x_{n-del(e)}\}$, where $\{x_n\}$ is the data sequence generated at the output of the actor $src(e)$.

A data flow actor is *enabled* for execution any time it has sufficient data on its incoming edges (i.e., in the associated FIFO queues) to perform its specified computation. An actor can execute (*fire*) at any time when it is enabled (*data-driven execution*). In general, the execution of an actor results in some number of tokens being removed (*consumed*) from each incoming edge and some number being placed (*produced*) on each outgoing edge. This production activity, in general, leads to the enabling of other actors.

The order in which actors execute is not part of a data flow specification and is constrained only by the simple principle of data-driven execution defined earlier. This is in contrast to many alternative programming models, such as those that underlie procedural languages, in which execution order is *overspecified* by the programmer [18]. The actor execution order for a data flow specification may be determined at compile time (if sufficient static information is available), at run time, or using a mixture of compile-time and run-time techniques.

2.2 Synchronous Data Flow

Synchronous data flow (SDF), introduced by Lee and Messerschmitt [19], is the simplest and, currently, the most popular form of data flow modeling for DSP design. SDF imposes the restriction that the number of data values produced by an actor onto each outgoing edge is constant; similarly, the number of data values consumed by an actor from each incoming edge is constant. Thus, an SDF edge e has two additional attributes: the number of data values produced onto e by each firing of the source actor, denoted prd(e), and the number of data values consumed from e by each firing of the sink actor, denoted cns(e).

Example 1

A simple example of an SDF abstraction is shown in Figure 2. Here, each edge is annotated with the numbers of data values produced and consumed by the source and sink actors, respectively. For example, prd((B, C)) = 1 and cns ((B, C)) = 2. The "2D" next to the edge (D, E) represents two units of delay. Thus, del ((D, E)) = 2.

The restrictions imposed by the SDF model offer a number of important advantages, including (1) *static scheduling*, which avoids the execution time and power consumption overhead and the unpredictability of dynamic scheduling approaches, and (2) decidability of key verification problems—in particular, determination of bounded memory requirements and deadlock avoidance. These two verification problems are critical in the development of DSP applications because DSP systems involve iterative operation on vast, often unbounded, sequences of input data. Not all SDF graphs permit *admissible* operation on unbounded input sets (i.e., operation without deadlock and without unbounded data accumulation on one or more edges). However, it can always be determined at compile time whether or not admissible operation is possible for a given SDF graph. In exchange for its strong advantages, the SDF model has limited expressive power—not all applications can be expressed in the model.

A necessary and sufficient condition for admissible operation to be possible for an SDF graph is the existence of a *valid schedule* for the graph, which is a finite sequence of actor firings that executes each actor at least once, fires actors only after they are enabled, and produces no net change in the number of tokens queued on each edge. SDF graphs for which valid schedules exist are called *consistent* SDF graphs.

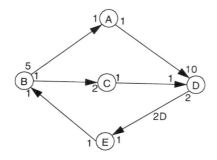

Figure 2   An example of an SDF graph.

Efficient algorithms have been developed by Lee and Messerschmitt [19] to determine whether or not a given SDF graph is consistent and to determine the minimum number of times that each actor must be fired in a valid schedule. We represent these minimum numbers of firings by a vector (called the *repetitions vector*) $\mathbf{q}_G$, indexed by the actors in G (we often suppress the subscript if G is understood). These minimum numbers of firings can be derived by finding the minimum positive integer solution to the *balance equations* for G, which specify that $\mathbf{q}$ must satisfy

$$\mathbf{q}(\text{src}(e)) \times \text{prd}(e) = \mathbf{q}(\text{snk}(e)) \times \text{cns}(e) \quad \text{for every edge } e \text{ in } G \qquad (6)$$

Associated with any valid schedule S, there is a positive integer $J(S)$ such that S fires each actor A exactly $(J(S) \times \mathbf{q}(A))$ times. This number $J(S)$ is referred to as the *blocking factor* of S.

Given a consistent SDF graph G, the *total number of samples exchanged* (per schedule iteration) on an SDF edge e in G, denoted $\text{TNSE}_G(e)$, is defined by the equal-valued products in the left-hand side and right-hand side of Eq. (6); that is,

$$\text{TNSE}_G(e) = \mathbf{q}(\text{src}(e)) \times \text{prd}(e) = \mathbf{q}(\text{snk}(e)) \times \text{cns}(e) \qquad (7)$$

Given a subset X of actors in G, the repetitions count of X, denoted $q_G(X)$, is defined by

$$q_G(X) \equiv gcd(\{\mathbf{q}_G(A) \mid A \in X\})$$

where *gcd* denotes the *greatest common divisor* operator.

Example 2

Consider again the SDF graph of Figure 2. The repetitions vector of this graph is given by

$$\mathbf{q}(A, B, C, D, E) = (10, 2, 1, 1, 2) \qquad (8)$$

Additionally, we have $\text{TNSE}_G((A, D)) = 10$ and $\text{TNSE}_G((B, C)) = 2$.

If a repetitions vector exists for an SDF graph but a valid schedule does not exist, then the graph is deadlocked. Thus, an SDF graph is consistent if and only if a repetitions vector exists and the graph is not deadlocked. For example, if we reduce the number of delays on the edge (D, E) in Figure 2 (without adding delay to any of the other edges), then the graph will become deadlocked.

In summary, SDF is currently the most widely used data flow model in commercial and research-oriented DSP design tools. Although SDF has limited expressive power, the model has proven to be of great practical value in the domain of signal processing and digital communication. SDF encompasses a broad and important class of applications, including modems, digital audio broadcasting systems, video encoders, multirate filter banks, and satellite receiver systems, just to name a few [2,19–23]. Commercial tools that employ SDF semantics include Simulink by The Math Works, SPW by Cadence, and ADS by Hewlett

Packard. SDF-based research tools include Gabriel [24] and several key domains in Ptolemy [25], from the University of California, Berkeley, and ASSIGN from Carnegie Mellon [26]. Except where otherwise noted, all of the cosynthesis techniques discussed in this chapter are applicable to SDF-based specifications.

2.3 Alternative Data Flow Models

To address the limited expressive power of SDF, a number of alternative data flow models have been investigated for the specification of DSP systems. These can be divided into three major groups: *the decidable data flow models*, which, like SDF, enable bounded memory and deadlock determination to be solved at compile time; *the dynamic data flow models*, in which there is sufficient dynamism and expressive power that the bounded memory and deadlock problems become undecidable; and the *data flow meta-models*, which are model-independent mechanisms for adding expressive power to broad classes of data flow-modeling approaches. Decidable data flow models include SDF; *cyclostatic data flow* [21] and *scalable synchronous data flow* [27], which we discuss in Sections 2.4 and 6, respectively; and *multidimensional synchronous data flow* [28] for expressing multidimensional DSP applications, such as those arising in image and video processing. Dynamic data flow models include *Boolean data flow and integer-controlled data flow* [29,30], and *bounded dynamic data flow* [31]. Meta-modeling techniques relevant to data flow include the *starcharts* approach [32], which provides flexible integration of finite-state machine and data flow models, and *parameterized data flow* [33,34], which provides a general mechanism for incorporating dynamic reconfiguration capabilities into arbitrary data flow models.

2.4 Cyclostatic Data Flow

Cyclostatic data flow (CSDF) and scalable synchronous data flow (described in Sec. 6) are presently the most widely used alternatives to SDF. In CSDF, introduced by Bilsen et al., the number of tokens produced and consumed by an actor is allowed to vary as long as the variation takes the form of a fixed, periodic pattern [21]. More precisely, each actor A in a CSDF graph has associated with it a fundamental period $\tau(A) \in Z^+$, which specifies the number of *phases* in one minimal period of the cyclic production/consumption pattern of A. For each incoming edge e of A, the scalar SDF attribute cns(e) is replaced by a $\tau(A)$-tuple $(C_{e,1}, C_{e,2}, \ldots, C_{e,\tau(A)})$, where each $C_{e,i}$ is a non-negative integer that gives the number of data values consumed from e by A in the ith phase of each period of A. Similarly, for each outgoing edge e, prd(e) is replaced by a $\tau(A)$-tuple $(P_{e,1}, P_{e,2}, \ldots, P_{e,\tau(A)})$, which gives the numbers of data values produced in successive phases of A.

Example 3

A simple example of a CSDF actor is a conventional downsampler actor from multirate signal processing. Functionally, a downsampler actor (with downsampling factor N) has one incoming edge and one outgoing edge and performs the function $y[i] = x[N(i - 1) + 1]$, where for $k \in Z^+$, $y[k]$ and $x[k]$ denote the kth data values produced and consumed, respectively, by the actor. Thus, for every input value that is copied to the output, $N - 1$ input values are discarded. This functionality can be specified by a CSDF actor that has N phases. A data value is consumed from the incoming edge for all N phases, resulting in the N-component *consumption tuple* $(1, 1, \ldots, 1)$; however, a data value is produced onto the outgoing edge only on the first phase, resulting in the *production tuple* $(1, 0, \ldots, 0)$.

   Like SDF, CSDF permits efficient verification of bounded memory requirements and deadlock avoidance [21]. Furthermore, static schedules can always be constructed for consistent CSDF graphs.
   A CSDF actor A can easily be converted into an SDF actor A' such that if identical sequences of input data values are applied to A and A', then identical output data sequences result. Such a functionally equivalent SDF actor A' can be derived by having each firing of A' implement one fundamental CSDF period of A [i.e., $\tau(A)$ successive phases of A]. Thus, for each incoming edge e' of A', the SDF parameters of e' are given by

$$\text{del}(e') = \text{del}(e); \qquad \text{prd}(e') = \sum_{i=1}^{\tau(A)} P_{e,i}$$

and similarly,

$$\text{cns}(e') = \sum_{i=1}^{\tau(A)} C_{e,i} \tag{9}$$

where e is the corresponding incoming edge of the CSDF actor A.
   Because any CSDF actor can be converted in this manner to a functionally equivalent SDF actor, it follows that CSDF does not offer increased expressive power at the level of individual actor functionality (input–output mappings). However, the CSDF model does offer increased flexibility in compactly and efficiently representing interactions between actors.

Example 4

As an example of increased flexibility in expressing actor interactions, consider the CSDF specification illustrated in Figure 3. This specification represents a recursive digital filter computation of the form

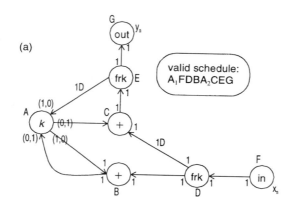

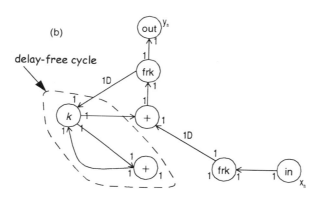

Figure 3 (a) An example that illustrates the compact modeling of resource sharing using CSDF. The actors labeled frk denote data flow "forks," which simply replicate their input tokens on all of their output edges. The top right portion of the figure gives a valid schedule for this CSDF specification. Here, A_1 and A_2 denote the first and second phases of the CSDF actor A, respectively. (b) The SDF version of the specification in (a). This graph is deadlocked due to the presence of a delay-free cycle.

$$y_n = k^2 y_{n-1} + k x_n + x_{n-1} \qquad (10)$$

In Figure 3, the two-phase CSDF actor labeled A represents a scaling (multiplication) by the constant factor k. In each of its two phases, actor A consumes a data value from one of its incoming edges, multiplies the data value by k, and produces the resulting value onto one of its outgoing edges. The CSDF specification of

Figure 3 thus exploits our ability to compute Eq. (10) using the equivalent formulation

$$y_n = k(ky_{n-1} + x_n) + x_{n-1} \tag{11}$$

which requires only addition actors and k-scaling actors. Furthermore, the two k-scaling operations contained in Eq. (11) are consolidated into a single CSDF actor (actor A).

Such consolidation of distinct operations from different data streams offers two advantages. First, it leads to more compact representations because fewer vertices are required in the CSDF graph. For large or complex applications, this can result in more intuitive representations and can reduce the time required to perform various analysis and synthesis tasks. Second, it allows a precise modeling of *resource sharing* decisions—prespecified assignments of multiple operations in a DSP application onto individual hardware resources (such as functional units) or software resources (such as subprograms)—within the framework of data flow. Such prespecified assignments may arise from constraints imposed by the designer and from decisions taken during synthesis or design space exploration.

Another advantage offered by CSDF that is especially relevant to cosynthesis tasks is that by decomposing actors into a finer level (phase-level) of specification granularity, basic behavioral optimizations such as *constant propagation and dead code elimination* [35,36] are facilitated significantly [37]. As a simple example of dead code elimination with CSDF, consider the CSDF specification shown in Figure 4a of a multirate finite impulse response (FIR) filtering system that is expressed in terms of basic multirate building blocks. From this graph, the equivalent "acyclic precedence graph" (APG) shown in Figure 4b, can be derived using concepts discussed in Refs. 19 and 21. In the CSDF APG, each actor corresponds to a single phase of a CSDF actor or a single firing of an SDF actor within a valid schedule. We will discuss the APG concept in more detail in Section 3.1.

From Figure 4b, it is apparent that the results of some computations (SDF firings or CSDF phases) are never needed in the production of any of the system outputs. Such computations correspond to dead code and can be eliminated during synthesis without compromising correctness. For this example, the complete set of subgraphs that correspond to dead code is illustrated in Figure 4b. Parks et al. show that such "dead subgraphs" can be detected with a straightforward algorithm [37].

Other advantages of CSDF include improved support for hierarchical specifications and more economical data buffering [21].

In summary, CSDF is a useful generalization of SDF that maintains the properties of efficient verification and static scheduling while offering a richer range of interactor communication patterns and improved support for basic behavioral optimizations. CSDF concepts were introduced in the GRAPE design

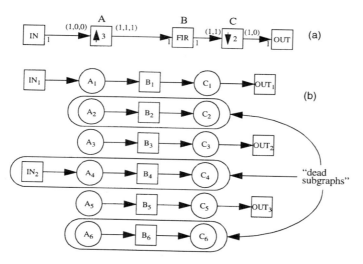

Figure 4   An example of efficient dead code elimination using CSDF.

environment [55], which is a research tool developed at K. U. Leuven, and are currently used in a number of commercial design tools such as DSP Canvas by Angeles Design Systems, and Virtuoso Synchro by Eonic Systems.

3   MULTIPROCESSOR IMPLEMENTATION OF DATA FLOW MODELS

A fundamental task in synthesizing hardware and software from a data flow specification is that of scheduling, which, as described in Section 2.2, refers to the process of determining the order in which actors will be executed. During co-synthesis, it is often desirable to obtain efficient, *parallel* implementations, which execute multiple actor firings simultaneously on different resources.

For this purpose, the class of "valid schedules" introduced in Section 2.2 is not sufficient; *multiprocessor schedules*, which consist of multiple firing sequences—one for each processing resource—are required. However, the consistency concepts developed in Section 2.2 are inherent to SDF specifications and apply regardless of whether or not parallel implementation is used. In particular, when performing static, multiprocessor scheduling of SDF graphs, it is still necessary to first compute the repetitions vector and to verify that the graph is deadlock-free, and the techniques for accomplishing these objectives are no different for the multiprocessor case.

However, there are a number of additional considerations that arise when attempting to construct and implement multiprocessor schedules. We elaborate on these in the remainder of this section.

3.1 Precedence Expansion Graphs

Associated with any connected, consistent SDF graph G, there is a unique directed graph, called its *equivalent acyclic precedence graph* (APG), that specifies the precedence relationships between distinct actor firings throughout an iteration of a valid schedule for G [19]. Cosynthesis algorithms typically operate on this APG representation because it fully exposes interfiring concurrency, which is hidden in the more compact SDF representation. The APG can thus be viewed as an intermediate representation when performing cosynthesis from an SDF specification.

Each vertex of the APG corresponds to an actor firing within a single iteration period of a valid schedule. Thus, for each actor A in an SDF graph, there are $q(A)$ corresponding vertices in the associated APG. For each $i = 1, 2, \ldots,$ $q(A)$, the vertex associated with the ith firing of A is often denoted as A_i. Furthermore, there is an APG edge directed from the vertex corresponding to firing A_i to the vertex corresponding to firing B_j if and only if at least one token produced by A_i is consumed by B_j.

Example 5

As a simple example, Figure 5 shows an SDF graph and its associated APG.

For an efficient algorithm that systematically constructs the equivalent APG from a consistent SDF graph, we refer the reader to Ref. 39. Similar techniques can be employed to map CSDF specifications into equivalent APG representations.

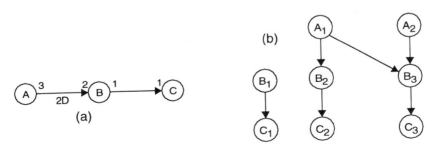

Figure 5 (a) An SDF graph and (b) its equivalent APG.

We refer to an APG representation of an SDF or CSDF application specification as a *data flow application graph* or, simply, an *application graph*. In other words, an application graph is an application specification in which each vertex represents *exactly* one firing within a valid schedule for the graph. Additionally, when the APG is viewed in isolation (i.e., independent of any particular SDF graph), each vertex in the APG may be referred to as an *actor* without ambiguity.

3.2 Multiprocessor Scheduling Models

Cosynthesis requires two central tasks: *allocation* of resources (e.g., programmable processors, FPGA devices, and so-called "algorithm-based" computing modules [40]), and *scheduling* of application actors onto the allocated resources. The scheduling task can be further subdivided into three main operations: assigning actors to processors, ordering actors on each processor, and determining the time at which each actor begins execution. Based on whether these scheduling operations are performed at run time or compile time, we can classify multiprocessor scheduling strategies into four categories: fully static, static assignment, fully dynamic, and self-timed scheduling [41]. In fully static scheduling, all three scheduling operations are performed at compile time; in static allocation, only the processor assignment is performed at compile time; and in the fully dynamic approach, all three operations are completed at run time. As we move from fully static to fully dynamic scheduling, we trade off simplicity and lower run-time cost for increased generality.

For DSP systems, an efficient and popular scheduling model is the *self-timed* model [41], where we obtain a fully static schedule, but we ignore the precise timing that such a strategy would enforce. Instead, processors synchronize with one another only based on interprocessor communication (IPC) requirements. Such a strategy retains much of the reduced overhead of fully static scheduling, offers robustness when actor execution times are not constant or precisely known, improves efficiency by eliminating extraneous synchronization requirements, eliminates the need for specialized synchronization hardware, and naturally supports asynchronous design [8,41]. The techniques discussed in this chapter are suitable for incorporation in the context of fully static or self-timed scheduling.

3.3 Scheduling Techniques

Numerous scheduling algorithms have been developed for multiprocessor scheduling of data flow application graphs. Two general categories of scheduling techniques that are frequently used in cosynthesis approaches are *clustering* and *list scheduling*.

Clustering algorithms for multiprocessor scheduling operate by incrementally constructing groupings, called *clusters*, of actors that are to be executed on the same processor. Clustering and list scheduling can be used in a complementary fashion. Typically, clustering is applied to focus the efforts of a list scheduling algorithm on effective processor assignments. When used efficiently, clustering can significantly enhance the results produced by list scheduling and a variety of other scheduling techniques.

In list scheduling, *a priority list L* of actors is constructed; a global time clock c_G is maintained; and each actor T is eventually mapped into a time interval $[x_T, y_T]$ on some processor (the time intervals for two distinct actors assigned to the same processor cannot overlap). The priority list L is a linear ordering $(v_1, v_2, \ldots, v_{|V|})$ of the actors in the input application graph $G = (V, E)$ ($V = \{v_1, v_2, \ldots, v_{|V|}\}$) such that for any pair of distinct actors v_i and v_j, v_i is to be given higher scheduling priority than v_j if and only if $i < j$. Each actor is mapped to an available processor as soon as it becomes the highest-priority actor—according to L—among all actors that are *ready*. An actor is ready if it has not yet been mapped but its predecessors have all been mapped, and all satisfy $y_T \leq t$, where t is the current value of c_G. For self-timed implementation, actors on each processor are ordered according to the order of their associated time intervals.

A wide variety of actor prioritization schemes for list scheduling can be specified in terms of a *parameterized longest path function*

$$\lambda_G(A, f_v, f_e) \tag{12}$$

where $G = (V, E)$ denotes the application graph that is being scheduled; $A \in V$ is any actor in G; $f_v: V \to \overline{Z}$ is a function that maps application graph actors into (extended) integers (vertex weights); and, similarly, $f_e: E \to \overline{Z}$ is a function that maps application graph edges into integers (edge weights). The value of $\lambda_G(A, f_v, f_e)$ is defined to be

$$\max\left(\left\{ \left| \sum_{i=1}^{n} f_v(\text{snk}(e_i)) + \sum_{i=1}^{n} f_e(e_i) + f_v(A) \right| \\ (e_1, e_2, \ldots, e_n) \text{ is a path in } G \text{ that originates at } A \right\}\right) \tag{13}$$

Under this formulation, the priority of an actor is taken to be the associated value of $\lambda_G(*, f_v, f_e)$; in other words, the priority list for list scheduling is constructed in decreasing order of the metric $\lambda_G(*, f_v, f_e)$.

Example 6

If actor execution times are constant, $f_v(A)$ is taken to be the execution time of A, and f_e is taken to be the *zero function on E* [$f(e') = 0$ for all $e' \in E$], then

$\lambda_G(*, f_v, f_e)$ gives the famous *Hu-level* priority function [42], which is the value of the longest-cumulative-execution-time path that originates at a given actor. For homogeneous communication networks, another popular priority function is obtained by taking $f_e(e')$ to be the interprocessor communication latency associated with edge e' [the communication latency if src(e') and snk(e') are assigned to different processors] and, again, taking $f_v(A)$ to be the execution time of A. In the presence of nondeterministic actor execution times, common choices for f_v include the average and worst-case execution times.

4 PARTITIONING INTO HARDWARE AND SOFTWARE

This section focuses on a fundamental component of the cosynthesis process: the partitioning of application graph actors into hardware and software. Because partitioning and scheduling are, in general, highly interdependent, these two tasks are usually performed jointly. The net result is an allocation (if applicable) of hardware and software processing resources and communication resources, an assignment of application graph actors to allocated resources, and a complete schedule for the derived allocation/assignment pair. Here, we examine three algorithms, ordered in increasing levels of generality, that address the partitioning problem.

4.1 GCLP

The *global criticality, local phase* (GCLP) *algorithm* [43], developed by Kalavade and Lee, gives an approach for combined hardware/software partitioning and scheduling for minimum latency. Input to the algorithm includes an application graph $G = (V, E)$, a target platform consisting of a programmable processor and a fabric for implementing custom hardware, and constraints on the latency and on the code size of the software component. Each actor $A \in V$ is characterized by its execution time $t_h(A)$ and area $a_h(A)$ if implemented in hardware, and by its execution time $t_s(A)$ and code size $a_s(A)$ if implemented in software. The GCLP algorithm attempts to compute a mapping of graph actors into hardware and software and a schedule for the mapped actors. The objective is to minimize the area of the custom hardware subject to the constraints on latency and software code size.

At each iteration i of the algorithm, a ready actor is selected for mapping and scheduling based on a dynamic priority function $P_i: V \to \aleph$ that takes into account the relative difficulty (*time criticality*) in achieving the latency constraint based on the partial schedule S_i constructed so far. Increasing levels of time criticality translate to increased affinity for hardware implementation in the computation P_i of actor priorities. Because it incorporates the structure of the entire application graph and current scheduling state S_i, this affinity for hardware implementation is called the *global criticality*. We denote the value of global criticality computed at algorithm iteration i by $C_g(i)$.

Once a ready actor A_i is chosen for scheduling based on global criticality considerations, the hardware and software mapping alternatives for A_i are taken into account, based on so-called *local phase* information, to determine the most attractive implementation target (hardware or software) for A_i, and A_i is scheduled accordingly.

The global criticality metric $C_g(i)$ is derived by determining a tentative implementation target for each unscheduled actor in an effort to efficiently extend the partial schedule S_i into a complete schedule. The goal in this rough, schedule extension step is to determine the most economical subset H_i of unscheduled actors to implement in hardware such that the latency constraint is achieved. This subset is iteratively computed based on an actor–priority function that captures the area/time trade-offs for each actor and on a fast scheduling heuristic that computes the overall latency for a given hardware/software mapping.

Given H_i, the global criticality at iteration i is computed as an estimate of the fraction of overall computation in the set U_i of unscheduled actors that is contained in the tentatively hardware-mapped subset H_i:

$$C_g(i) = \frac{\sum_{A \in H_i} \text{ElemOps}(A)}{\sum_{A \in U_i} \text{ElemOps}(A)} \tag{14}$$

where $\text{ElemOps}(A)$ denotes the number of elementary operations (e.g., addition, multiplication, etc.) within actor A.

Once $C_g(i)$ is computed, the hardware mapping H_i is discarded and $C_g(i)$ is loosely interpreted as an *actor-invariant probability* that any given actor will be implemented in hardware. This probabilistic interpretation is applied to compute "critical path lengths" in the application graph, in which the implementation targets, and hence the execution times, of unscheduled actors are not yet known. More specifically, the actor that is selected for mapping and scheduling at algorithm iteration i is chosen to be one (ties are broken arbitrarily) that maximizes

$$\lambda_G(A, \tau_i, \varepsilon_0) \tag{15}$$

over all $A \in \text{Ready}(S_i)$, where λ_G is the parameterized longest path function defined by Eq. (13); $\tau_i: V \to \aleph$ is defined by

$$\tau_i(X) = C_g(i)t_h(X) + (1 - C_g(i))t_s(X); \tag{16}$$

$\varepsilon_0: E \to \{0\}$ is the zero function on E; and $\text{Ready}(S_i)$ is the set of application graph actors that are ready at algorithm iteration i. The "execution time estimate" given in Eq. (16) can be interpreted loosely as the expected execution time of actor X if one wishes to extend the partial schedule S_i into an economical implementation that achieves the given latency constraint.

4.1.1 Hardware/Software Selection Threshold

In addition to determining [via Eq. (16)] the actor A_i that is to be scheduled at algorithm iteration i, the global criticality $C_g(i)$ is used to determine whether A_i should be implemented in hardware or software. In particular, an actor-dependent cutoff point threshold(A_i) is computed such that if $C_g(i) \geq$ threshold(A_i), then A_i is mapped into hardware or software based on the alternative that results in the earliest completion time for A_i (based on the partial schedule S_i), whereas if $C_g(i) <$ threshold(A_i), then the mapping for A_i is chosen to be the one that results in the leanest resource consumption.

The objective function selection threshold associated with an actor A_i is computed as

$$\text{threshold}(A_i) = 0.5 + \text{LocalPhaseDelta}(A_i) \tag{17}$$

where LocalPhaseDelta (A_i) measures aspects of the specific hardware/software trade-offs associated with actor A_i. More specifically, this metric incorporates the classification of A_i as either an *extremity* actor, a *repeller* actor, or a "normal" actor. An extremity actor is either a software extremity or a hardware extremity. Intuitively, a software extremity is an actor whose software execution time (SET) is one of the highest SETs among all actors, but whose hardware implementation area (HIA) is *not* among the highest HIAs. Similarly, a hardware extremity is an actor whose HIA is one of the highest HIAs, but whose SET is *not* among the highest SETs. The precise methods to compute thresholds that determine the classes of "highest" SET and HIA values are parameters of the GCLP framework that are to be configured by the tool developer or the user.

An actor is a repeller with respect to software (hardware) implementation if it is not an extremity actor and its functionality contains components that are distinguishably ill-suited to efficient software (hardware) implementation. For example, *the bit-level instruction mix*, defined as the overall proportion of bit-level operations, has been identified as an actor property that is useful in identifying software repellers (*a software repeller property*). Similarly, the proportion of memory-intensive instructions is a hardware repeller property. For each such repeller property of a given repeller actor, a numeric estimate is computed to characterize the degree to which the property favors software or hardware implementation for the actor.

The LocalPhaseDelta value in Eq. (17) is set to zero for normal actors (i.e., actors that are neither extremity nor repeller actors). For extremity actors, the value is determined as a function of the SETs and HIAs, and for repeller actors, it is computed as

$$\text{LocalPhaseDelta}(A_i) = \frac{1}{2}(\phi_h - \varphi_s) \tag{18}$$

where φ_h and φ_s represent normalized, weighted sums of contributions from individual hardware and software repeller properties, respectively. Thus, for example, if the hardware repeller properties of actor A_i dominate ($\varphi_h > \varphi_s$), it becomes more likely [from Eqs. (17) and (18)] that $C_g(i) <$ threshold(A_i) and, thus, that A_i will be mapped to software (assuming that the communication and code size costs associated with software mapping are not excessive).

The overall appeal of the GCLP algorithm stems from its ability to integrate global, application- and partial-schedule-level information with the actor-specific, heterogeneous-mapping metrics associated with the local phase concept. Also, the scheduling, estimation, and mapping heuristics within the GCLP algorithm consider area and latency overheads associated with communication between hardware and software. Thus, the algorithm jointly considers actor execution times, hardware and software capacity costs, and both temporal and spacial costs associated with interprocessor communication.

4.1.2   Cosynthesis for Multifunction Applications

Kalavade and Subrahmanyam have extended the GCLP algorithm to handle cosynthesis involving multiple applications that are operated in a time-multiplexed manner [44]. Such *multifunction* systems arise commonly in embedded applications. For example, a video encoding system may have to be designed to support a variety of formats, such as MPEG2, H.261, and JPEG, based on the different modes of operation available to the user.

The *multiapplication codesign problem* is a formulation of multifunction cosynthesis in which the objective is to exploit similarities between distinct system functions to streamline the result of synthesis. An instance of this problem can be viewed as a finite set of inputs to the original GCLP algorithm described earlier in this section. More precisely, an instance of multiapplication codesign consists of a set of application graphs $appset = \{G_1, G_2, \ldots, G_N\}$, where each $G_i = (V_i, E_i)$ has an associated latency constraint L_i. Furthermore, if we define $V_{appset} = (V_1 \cup V_2 \cup \cdots \cup V_N)$, then each actor $A \in V_{appset}$ is characterized by its *node type* type(A), execution time $t_h(A)$ and area $a_h(A)$ if implemented in hardware, and execution time $t_s(A)$ and code size $a_s(A)$ if implemented in software. The objective is to construct an assignment of actors in V_{appset} into hardware and software, and schedules for all of the application graphs in *appset* such that the schedule for each G_i satisfies its associated latency constraint L_i, and overall hardware area is minimized. An underlying assumption in this codesign problem is that at any given time during operation, at most one of the application graphs in *appset* may be active.

The node-type attribute specifies the function class of the associated actor and is used to identify opportunities for resource sharing across multiple actors within the same application, as well as across actors in different applications. For example, if two application graphs each contain a DCT module (an actor whose node type is that of a DCT) and one of these is mapped to hardware, then it may be profitable to map the other DCT actor into hardware as well, especially because both DCT actors will never be active at the same time.

4.1.3 Modified Threshold Adjustment

Kalavade's "multifunction extension" to GCLP, which we call GCLP-MF, retains the global criticality concept and the threshold-based approach to mapping actors into hardware and software. However, the metrics associated with local phase computation (threshold adjustment) are replaced with a number of alternative metrics, called *commonality measures*, that take into account characteristics that are relevant to the multifunction case. These metrics are consistently normalized to keep their values within predictable and meaningful ranges.

Recall that higher values of the GCLP threshold favor software implementation, whereas lower values favor hardware implementation, and the threshold in GCLP is computed from Eq. (17) as the sum of 0.5 and an adjustment term, called the local phase. In GCLP, this local phase adjustment term is replaced by an alternative function that incorporates reuse of node types across different actors and applications, and actor-specific performance-area trade-offs. Type reuse is quantified by a *type repetitions* metric, denoted R, which gives the total number of actor instances of a given type over all application graphs in *appset*. In other words, for a given node type θ,

$$R(\theta) = \sum_{(V, E) \in \text{appset}} |\{A \in V | (\text{type}(A) = \theta)\}| \tag{19}$$

and the normalized form of this metric, which we denote R_N, is defined by normalizing to values restricted within [0, 1]:

$$R_N(\theta) = \frac{R(\theta)}{\max(\{R(\text{type}(A)) | (A \in V_{\text{appset}})\})} \tag{20}$$

Performance-area trade-off information is quantified by a metric T that measures the speedup in moving an actor implementation from software to hardware relative to the required hardware area:

$$T(A) = \frac{t_s(A) - t_h(A)}{a_h(A)} \quad \text{for each } A \in v_{\text{appset}} \tag{21}$$

The normalized form of this metric, T_N, is defined in a fashion analogous to Eq. (20) to again obtain a value within [0, 1].

4.1.4 GCLP-MF Algorithm Versions

Two versions of GCLP-MF have been proposed. In the first version, which we call GCLP-MF-A, the normalized commonality metrics R_N and T_N are combined into a composite metric κ, based on user-defined weighting factors α_1 and α_2:

$$\kappa(A) = \alpha_1 R_N(\text{type}(A)) + \alpha_2 T_N(A) \quad \text{for each } A \in V_{\text{appset}} \tag{22}$$

This composite metric, in turn, is mapped into a [0, 0.5]-normalized form by applying a formula analogous to Eq. (20) and then multiplying by 0.5. The resulting normalized, composite metric, which we denote by κ_N, becomes the threshold adjustment value for GCLP-MF-A. More specifically, in GCLP-MF-A, the hardware/software mapping threshold is computed as

$$\text{threshold}(A) = 0.5 - \kappa_N(A) \tag{23}$$

This threshold value, which replaces the original GCLP threshold expression of Eq. (17), is compared against an actor's application-specific global criticality measure during cosynthesis. Intuitively, this threshold systematically favors hardware implementation for actor types that have relatively high type-repetition counts, and for actors that deliver large hardware versus software performance gains with relatively small amounts of hardware area overhead.

The GCLP-MF-A algorithm operates by applying to each member of *appset* the original GCLP algorithm with the threshold computation of Eq. (17) replaced by that of Eq. (23).

The second version, GCLP-MF-B, attempts to achieve some amount of "interaction" across cosynthesis decisions of different application graphs in appset rather than processing each application in isolation. In particular, the composite adjustment term (22) is discarded, and instead, a mechanism is introduced to allow cosynthesis decisions for the most difficult (from a synthesis perspective) applications to influence those that are made for less difficult applications. The difficulty of an application graph $G_i \in$ *appset* is estimated by its *criticality*, which is defined to be the sum of the software execution times divided by the latency constraint:

$$\text{criticality}(G_i) = = \frac{\displaystyle\sum_{v \in V_i} t_s(v)}{L_i} \tag{24}$$

Intuitively, an application with high criticality requires a large amount of hardware area to satisfy its latency constraint and thus makes it more difficult to meet the minimization objective of cosynthesis.

Version GCLP-MF-B operates by processing application graphs in decreasing order of their criticality, keeping track of interapplication resource-sharing possibilities throughout the cosynthesis process, and systematically incorporating these possibilities into the hardware/software selection threshold. Resource-sharing information is effectively stored as an actor-indexed array S of three-valued "sharing state" elements. For a given actor A, $S[A]$ = NULL indicates that no actor of type type(A) has been considered in a previous mapping step; $S[A]$ = HW indicates that a type(A) actor has previously been considered and has been mapped into hardware; and $S[A]$ = SW indicates a previous software mapping decision for type(A).

Like GCLP-MF-A, the GCLP-MF-B algorithm applies the original GCLP algorithm to each application graph separately with a modification of the hardware/software threshold function [17]. Specifically, the threshold in GCLP-MF-B is computed as

$$\text{threshold } (A) = \begin{cases} 0.5 - T_N(A) & \text{if } (S[A] = \text{NULL}) \\ 0.5 - R_N(A) & \text{if } (S[A] = \text{HW}) \\ 0.5 + R_N(A) & \text{if } (S[A] = \text{SW}) \end{cases} \qquad (25)$$

Thus, previous mapping decisions (from equal- or higher-criticality applications), together with commonality metrics, are used to determine whether or not a given actor is mapped into hardware or software.

Experimental results have shown that for multifunction systems, both versions of GCLP-MF significantly outperform isolated applications of the original GCLP algorithm to the application graphs in *appset* and that version B, which incorporates the commonality metrics used in version A in addition to the shared mapping state S, outperforms version A.

4.2 COSYN

Optimal or nearly optimal hardware/software cosynthesis solutions are difficult to achieve because there are numerous relevant implementation considerations and constraints. The COSYN algorithm [45], developed by Dave et al., takes numerous elements of this complexity into account. The design considerations and objectives addressed by the algorithm include allowing arbitrary, possibly heterogeneous collections of processors and communication links, intraprocessor concurrency (e.g., in FPGAs and ASICs), pre-emptive versus non-pre-emptive scheduling, actor duplication on multiple processors to alleviate communication bottlenecks, memory constraints, average, quiescent and peak power dissipation in processing elements and communication links, latency (in the form of actor deadlines), throughput (in the form of subgraph initiation rates), and overall dollar cost, which is the ultimate minimization objective.

4.2.1 Algorithm Flow

Input to the COSYN algorithm includes an application graph $G = (V, E)$ that may consist of several independent subgraphs that operate at different rates (periods) and with different deadlines; a library of processing elements $R = \{r_1, r_2, \ldots, r_m\}$; a set of communication resources ("links") $C = \{c_1, c_2, \ldots, c_n\}$; an actor execution time function $t_e: V \times R \to \overline{\aleph}$, which specifies the execution time of each actor on each candidate processing resource; a communication time function $t_c: E \times C \to \overline{\aleph}$, which gives the latency of communication of each edge on each candidate communication resource; and a deadline function

$$\text{deadline: } V \to \overline{\aleph}$$

which specifies an optional maximum allowable completion time for each actor. Under this notation, an infinite value of $t_e(t_c)$ indicates an incompatibility between the associated actor/resource (edge/resource) pair, and, similarly, deadline(v) = ∞ if there is no deadline specified for actor v. The overall flow of the COSYN algorithm is as follows:

```
function COSYN
     FormClusters(G) → Cluster set X
     unallocated = X
     for i = 1, 2, . . . ., |X|
         ComputeClusterPriorities (unallocated)
         Select a maximum priority cluster Cᵢ ∈ unallocated
         Evaluate possible allocations for Cᵢ and select best one
         unallocated = unallocated − {Cᵢ}
     end for
end function
```

In the initial *FormClusters* phase, the application graph is analyzed to identify subgraphs that are to be grouped together during the allocation and assignment exploration phases. After clusters have been formed, they are examined— one by one—and allocated by exploring their respective ranges of possible allocations and selecting the ones that best satisfy certain criteria that relate to the given performance and cost objectives. As individual allocation decisions are made, execution times of actors in the associated clusters become fixed, and this information is used to re-evaluate cluster priorities for future cluster selection decisions and to re-evaluate actor edge priorities during scheduling (to evaluate candidate allocations). Thus, cluster selection and scheduling decisions are computed dynamically based on all previously committed allocations.

4.2.2 Cluster Formation

Clustering decisions during the *FormCluster* phase are guided by a metric that prioritizes actors based on deadline- and communication-conscious critical path analysis. Like cluster selection and allocation decisions, actor priorities for clustering are dynamically evaluated based on all previous clustering operations. The priority of an actor for clustering is computed as

$$\lambda_G(A, f_t, f_c) \tag{26}$$

where the execution time contribution function $f_t: V \to \overline{Z}$, is given as the worst-case execution time offset by the actor deadline:

$$f_t(v) = \max(\{t_e(v, r_i) | (r_i \in R) \quad \text{and} \quad (t_e(v, r_i) < \infty)\}) - \text{deadline}(v) \tag{27}$$

and the communication time contribution function $f_c: E \to \aleph$ is given as the worst-case communication cost, based on all previous clustering decisions:

$$f_c(e, c) = \begin{cases} 0 & \text{if } (e \in subsumed) \\ \max(t_c(e, c_i) | (c_i \in C) \quad \text{and} \quad (t_c(e, c_i) < \infty)\}) & \text{otherwise} \end{cases} \tag{28}$$

Here, *subsumed* denotes the set of edges in E that have been "enclosed" by the clusters created by all previous clustering operations; that is, the set of edges e such that $src(e)$ and $snk(e)$ have already been clustered and both belong to the same cluster.

At each clustering step, an unclustered actor A that maximizes $\lambda_G(*, f_t, f_c)$ is selected, and based on certain compatibility criteria, A is first either merged into the cluster of a predecessor actor or inserted into a new cluster, and then the resulting cluster may be further expanded to contain a successor of A.

4.2.3 Cluster Allocation

After clustering is complete, we have a disjoint set of clusters $X = \{X_1, X_2, \ldots, X_p\}$, where each X_i represents a subset of actors that are to be assigned to the same physical processing element. Clusters are then selected one at a time, and for each selected cluster, the possible allocations are evaluated by scheduling. At each cluster selection step, a cluster with maximal priority (among all clusters that have not been selected in previous steps) is selected, where the priority of a cluster is simply taken to be the priority of its highest-priority actor, and actor priorities are determined using an extension of Eq. (26) that takes into account the effects of any previously committed allocation decisions. More precisely, we suppose that for each edge $e \notin subsumed$, asgn$(e) =$ NULL if e has not yet been assigned to a communication resource; otherwise, asgn$(e) \in C$ gives the resource type of the communication link to which e has been assigned. Similarly, we allow a minor abuse of notation and suppose that for each actor A, asgn$(A) =$ NULL if e has not yet been assigned to a processing element (i.e., the enclosing cluster has not yet been allocated); otherwise, $\text{asgn}(A) \in R$ gives the resource type of the processing element to which e has been assigned. Actor priority throughout the cluster allocation phase of COSYN is then computed as

$$\lambda_G(A, g_t, g_c) \qquad (29)$$

where $g_t: V \to \overline{Z}$ is defined by

$$g_t(v) = \begin{cases} f_t(v) & \text{if } (\text{asgn}(v) = \text{NULL}) \\ t_e(v, \text{asgn}(v)) - \text{deadline}(v) & \text{otherwise} \end{cases} \qquad (30)$$

and, similarly, $g_c: E \to \aleph$ is defined by

$$g_c(e) = \begin{cases} f_c(e) & \text{if } (\text{asgn}(e) = \text{NULL}) \\ t_c(e, \text{asgn}(e)) & \text{otherwise} \end{cases} \qquad (31)$$

In other words, if an actor or edge x has been assigned to a resource r, x is modeled with the latency of x on the resource type associated with r; otherwise, the worst case latency is used to model x.

As clusters are allocated, the values $\{\text{asgn}(x)|x \in (E \cup V)\}$ change, in general, and thus, for improved accuracy, actor priorities are reevaluated—using Eqs. (29)–(31)—during subsequent cluster allocation steps.

4.2.4 Allocation Selection

After a cluster is selected for allocation, candidate allocations are evaluated by scheduling and *finish-time estimation*. During scheduling, actors and edges are processed in an order determined by their priorities, and considerations such as overlapped versus nonoverlapped communication and actor preemption are taken into account at this time. Once scheduling is complete, the best- and worst-case finish times of the actors and edges in the application graph are estimated—based on their individual best- and worst-case latencies—to formulate an overall evaluation of the candidate allocation.

The best- and worst-case latencies associated with actors and edges are determined in a manner analogous to the "allocation-conscious" priority contribution values $g_t(*)$ and $g_c(*)$ computed in Eqs. (30) and (31). For each actor $v \in V$, the best case latency is defined by

$$t_{\text{best}}(v) = \begin{cases} \min(\{t_e(v, r_i)|(r_i \in R)\}) & \text{if } (\text{asgn}(v) = \text{NULL}) \\ t_e(v, \text{asgn}(v)) & \text{otherwise} \end{cases} \qquad (32)$$

and, similarly, the best-case latency for each edge $e \in E$ is defined by

$$t_{\text{best}}(e) = \begin{cases} \min(\{t_c(e, c_i)|(c_i \in C)\}) & \text{if asgn}(e) = \text{NULL} \\ t_c(e, \text{asgn}(e)) & \text{otherwise} \end{cases} \tag{33}$$

The worst-case latencies, denoted $t_{\text{worst}}(v)$ and $t_{\text{worst}}(e)$, are defined (using the same minor abuse of notation) in a similar fashion.

From these best- and worst-case latencies, allocation-conscious best- and worst-case *finish*-time estimates F_{best} and F_{worst} of each actor and each edge are computed by

$$F_{\text{best}}(v) = \max(\{F_{\text{best}}(e_{\text{in}}) + t_{\text{best}}(v)|e_{\text{in}} \in \text{in}(v)\}), \text{ and} \tag{34}$$

$$F_{\text{worst}}(v) = \max(\{F_{\text{worst}}(e_{\text{in}}) + t_{\text{worst}}(v)|e_{\text{in}} \in \text{in}(v)\}) \quad \text{for } v \in V; \tag{35}$$

$$F_{\text{best}}(e) = F_{\text{best}}(\text{src}(e)) + t_{\text{best}}(e), \text{ and} \tag{36}$$

$$F_{\text{worst}}(e) = F_{\text{worst}}(\text{src}(e)) + t_{\text{worst}}(e) \quad \text{for } e \in E \tag{37}$$

The worst-case and best-case finish times, as computed by Eqs. (34–37), are used in evaluating the quality of a candidate allocation. Let $V_{\text{deadline}} \subseteq V$ denote the subset of actors for which deadlines are specified; let α denote the set of candidate allocations for a selected cluster; and let $\alpha' \subseteq \alpha$ be the set of candidate allocations for which all actors in V_{deadline} have their corresponding deadlines satisfied in the best case (i.e., according to $\{F_{\text{best}}(v)\}$). If $\alpha' \neq \varnothing$, then an allocation is chosen from the subset α' that *maximizes* the sum

$$\sum_{v \in V_{\text{deadline}}} F_{\text{worst}}(v) \tag{38}$$

of worst-case finish times over all actors for which prespecified deadlines exist. On the other hand, if $\alpha' = \varnothing$, then an allocation is chosen from α that *maximizes* the sum

$$\sum_{v \in V_{\text{deadline}}} F_{\text{best}}(v) \tag{39}$$

of best-case finish times over all actors for which deadlines exist. In both cases, the maxima over the respective sets of sums are taken because they ultimately lead to final allocations that have lower overall dollar cost [45].

4.2.5 Accounting for Power Consumption

A "low-power version" of the COSYN algorithm, called COSYN-LP, has been developed to minimize power consumption along with overall dollar cost. In addition to the algorithm inputs defined in Section 4.2.1, COSYN-LP also em-

ploys *average power dissipation functions* p_e: $V \times R \to \overline{\aleph}$ and p_c: $E \times C \to$
$\overline{\aleph}$. The value of $p_e(v, r_i)$ gives an estimate of the average power dissipated while
actor v executes on processing resource r_i, and, similarly, the value of $p_c(e, c_i)$
estimates the average power dissipated when edge e executes on communication
resource c_i. Again, infinite values in this notation correspond to incompatibility
relationships between operations (actors or edges) and resource types. Similar
functions are also defined for peak (maximum instantaneous) power consumption
and quiescent power consumption (power consumption during periods of inactiv-
ity) of resources for processing and communication.

The COSYN-LP algorithm incorporates modifications to the clustering and
allocation evaluation phases that take actor and edge power consumption information into account. For example, the cluster formation process is modified to
use the following power-oriented actor priority function:

$$\lambda_G(A, \rho_t, \rho_c) \tag{40}$$

Here, ρ_t: $V \to \aleph$ is defined by

$$\rho_t(v) = t_e(v, r_{\text{worst}}(v)) \times p_e(v, r_{\text{worst}}(v)) \tag{41}$$

where $r_{\text{worst}}(v)$ is a processing resource type that maximizes the execution time
$t_e(v, *)$ of v; similarly, ρ_c: $E \to \aleph$ is defined by

$$\rho_c(e) = t_c(e, c_{\text{worst}}(e)) \times p_c(e, c_{\text{worst}}(e)) \tag{42}$$

where $c_{\text{worst}}(e)$ is a communication resource type that maximizes the communication latency $t_c(e, *)$ of e. There is slight ambiguity here because there may be
more than one processing (communication) resource that maximize the latency
for a given actor (edge); tie breaking in such cases can be performed arbitrarily
(it is not specified as part of the algorithm).

Thus, in COSYN-LP, priorities for cluster formation are computed on the
basis of average power dissipation based on worst-case execution times.

In a similar manner, the average power dissipation metrics along with the
peak and quiescent power metrics are incorporated into the cluster allocation
phase of COSYN-LP. For details, we refer the reader to Ref. 45.

4.3 CodeSign

As part of the CodeSign project at ETH Zurich, Blickle et al. have developed
a search technique for hardware/software cosynthesis [46] that is based on the
framework of *evolutionary algorithms*. In evolutionary algorithms, complex
search spaces are explored by encoding candidate solutions as ''chromosomes''
and evolving ''populations'' of these chromosomes by applying the principles
of *reproduction* (retention of chromosomes in a population), *crossover* (derivation of new chromosomes from two or more ''parent'' chromosomes), *mutation*

(modification of individual chromosomes), and *fitness* (metrics for evaluating the quality of chromosomes) [47]. These principles incorporate probabilistic techniques to derive new chromosomes from an existing population and to replace portions of a population with selected, newly derived chromosomes.

4.3.1 Specifications

A key innovation in the CodeSign approach is a novel formulation of joint allocation, assignment, and scheduling as mappings between sequences of graphs and "activations" of vertices and edges in these graphs. This formulation is intuitively appealing and provides a natural encoding structure for embedding within the framework of evolutionary algorithms.

The central data structure that underlies the CodeSign cosynthesis formulation is the *specification*. A CodeSign specification can be viewed as an ordered pair $S = (H_S, M_S)$, where $H_s = \{G_1, G_2, \ldots, G_N\}$; each G_i is a directed graph (called a "dependence graph") (V_i, E_i); and each M_i is a set of *mapping edges* that connect vertices in successive dependence graphs (i.e., for each $e \in M_i$, $src(e) \in V_i$ and $snk(e) \in V_{i+1}$). If the specification in question is understood, we write

$$V_H = \bigcup_{i=1}^{N} V_i, \quad E_H = \bigcup_{i=1}^{N} E_i, \quad E_M = \bigcup_{i=1}^{N-1} M_i \qquad (43)$$

Thus, V_H and E_H denote the sets of all dependence graph vertices and edges, respectively, and E_M denotes the set of all mapping edges. The *specification graph* of S is the graph $G_S = (V_S, E_S)$ obtained by integrating all of the dependence graphs and mapping edges: $V_S = V_H$ and $E_S = (E_H \cup E_M)$.

The "top-level" dependence graph (the *problem graph*) G_1 gives a behavioral specification of the application to be implemented. In this sense, it is similar to the application graph concept defined in Section 3.1. However, it is slightly different in its incorporation of special *communication vertices* that explicitly represent interactor communication and are ultimately mapped onto communication resources in the target architecture [46].

The remaining dependence graphs $G_2, G_3, \ldots, G_N$ specify different levels of abstraction or refinement during implementation. For example, a dependence graph could specify an architectural description consisting of available resources for computation and communication (*architecture graph*) and another dependence graph could specify the decomposition of a target system into integrated circuits and off-chip buses (*chip graph*). Due to the general nature of the CodeSign specification formulation, there is full flexibility to define alternative or additional levels of abstraction in this manner.

Dependence graph edges specify connectivity between modules within the same level of abstraction, and mapping edges specify *compatibility* relationships between successive abstraction levels in a specification; that is, $e \in E_M$ indicates that $src(e)$ "can be implemented by" $snk(e)$.

Example 7

Figure 6a provides an illustration of a CodeSign specification for hardware/software cosynthesis onto an architecture that consists of a programmable processor resource P_S, a resource for implementing custom hardware P_H, and a bidirectional bus B that connects these two processing resources. The v_i's denote problem graph actors and the c_i's denote communication vertices. Here, only hardware implementation is allowed for v_2 and v_5, only software implementation is allowed for v_3, and v_1 and v_4 may each be mapped to either hardware or software. Thus, for example, there is no edge connecting v_2 or v_5 with the vertex P_S associated with the programmable processor. In general, communication vertices can be mapped either to the bus B (if the source and sink vertices are mapped to different processing resources) or *internally* to either the hardware (P_H) or software (P_S) resource (if the source and sink are mapped to the same processing resource). However, mapping restrictions of the problem graph actors may limit the possible mapping targets of a communication vertex. For example, because v_2 and v_3 are restricted respectively to hardware and software implementations, communication vertex c_2 must be mapped to the bus B. Similarly, c_3 can be mapped to P_s or B, but not to P_H. The set of mapping edges for this example is given by

$$E_M = \{(v_1, P_H), (v_1, P_S), (v_2, P_H), (v_3, P_S), (v_4, P_H), (v_4, P_S), \tag{44}$$
$$(v_5, P_H), (c_1, B), (c_1, P_S), (c_2, B), (c_3, B), (c_3, P_S), (c_4, B)\}$$

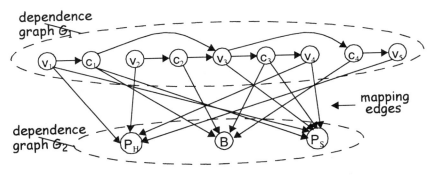

Figure 6 An illustration of a specification in CodeSign.

4.3.2 Activation Functions

Allocations and assignments of specification graphs are formulated in terms of *activation functions*. An activation function for a specification graph G_S is any function $a: (V_S \cup E_S) \rightarrow \{0, 1\}$ that maps vertices and edges of G_S into binary numbers. If $x \in (V_S \cup E_S)$ is a vertex or a dependence graph edge, then $a(x) = 1$ is equivalent to the *use* or *instantiation* of x in the associated allocation. On the other hand, if x is a mapping edge, then $a(x) = 1$ if and only if $src(x)$ is implemented by $snk(x)$ according to the associated assignment.

Thus, an activation function uniquely determines an allocation and assignment for the associated specification. The allocation associated with an activation function a can be expressed in precise terms by

$$\alpha(a) = \{x \in (V_H \cup E_H) | a(x) = 1\} \tag{45}$$

and, similarly, the assignment associated with a is defined by

$$\beta(a) = \{e \in E_M | a(e) = 1\} \tag{46}$$

We say that $x \in (V_S \cup E_S)$ is activated if $a(x) = 1$. The allocation and assignment associated with an activation function a are *feasible* if for each activated mapping edge $e \in \beta(a)$, the source and sink vertices are activated [i.e., $src(e), snk(e) \in \alpha(a)$]; for each activated vertex $v \in \alpha(a)$, there exists exactly one activated, output mapping edge mapping(v) [i.e., $|(\text{out}(v) \cap \beta(a))| = 1$]; and for each activated dependence graph edge $e \in \alpha(a)$, either

$$\text{mapping}(src(e)) = \text{mapping}(snk(e))$$

or

$$(\text{mapping}(src(e)), \text{mapping}(snk(e))) \in \alpha(a) \tag{47}$$

This last condition, Eq. (47), simply states that $src(e)$ and $snk(e)$ must either be assigned to the same vertex in the succeeding dependence graph or there must be an activated edge that provides the appropriate communication between the distinct vertices to which $src(e)$ and $snk(e)$ are mapped.

4.3.3 Evolutionary Algorithm Approach

The overall approach in the CodeSign synthesis algorithm is to encode allocation and assignment information in the chromosome data structure of the evolutionary algorithm and to use a deterministic heuristic for scheduling, because effective deterministic techniques exist for computing schedules given prespecified allocations and assignments [36].

Decoding of a chromosome (e.g., to evaluate its fitness) begins by interpreting the allocation (activation) status (0 or 1) of each specification graph vertex that is given in the chromosome. Some allocation obtained in this way may be "incomplete" in the sense that there may be some functional vertices for which no compatible resources are instantiated. Such incompleteness in allocations is "repaired" by activating additional vertices based on a *repair allocation priority list*, which is also a component of the chromosome due to the relatively large impact of resource activation decisions on critical implementation metrics, such as performance and area. This priority list specifies the order in which vertices will be considered for activation during repair of allocation incompleteness.

After a chromosome has been converted into its associated allocation and incompleteness of the allocation has been repaired, the assignment information from the chromosome is decoded. The coding convention for assignment information has been carefully devised to be orthogonal to the allocation encoding, so that the process of interpreting assignment information is independent of the given allocation. This independence between the interpretation of allocation and assignment information is important in facilitating efficient evolution of the chromosome population [46].

Like allocation repair information, assignment information is encoded in the form of priority lists: Each dependence graph vertex has an associated priority list $L_\beta(v)$ of its outgoing mapping edges (out$(v) \cap E_M$). These priority lists are interpreted by examining each allocated vertex v and activating the first member of $L_\beta(v)$ that does not conflict with the requirements of a feasible allocation/ assignment that were discussed in Section 4.3.2

It is possible that a feasible allocation/assignment does not result from the decoding of a particular chromosome. Indeed, Blickle et al. has shown that the problem of determining a feasible allocation/assignment is computationally intractable [46] so straightforward techniques—such as applying the decoding process to random chromosomes—cannot be relied upon to consistently achieve feasiblity.

If such infeasibility is determined during the decoding process, then a significant *penalty* is incorporated into the fitness of the associated chromosome. Otherwise, the decoded allocation and assignment are scheduled using a deterministic scheduling heuristic, and the resulting schedule, along with the assignment and allocation, are assessed in the context of the designer's optimization constraints and objectives to determine the chromosome fitness.

In summary, the CodeSign cosynthesis algorithm incorporates a novel *specification graph* data structure and an evolutionary algorithm formulation that encodes allocation and assignment information in terms of specification graph concepts. Due to space limitations, we have suppressed several interesting details of the complete synthesis algorithm, including mechanisms for promoting resource sharing, and details of the scheduling heuristic. The reader is encouraged to consult Ref. 46 for a comprehensive discussion.

5 SYNCHRONIZATION OPTIMIZATION

In Section 3.2, we discussed the utility of self-timed multiprocessor implementation strategies in the design of efficient and robust parallel processing engines for DSP. For self-timed DSP multiprocessors, an important consideration in addition to hardware/software partitioning and the associated scheduling task is *synchronization* to ensure the integrity of *interprocessor communication operations* associated with data flow edges whose source and sink actors are mapped to different processing elements. Because cost is often a critical constraint, embedded multiprocessors must often use simple communication topologies and limited, if any, hardware support for synchronization. A variety of efficient techniques have been developed to optimize synchronization for such cost-constrained, self-timed multiprocessors [8,48,49]. Such techniques can significantly reduce the execution time and power consumption overhead associated with synchronization and can be used as postprocessing steps to any of the partitioning algorithms discussed in Section 4, as well as to a wide variety of multiprocessor scheduling algorithms for data flow graphs, such as those described in Refs. 39, 50, and 51.

In this section, we present an overview of these approaches to synchronization optimization. Specifically, we discuss two closely related graph-theoretic models, the *IPC graph* G_{ipc} [52] and the *synchronization graph* G_s [48], that are used to model the self-timed execution of a given parallel schedule for an application graph, and we discuss the application of these models to the systematic streamlining of synchronization functionality.

Given a self-timed multiprocessor schedule for an application graph G, we derive G_{ipc} and G_s by first instantiating a vertex for each actor, connecting an edge from each actor to the actor that succeeds it on the same processor, and adding an edge that has unit delay from the last actor on each processor to the first actor on the same processor. Also, for each edge (x, y) in G that connects actors that execute on different processors, an *IPC edge* is instantiated in G_{ipc} from x to y. Figure 7c shows the IPC graph that corresponds to the application

(a)

(b)

Processor	Actor ordering
Proc. 1	B, D, F
Proc. 2	A, C, E

(c)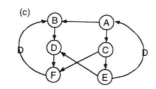

Figure 7   An illustration of a self-timed schedule and its associated IPC graph.

graph of Figure 7a and the processor assignment and actor ordering of Figure 7b.

Each edge in G_{ipc} and G_s is either an *intraprocessor edge* or an *interprocessor edge*. Intraprocessor edges model the ordering (specified by the given parallel schedule) of actors assigned to the same processor; interprocessor edges in G_{ipc}, called *IPC edges*, connect actors assigned to distinct processors that must communicate for the purpose of data transfer; and interprocessor edges in G_s, called *synchronization edges*, connect actors assigned to distinct processors that must communicate for synchronization purposes.

Each edge e in G_{ipc} represents the *synchronization constraint*

$$\text{start}(\text{snk}(e), k) \geq \text{end}(\text{src}(e), k - \text{del}(e)) \quad \text{for all } k \tag{48}$$

where start (v, k) and end (v, k) respectively represent the times at which firing k of actor v begins execution and completes execution.

Initially, the synchronization graph G_s is identical to G_{ipc}. However, various transformations can be applied to G_s in order to make the overall synchronization structure more efficient. After all transformations on G_s are complete, G_s and G_{ipc} can be used to map the given parallel schedule into an implementation on the target architecture. The IPC edges in G_{ipc} represent buffer activity and are implemented as buffers in shared memory, whereas the synchronization edges of G_s represent synchronization constraints, and are implemented by updating and testing flags in shared memory. If there is an IPC edge as well as a synchronization edge between the same pair of actors, then a synchronization protocol is executed before the buffer corresponding to the IPC edge is accessed to ensure sender–receiver synchronization. On the other hand, if there is an IPC edge between two actors in the IPC graph but there is no synchronization edge between the two, then no synchronization needs to be done before accessing the shared buffer. If there is a synchronization edge between two actors but no IPC edge, then no shared buffer is allocated between the two actors; only the corresponding synchronization protocol is invoked.

Any transformation that we perform on the synchronization graph must respect the synchronization constraints implied by G_{ipc}. If we ensure this, then we only need to implement the synchronization edges of the optimized synchronization graph. If $G_1 = (v, E_1)$ and $G_2 = (V, E_2)$ are synchronization graphs with the same vertex-set and the same set of intraprocessor edges (edges that are not synchronization edges), we say that G_1 *preserves* G_2 if for all $e \in E_2$ such that $e \notin E_1$, we have $\rho_{G_1}(\text{src}(e), \text{snk}(e)) \leq \text{del}(e)$, where $\rho_G(x, y) \equiv \infty$ if there is no path from x to y in the synchronization graph G, and if there is a path from x to y, then $\rho_G(x, y)$ is the minimum over all paths p directed from x to y of the sum of the edge delays on p. The following theorem (developed in Ref. 48) underlies the validity of a variety of useful synchronization graph transformations, which we discuss in Sections 5.1–5.4.

THEOREM 1 *The synchronization constraints (as specified by Ref. 52) of G_1 imply the constraints of G_2 if G_1 preserves G_2.*

5.1 Removal of Redundant Synchronization Edges

A synchronization edge is *redundant* in a synchronization graph G if its removal yields a graph that preserves G. Equivalently, a synchronization edge e is redundant if there is a path $p \neq (e)$ from src(e) to snk(e) such that $\delta(p) \leq$ del(e), where $\delta(p)$ is the sum of the edge delays on path p. Thus, the synchronization function associated with a redundant synchronization edge "comes for free" as a by-product of other synchronizations.

Example 8

Figure 8 shows an example of a redundant synchronization edge. The dashed edges in this figure are synchronization edges. Here, before executing actor D, the processor that executes $\{A, B, C, D\}$ does not need to synchronize with the processor that executes $\{E, F, G, H\}$ because due to the synchronization edge x_1, the corresponding firing of F is guaranteed to complete before each firing of D is begun. Thus, x_2 is redundant.

The following result establishes that the order in which we remove redundant synchronization edges is not important.

THEOREM 2 [48] *Suppose $G_s = (V, E)$ is a synchronization graph, e_1 and e_2 are distinct redundant synchronization edges in G_s, and $\tilde{G}_s = (V, E - \{e_1\})$. Then, e_2 is redundant in $\tilde{G}_s$.*

Theorem 2 tells us that we can avoid implementing synchronization for all redundant synchronization edges because the "redundancies" are not interdepen-

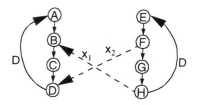

Figure 8 An example of a redundant synchronization edge.

dent. Thus, an optimal removal of redundant synchronizations can be obtained by applying a straightforward algorithm that successively tests the synchronization edges for redundancy in some arbitrary sequence and removes each of the edges that are found to be redundant. Such testing and removal of redundant edges can be performed in $O(|V|^2 \log_2(|E|) + |V||E|)$ time.

Example 9

Figure 9a shows a synchronization graph that arises from a two-processor schedule for a four-channel multiresolution quadrature mirror filter (QMF) bank, which has applications in signal compression. As in Figure 8, the dashed edges are synchronization edges. If we apply redundant synchronization removal to the synchronization graph of Figure 9a, we obtain the synchronization graph in Figure 9b; the edges (A_1, B_2), (A_3, B_1), (A_4, B_1), (B_2, E_1), and (B_1, E_2) are detected to be redundant; and the number of synchronization edges is reduced from 8 to 3 as a result.

5.2 Resynchronization

The goal of resynchronization is to introduce new synchronizations in such a way that the number of original synchronizations that become redundant exceeds

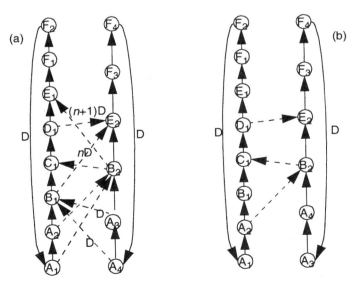

Figure 9 An example of redundant synchronization removal.

the number of new synchronizations that are added and, thus, the net synchronization cost is reduced. To ensure that the serialization introduced by resynchronization does not degrade the throughput, the new synchronizations are restricted to
lie outside the SCCs of the synchronization graph (*feedforward resynchronization*) [8].

Resynchronization of self-timed multiprocessors has been studied in two
contexts [49]. In *maximum-throughput resynchronization*, the objective is to
compute a resynchronization that minimizes the total number of synchronization
edges over all synchronization graphs that preserve the original synchronization
graph. It has been shown that optimal resynchronization is NP-complete. However, a broad class of synchronization graphs has been identified for which
optimal resynchronization can be performed by an efficient, polynomial-time
algorithm. A heuristic for general synchronization graphs called Algorithm
Global-resynchronize has also been developed that works well in practice.

Effective resynchronization improves the throughput of a multiprocessor

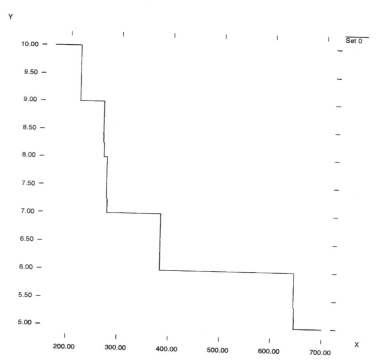

Figure 10   An illustration of resynchronization. The vertical axis gives the number of
synchronization edges, and the horizontal axis gives the latency constraint.

implementation by reducing the rate at which synchronization operations must be performed. However, because additional serialization is imposed by the new synchronizations, resynchronization can produce a significant increase in latency. In *latency-constrained resynchronization,* the objective is to compute a resynchronization that minimizes the number of synchronization edges over all valid resynchronizations that do not increase the latency beyond a prespecified upper bound on the tolerable latency. Latency-constrained resynchronization is intractable even for the very restricted subclass of synchronization graphs in which each SCC contains only one actor, and all synchronization edges have zero delay. However, an algorithm has been developed that computes optimal latency-constrained resynchronizations for two-processor systems in $O(N^2)$ time, where N is the number of actors. Also, an efficient extension of Algorithm Global-resynchronize, called Algorithm *Global-LCR*, has been developed for latency-constrained resynchronization of general synchronization graphs.

Figure 10 illustrates the results delivered by Global-LCR when it is applied to a six-processor schedule of a synthesizer for plucked-string musical instruments in 11 voices. The plot in Figure 10 shows how the number of synchronization edges in the result computed by Global-LCR changes as the latency constraint varies. The alternative synchronization graphs represented in Figure 10 offer a variety of latency/throughput trade-off alternatives for implementing the given schedule. The (rightmost) extreme of these trade-off points offers 22–27% improvement in throughput and 32–37% reduction in the average rate at which shared memory is accessed, depending on the access time of the shared memory. Because accesses to shared memory typically require significant amounts of energy, this reduction in the average rate of shared memory accesses is especially useful when low power consumption is an important implementation issue.

5.3 Feed-Forward and Feedback Synchronization

In general, self-timed execution of a multiprocessor schedule can result in unbounded data accumulation on one or more more IPC edges. However, the following result states that each *feedback edge* (an edge that is contained in an SCC) has a bounded buffering requirement. This result emerges from the theory of *timed marked graphs*, a family of computation structures to which synchronization graphs belong.

THEOREM 3 *Throughout the self-timed execution of an IPC graph G_{ipc}, the number of tokens on a feedback edge e of G_{ipc} is bounded; an upper bound is given by*

$$\min(\{\delta(C)|(e \in \text{edges}(C))\}) \tag{49}$$

*where δ(C) denotes the sum of the edge delays in cycle C. The constant bound
specified by the term [49] is called the* self-timed buffer bound *of that edge.*

A *feed-forward edge* (an edge that is not contained in an SCC), however,
has no such bound on the buffer size.

Based on Theorem 3, two efficient protocols can be derived for the implementation of synchronization edges. Given an IPC graph (V, E) and an IPC edge
$e \in E$, if e is a feed-forward edge, then we can apply a synchronization protocol
called *unbounded buffer synchronization* (UBS), which guarantees that snk(e)
never attempts to read data from an empty buffer (to prevent underflow) and
src(e) never attempts to write data into the buffer unless the number of tokens
already in the buffer is less than some prespecified limit, which is the amount
of memory allocated to that buffer (to prevent overflow). If e is a feedback edge,
then we use a simpler protocol, called *bounded buffer synchronization* (BBS),
which only explicitly ensures that overflow does not occur. The simpler BBS
protocol requires only half of the run-time overhead that is incurred by UBS.

5.4 Implementation Using Only Feedback Synchronization

One alternative to implementing UBS for a feedforward edge e is to add synchronization edges to G_s so that e becomes encapsulated in an SCC, and then implement e using BBS, which has lower cost. An efficient algorithm, called Convert-to-SC-graph, has been developed to perform this graph transformation in such a
way that the net synchronization cost is minimized and the impact on the self-timed buffer bounds of the IPC edges is optimized. Convert-to-SC-graph effectively "chains together" the source SCCs, chains together the sink SCCs, and
then connects the first SCC of the "source chain" to the last SCC of the sink
chain with an edge. Depending on the structure of the original synchronization
graph, Convert-to-SC-graph can reduce the overall synchronization cost by up
to 50%.

Because conversion to a strongly connected graph must introduce one or
more new cycles, it may be necessary to insert delays on the edges added by
Convert-to-SC-graph. These delays may be needed to avoid deadlock and to ensure that the serialization introduced by the new edges does not degrade the
throughput. The location (edge) and magnitude of the delays that we add are
significant because (from Theorem 3) they affect the self-timed buffer bounds
of the IPC edges, which, in turn, determine the amount of memory that we allocate for the corresponding buffers.

A systematic technique has been developed, called Algorithm *Determine
Delays*, that efficiently inserts delays on the new edges introduced during the
conversion to a strongly connected synchronization graph. For a broad class of
practical synchronization graphs—those synchronization graphs that contain only
one source SCC or only one sink SCC—Determine Delays computes a solution
(placement of delays) that minimizes the sum of the resulting self-timed buffer

bounds. For general synchronization graphs, Determine Delays serves as an efficient heuristic.

6 BLOCK PROCESSING

Recall from Section 2.2 that DSP applications are characterized by groups of operations that are applied repetitively on large, often unbounded, data streams. *Block processing* refers to the uninterrupted repetition of the same operation (e.g., data flow graph actor) on two or more successive elements from the same data stream. The *scalable synchronous data flow* (SSDF) model is an extension of SDF that enables software synthesis of *vectorized* implementations, which exploit the opportunities for efficient block processing and, thus, form an important component of the cosynthesis design space. The internal specification of an SSDF actor A assumes that the actor will be executed in groups of ($N_v(A)$ successive firings, which operate on $N_v(A) \times \mathrm{cns}(e)$)-unit blocks of data at a time from each incoming edge e. Block processing with well-designed SSDF actors reduces the rate of interactor context switching and context switching between successive code segments within complex actors, and it may improve execution efficiency significantly on deeply pipelined architectures.

At the Aachen University of Technology, as part of the COSSAP [27] software synthesis environment for DSP (now developed by Synopsys), Ritz et al. investigated the optimized compilation of SSDF specifications [53]. This work has targeted the minimization of the context-switch overhead, or the average rate at which *actor activations* occur. An actor activation occurs whenever two distinct actors are invoked in succession. Activation overhead includes saving the contents of registers that are used by the next actor to invoke, if necessary, and loading state variables and buffer pointers into registers.

For example, the schedule

$$(2(2B)(5A))(5C) \tag{50}$$

results in five activations per schedule period. Parenthesized terms in Eq. (50) represent *schedule loops*, which are repetitive firing patterns that are to be translated into loops in the target code. More precisely, a parenthesized term of the form $(nT_1T_2 \ldots T_n)$ specifies the successive repetition n times of the subschedule $T_1T_2 \ldots T_n$. Schedules that contain only one appearance of each actor, such as the schedule of Eq. (50), are referred to as *single appearance schedules*. Because of their code size optimality and because they have been shown to satisfy a number of useful formal properties [2], single appearance schedules have been the focus of a significant component of work in DSP software synthesis.

Ritz estimates the average rate of activations for a valid schedule S as the number of activations that occur in one iteration of S divided by the blocking factor $J(S)$. This quantity is denoted by $N'_{act}(S)$. For example, suppose we have an SDF graph for which $\mathbf{q}(A, B, C) = (10, 4, 5)$. Then,

$$N'_{act}((2(2B)(5A))(5C)) = 5$$

and

$$N'_{act}((4(2B)(5A))(10C)) = 9/2 = 4.5 \tag{51}$$

If, for each actor, each firing takes the same amount of time and if we ignore the time spent on computation that is not directly associated with actor firings (e.g., schedule loops), then $N'_{act}(S)$ is directly proportional to the number of actor activations per unit time. In practice, these assumptions are seldom valid; however, $N'_{act}(S)$ gives a useful estimate and means for comparing schedules. For consistent acyclic SDF graphs, clearly N'_{act} can be made arbitrarily small by increasing the blocking factor sufficiently; thus, the extent to which the activation rate can be minimized is limited by the SCCs.

Ritz's algorithm for vectorization, which we call *complete hierarchization vectorization* (CHV), attempts to find a valid single appearance schedule that minimizes N'_{act} over all valid single-appearance schedules. Minimizing the number of activations does not imply minimizing the number of appearances and, thus, the primary objective of CHV is, implicitly, code size minimization. As a simple example, consider the SDF graph in Figure 11. It can be verified that for this graph, the lowest value of N'_{act} that is obtainable by a valid single-appearance schedule is 0.75, and one valid single-appearance schedule that achieves this minimum rate is $(4B)(4A)(4C)$. However, valid schedules exist that are not single-appearance schedules and that have values of N'_{act} below 0.75; for example, the valid schedule $(4B)(4A)(3B)(3A)(7C)$ contains two appearances of A and B and satisfies $N'_{act} = 5/7 = 0.71$.

In the CHV approach, the *relative vectorization degree* of a simple cycle C in a consistent, connected SDF graph $G = (V, E)$ is defined by

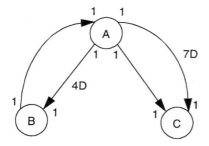

Figure 11 This example illustrates that minimizing actor activations does not imply minimizing actor appearances.

$$N_G(C) \equiv \max(\{\min(\{D_G(\alpha')|\alpha' \in \text{parallel}(\alpha)\})|\alpha \in \text{edges}(C)\}) \quad (52)$$

where

$$D_G(\alpha) \equiv \frac{\text{del}(\alpha)}{\text{TNSE}_G(\alpha)} \quad (53)$$

is the delay on edge α normalized by the total number of tokens consumed by $\text{snk}(\alpha)$ in a minimal schedule period of G, and

$$\text{parallel}(\alpha) \equiv \{\alpha' \in E|\text{src}(\alpha') = \text{src}(\alpha) \text{ and } \text{snk}(\alpha') = \text{snk}(\alpha)\} \quad (54)$$

is the set of edges with the same source and sink as α. For example, if G denotes the graph in Figure 11 and χ denotes the cycle whose associated *vertices* set contains A and C, then $D_G(\chi) = (7/1) = 7$.

Given a strongly connected SDF graph, a valid single-appearance schedule that minimizes N'_{act} can be constructed from a *complete hierarchization*, which is a cluster hierarchy such that only connected subgraphs are clustered, all cycles at a given level of the hierarchy have the same relative vectorization degree, and cycles in higher levels of the hierarchy have strictly higher relative vectorization degrees than cycles in lower levels [53].

Example 10

Figure 12 depicts a complete hierarchization of an SDF graph. Figure 12a shows the original SDF graph; here, q $(A, B, C, D) = (1, 2, 4, 8)$. Figure 12b shows

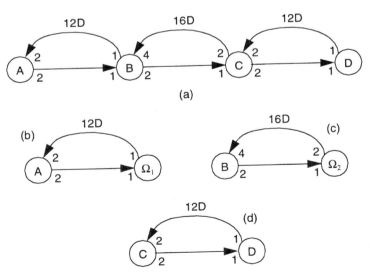

Figure 12 A complete hierarchization of a strongly connected SDF graph.

the top level of the cluster hierarchy. The hierarchical actor Ω_1 represents subgraph($\{B, C, D\}$) and this subgraph is decomposed as shown in Figure 12c, which gives the next level of the cluster hierarchy. Finally, Figure 12d, shows that subgraph($\{C, D\}$) corresponds to Ω_2 and is the bottom level of the cluster hierarchy. Now, observe that the relative vectorization degree of the simple cycle in Figure 12c with respect to the original SDF graph is $\lfloor 16/8 \rfloor = 2$, while the relative vectorization degree of the simple cycle in Figure 12b is $\lfloor 12/2 \rfloor = 6$; and the relative vectorization degree of the simple cycle in Figure 12d is $\lfloor 12/8 \rfloor = 1$. Thus, we see that the relative vectorization degree decreases as we descend the hierarchy and, thus, the hierarchization depicted in Figure 12 is complete.

The hierarchization step defined by each of the SDF graphs in Figures 12b–12d is called a *component* of the overall hierarchization.

The CHV technique constructs a complete hierarchization by first evaluating the relative vectorization degree of each simple cycle, determining the maximum vectorization degree, and then clustering the graphs associated with the simple cycles that do not achieve the maximum vectorization degree. This process is then repeated recursively on each of the clusters until no new clusters are produced. In general, this bottom-up construction process has unmanageable complexity; however, this normally does not create problems in practice because the SCCs of useful signal processing systems are often small, particularly in large-grain descriptions.

Once a complete hierarchization is constructed, CHV constructs a schedule ''template''—a sequence of loops whose iteration counts are to be determined later. For a given component Π of the hierarchization, if v_Π is the vectorization degree associated with Π, then all simple cycles in Π contain at least one edge α for which $D_G(\alpha) = v_\Pi$. Thus, if we remove from Π all edges in the set $\{\alpha | D_G(\alpha) = v_\Pi\}$, the resulting graph is acyclic, and if $F_{\Pi,1}, F_{\Pi,2}, \ldots, F_{\Pi,n_\Pi}$ is a topological sort of this acyclic graph, then valid schedules exist for Π that are of the form

$$T_\Pi \equiv (i_\Pi (i_{\Pi,1} F_{\Pi,1})(i_{\Pi,2} F_{\Pi,2}) \ldots (i_{\Pi,n_\Pi} F_{\Pi,n_\Pi})). \qquad (55)$$

This is the subschedule template for Π.

Here, each $F_{\Pi,j}$ is a vertex in the hierarchical SDF graph G_Π associated with Π. Thus, each $F_{\Pi,j}$ is either a *base block*—an actor in the original SDF graph G—or a hierarchical actor that represents the execution of a valid schedule for the corresponding subgraph of G. Now, let A_Π denote the set of actors in G that are contained in G_Π and in all hierarchical subgraphs nested within G_Π, and let $k_\Pi \equiv \gcd(\{i_{\Pi,j} | 1 \leq j \leq n_\Pi\})$. Thus, we have

$$i_{\Pi,j} = k_\Pi \mathbf{q}_{G_\Pi}(F_{\Pi,j}), \quad j = 1, 2, \ldots, n_\Pi \qquad (56)$$

The number of activations that T_Π contributes to N'_{act} is given by $((|B_\Pi| q_G(A_\Pi))/k_\Pi)$, where B_Π is the set of base blocks in G_Π [53]. Thus, if H

denotes the set of hierarchical components in the given complete hierarchization, then

$$N'_{act} = \sum_{\Pi \in H} \frac{|B_\Pi| q_G(A_\Pi)}{k_\Pi} \tag{57}$$

In the CHV approach, an exhaustive search over all i_Π and k_Π is carried out to minimize Eq. (57). The search is restricted by constraints derived from the requirement that the resulting schedule for G be valid. As with the construction of complete hierarchizations, the simplicity of SCCs in many practical applications often permits this expensive evaluation scheme.

Joint optimization of vectorization and buffer memory cost is developed in Ref. 22 and adaptations of the retiming transformation to improve vectorization for SDF graphs is addressed in Refs. 38 and 54.

7 SUMMARY

In this chapter, we have reviewed techniques for mapping high-level specifications of DSP applications into efficient hardware/software implementations. Such techniques are of growing importance in DSP design technology due to the increased use of heterogeneous multiprocessor architectures in which processing components, such as the ones discussed in Chapters 1–5, incorporate varying degrees and forms of programmability. We have discussed specification models based on coarse-grain data flow principles that expose valuable application structure during cosynthesis. We then developed a number of systematic techniques for partitioning coarse-grain data flow specifications into the hardware and software components of heterogeneous architectures for embedded multiprocessing. Synchronization between distinct processing elements in a partitioned specification was then discussed, and in this context, we examined a number of complementary strategies for reducing the execution-time and power consumption penalties associated with synchronization. We also reviewed techniques for effectively incorporating block processing optimization into the software component of a hardware/software implementation to improve system throughput.

Given the vast design spaces in hardware/software implementation and the complex range of design metrics (e.g., latency, throughput, peak and average power consumption, memory requirements, memory partitioning efficiency, and overall dollar cost), important areas for further research include developing and precisely characterizing a better understanding of the interactions between different implementation metrics during cosynthesis; of relationships between various classes of architectures and the predictability and efficiency of implementations with respect to different implementation metrics; and of more powerful modeling techniques that expose additional application structure in innovative ways, and handle dynamic application behavior (such as the dynamic data flow models and data flow meta-models mentioned in Sec. 2.3). We expect all three of these directions to be highly active areas of research in the coming years.

REFERENCES

1. M Ade, R Lauwereins, JA Peperstraete. Data memory minimisation for synchronous data flow graphs emulated on DSP–FPGA targets. Proceedings of the Design Automation Conference, 1997, pp 64–69.
2. SS Bhattacharyya, PK Murthy, EA Lee. Software Synthesis from Dataflow Graphs. Boston, MA: Kluwer Academic, 1996.
3. B Jacob. Hardware/software architectures for real-time caching. Proceedings of the International Workshop on Compiler and Architecture Support for Embedded Systems, 1999.
4. Y Li, W Wolf. Hardware/software co-synthesis with memory hierarchies. Proceedings of the International Conference on Computer-Aided Design, 1998, pp 430–436.
5. S Wuytack, J-P Diguet, FVM Catthoor, HJ De Man. Formalized methodology for data reuse exploration for low-power hierarchical memory mappings. IEEE Trans VLSI Syst 6:529–537, 1998.
6. P Marwedel, G Goossens, eds. Code Generation for Embedded Processors. Boston: Kluwer Academic, 1995.
7. YTS Li, S Malik. Performance analysis of embedded software using implicit path enumeration. IEEE Trans Computer-Aided Design 16:1477–1487, 1997.
8. S Sriram, SS Bhattacharyya. Embedded Multiprocessors: Scheduling and Synchronization. New York: Marcel Dekker, 2000.
9. TY Yen, W Wolf. Performance estimation for real-time distributed embedded systems. IEEE Trans Parallel Distrib Syst 9:1125–1136, 1998.
10. G De Micheli, M Sami. Hardware–software Co-design. Boston: Kluwer Academic, 1996.
11. P Paulin, C Liem, T May, S Sutarwala. DSP design tool requirements for embedded systems: A telecommunications industrial perspective. J VLSI Signal Process 9(1–2):23–47, January 1995.
12. R Ernst, J Henkel, T Benner. Hardware–software cosynthesis for microcontrollers. IEEE Design Test Computers Mag 10(4):64–75, 1993.
13. DE Thomas, JK Adams, H Schmitt. A model and methodology for hardware/software codesign. IEEE Design Test Computers Mag 10:6–15, 1993.
14. F Balarin. Hardware-Software Co-Design of Embedded Systems: The Polis Approach. Boston: Kluwer Academic, 1997.
15. EA Lee. Embedded software—An agenda for research. Technical Report. Electronics Research Laboratory, University of California at Berkeley UCB/ERL M99/63, December 1999.
16. TH Cormen, CE Leiserson, RL Rivest. Introduction to Algorithms. Cambridge, MA: MIT Press, 1992.

17. DB West. Introduction to Graph Theory. Englewood Cliffs, NJ: Prentice-Hall, 1996.

18. AL Ambler, MM Burnett, BA Zimmerman. Operational versus definitional: A perspective on programming paradigms. IEEE Computer Mag 25:28–43, 1992.

19. EA Lee, DG Messerschmitt. Synchronous dataflow. Proc IEEE 75:1235–1245, 1987.

20. M Ade, R Lauwereins, JA Peperstraete. Buffer memory requirements in DSP applications. Proceedings of the International Workshop on Rapid System Prototyping, 1994, pp 198–223.

21. G Bilsen, M Engels, R Lauwereins, JA Peperstraete. Cyclo-static dataflow. IEEE Trans Signal Process 44:397–408, 1996.

22. S Ritz, M Willems, H Meyr. Scheduling for optimum data memory compaction in block diagram oriented software synthesis. Proceedings of the International Conference on Acoustics, Speech, and Signal Processing, 1995.

23. PP Vaidyanathan. Multirate Systems and Filter Banks. Englewood Cliffs, NJ: Prentice-Hall, 1993.

24. EA Lee, WH Ho, E Goei, J Bier, SS Bhattacharyya. Gabriel: A design environment for DSP. IEEE Tran Acoust Speech Signal Process 37(11):1531–1562, 1989.

25. JT Buck, S Ha, EA Lee, DG Messerschmitt. Ptolemy: A framework for simulating and prototyping heterogeneous systems. Int J Computer Simul, January 1994.

26. DR O'Hallaron. The ASSIGN parallel program generator. Technical Report. School of Computer Science, Carnegie Mellon University, May 1991.

27. S Ritz, M Pankert, H Mey. High level software synthesis for signal processing systems. Proceedings of the International Conference on Application Specific Array Processors, 1992.

28. EA Lee. Representing and exploiting data parallelism using multidimensional dataflow diagrams. Proceedings of the International Conference on Acoustics, Speech, and Signal Processing, 1993, pp 453–456.

29. JT Buck. Static scheduling and code generation from dynamic dataflow graphs with integer-valued control systems. Proceedings of the IEEE Asilomar Conference on Signals, Systems, and Computers, 1994.

30. JT Buck, EA Lee. Scheduling dynamic dataflow graphs using the token flow model. Proceedings of the International Conference on Acoustics, Speech, and Signal Processing, 1993.

31. M Pankert, O Mauss, S Ritz, H Meyr. Dynamic data flow and control flow in high level DSP code synthesis. Proceedings of the International Conference on Acoustics, Speech, and Signal Processing, 1994.

32. A Girault, B Lee, EA Lee. Hierarchical finite state machines with multiple concurrency models. IEEE Trans Computer-Aided Design Integrated Circuits Syst 18(6): 742–760, 1999.

33. B Bhattacharya, SS Bhattacharyya. Parameterized dataflow modeling of DSP systems. Proceedings of the International Conference on Acoustics, Speech, and Signal Processing, 2000.

34. B Bhattacharya, SS Bhattacharyya. Quasi-static scheduling of re-configurable dataflow graphs for DSP systems. Proceedings of the International Workshop on Rapid System Prototyping, 2000.

35. AV Aho, R Sethi, JD Ullman. Compilers Principles, Techniques, and Tools. Reading, MA: Addison-Wesley, 1988.

36. G De Micheli. Synthesis and Optimization of Digital Circuits. New York: McGraw-Hill, 1994.

37. TM Parks, JL Pino, EA Lee. A comparison of synchronous and cyclo-static dataflow. Proceedings of the IEEE Asilomar Conference on Signals, Systems, and Computers, 1995.

38. KN Lalgudi, MC Papaefthymiou, M Potkonjak. Optimizing systems for effective block-processing: The k-delay problem. Proceedings of the Design Automation Conference, 1996, pp 714–719.

39. GC Sih. Multiprocessor scheduling to account for interprocessor communication. PhD thesis, University of California at Berkeley, 1991.

40. KJR Liu, A Wu, A Raghupathy, J Chen. Algorithm-based low-power and high-performance multimedia signal processing. Proc IEEE 86:1155–1202, 1998.

41. EA Lee, SHa. Scheduling strategies for multiprocessor real time DSP. Global Telecommunications Conference, 1989.

42. TC Hu. Parallel sequencing and assembly line problems. Oper Res 9, 1961.

43. A Kalavade, EA Lee. A global critically/local phase driven algorithm for the constrained hardware/software partitioning problem. Proceedings of the International Workshop on Hardware/Software Co-Design, 1994, pp 42–48.

44. A Kalavade, PA Subrahmanyam. Hardware/software partitioning for multifunction systems. IEEE Trans Computer-Aided Design 17:819–837, 1998.

45. BP Dave, G Lakshminarayana, NK Jha. COSYN: Hardware–soft ware co-synthesis of embedded systems. Proceedings of the Design Automation Conference, 1997.

46. T Blickle, J Teich, L Thiele. System-level synthesis using evolutionary algorithms. J Design Automat Embed Syst 3(1):23–58, 1998.

47. T Back, U Hammel, HP Schwefel. Evolutionary computation: Comments on the history and current state. IEEE Trans Evolut Comput 1:3–17, 1997.

48. SS Bhattacharyya, S Sriram, EA Lee. Optimizing synchronization in multiprocessor DSP systems. IEEE Trans Signal Process 45, 1997.

49. SS Bhattacharyya, S Sriram, EA Lee. Resynchronization for multiprocessor DSP systems. IEEE Trans Circuits Syst: Fundam Theory Applic 47(11):1597–1609, 2000.

50. SMH De Groot, S Gerez, O Herrmann. Range-chart-guided iterative data-flow graph scheduling. IEEE Trans Circuits Syst: Fundam Theory Applic 39(5):351–364, May 1992.

51. P Hoang, J Rabaey. Hierarchical scheduling of DSP programs onto multiprocessors for maximum throughput. Proceedings of the International Conference on Application Specific Array Processors, 1992.

52. S Sriram, EA Lee. Determining the order of processor transactions in statically scheduled multiprocessors J VLSI Signal Process 15(3):207–220, 1997.

53. S Ritz, M Pankert, H Meyr. Optimum vectorization of scalable synchronous dataflow graphs. Proceedings of the International Conference on Application Specific Array Processors, 1993.

54. V Zivojnovic, S Ritz, H Meyr. Retiming of DSP programs for optimum vectorization. Proceedings of the International Conference on Acoustics, Speech, and Signal Processing, 1994.

55. R Lauwereins, M Engels, M Ade, JA Peperstraete. GRAPE-II: A system-level prototyping environment for DSP applications. IEEE Computer Mag 28(2):35–43, 1995.

9
Data Transfer and Storage Architecture Issues and Exploration in Multimedia Processors

Francky Catthoor, Koen Danckaert, Chidamber Kulkarni, and Thierry Omnès
IMEC, Leuven, Belgium

1 INTRODUCTION

Storage technology "takes the center stage" [1] in more and more systems because of the eternal push for more complex applications with especially larger and more complicated data types. In addition, the access speed, size, and power consumption associated with this storage form a severe bottleneck in these systems (especially in an embedded context). In this chapter, several building blocks for memory storage will be investigated, with the emphasis on internal architectural organization. After a general classification of the memory hierarchy components in Section 2, cache architecture issues will be treated in Section 3, followed by main memory organization aspects in Section 4. The main emphasis will lie on modern multimedia and telecom oriented processors, both of the microprocessor and DSP type.

Apart from the storage architecture itself, the way data are mapped to these architecture components are as important for a good overall memory management solution. Actually, these issues are gaining in importance in the current age of deep submicron technologies where technology and circuit solutions are not sufficient on their own to solve the system design bottlenecks. Therefore, the last three sections are devoted to different aspects of data transfer and storage exploration: source code transformations (Sec. 5), task versus data parallelism exploitation (Sec. 6), and memory data layout organization (Sec. 7). Realistic multimedia and telecom applications will be used to demonstrate the impressive effects of such techniques.

2 HIERARCHICAL MEMORY ORGANIZATION IN PROCESSORS

The goal of a storage device is, in general, to store a number of n-bit data words for a short or long term. These data words are sent to processing units (processors) at the appropriate point in time (cycle) and the results of the operations are then written back in the storage device for future use. Due to the different characteristics of the storage and access, different styles of devices have been developed.

2.1 General Principles and Storage Classification

A very important difference can be made between memories (for frequent and repetitive use) for short-term and long-term storage. The former are, in general, located very close to the operators and require a very small access time. Consequently, they should be limited to a relatively small capacity (<32 words typically) and are usually taken up in feedback loops over the operators (e.g., RA–EXU–TRIA–BusA–RA in Fig. 1). The devices for longer-term storage are, in general, meant for (much) larger capacities (from 64 to 16M words) and take a separate cycle for read or write access (Fig. 2). Both categories will be described in more detail in the following subsections. A full memory hierarchy chain is illustrated in Figure 3 (for details, see Sec. 4.3).

Six other important distinctions can be made using the treelike "genealogy" of storage devices presented in Figure 4:

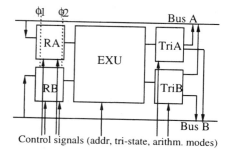

Figure 1 Two register files RA and RB in feedback loop of a data path.

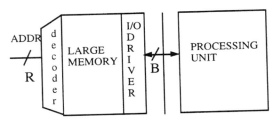

Figure 2 Large-capacity memory communicating with processing unit.

1. **Read-only or read/write (R/W) access**: Some memories are used
    only to store constant data (such as ROMs). Good alternatives for this
    are, for example, programmable logic arrays (PLA) or multilevel logic
    circuits, especially when the amount of data is relatively small. In most
    cases, data need to be overwritable at high speeds, which means that
    read and write are treated with the same priority (R/W access) such
    as in random-access memories or RAMs. In some cases, the ROMs
    can be made "electrically alterable" (=Write-few) with high energies
    (EAROM) or "programmable" by means of, for example, fuses
    (PROM). Only the R/W memories will be discussed later.

2. **Volatile or not**: For R/W memories, usually, the data are removed
    once the power goes down. In some cases, this can be avoided but
    these nonvolatile options are expensive and slow. Examples are mag-

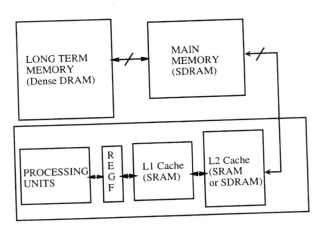

Figure 3 Typical memory hierarchy in processors.

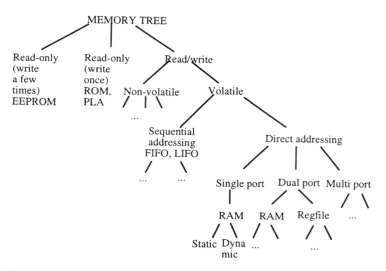

Figure 4 Storage classification.

netic media and tapes which are intended for slow access of mass data. We will restrict ourselves to the most common case on the chip, namely volatile.

3. **Address mechanism**: Some devices require only sequential addressing, such as the first-in first-out (FIFO) queue, first-in last-out (FILO), or stack structures discussed in Section 2.3, which put a severe restriction on the order in which the data are read out. Still, this restriction is acceptable for many applications. A more general but still sequential access order is available in a pointer-addressed memory (PAM). In the PAM, the main limitation is that each data value is both written and read once in any statically predefined order. However, in most cases the address sequence should be random (including repetition). Usually, this is implemented with a direct addressing scheme (typically called a random-access memory or RAM). Then, an important requirement is that in this case, the access time should be independent of the address selected. In many programmable processors, a special case of random-access-based buffering is realized, exploiting comparisons of tags and usually also including (full or partial) associativity (in a so-called cache buffer).

4. **Number of independent addresses and corresponding gateways (buses) for access**: This parameter can be one (single port), two (dual port), or even more (multiport). Any of these ports can be for reading

only, writing only, or R/W. Of course, the area occupied will increase considerably with the number of ports.

5. **Internal organization of the memories**: The memory can be meant of capacities which remain small or which can become large. Here, a trade-off is usually involved between speed and area efficiency. The register files in Section 2.2 constitute an example of the fast small-capacity organizations which are usually also dual ported or even multiported. The queues and stacks in Section 2.3 are meant for medium-sized capacities. The RAMs in Section 4 can become extremely large (up to 256 Mbit for the state of the art) but are also much slower in random access.

6. **Static or dynamic**: For R/W memories, the data can remain valid as long as VDD is on (static cell) or the data should be refreshed about every millisecond (dynamic cell). Circuit-level issues are discussed in overview articles like Ref. 2 for SRAMs and Refs. 3 and 4 for DRAMs.

In the following subsections, the most important read/write-type memories and their characteristics will be investigated in more detail.

2.2 Register File and Local Memory Organization

In this subsection, we discuss the register file and local memory organization. An illustrative organization for a dual-port register file with two addresses where the separate read and write addresses are generated from an instruction ROM is shown in Figure 5. In this case, two buses (A and B) are used but only in one direction so the write and read addresses directly control the port access. In general, the number of addresses can be less than the number of port(s) and the buses

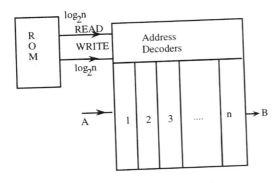

Figure 5 Regfile with both R and W address.

can be bidirectional. Additional control signals decide whether to write or read and for which port the address applies. The register file of Figure 5 can be used very efficiently in the feedback loop of a data path, as already illustrated in Figure 1. In general, it is used only for the storage of temporary variables in the application running on the data path (sometimes also referred to as execution unit). This is true both for most modern general-purpose reduced instruction-set computers (RISCs) and especially for modern multimedia-oriented signal processors which have regfiles up to 128 locations. For multimedia-oriented very long instruction word (VLIW) processors or modern superscalar processors, regfiles with a very large access bandwidth are provided, up to 17 ports (see, e.g., Ref. 5). Application-specific instruction set processors (ASIPs) and custom instruction set processors make heavy use of regfiles for the same purpose. It should be noted that although it has the clear advantage of very fast access, the number of data words to be stored should be minimized as much as possible due to the relatively area-intensive structure of such register files (both due to the decoder and the cell area overhead). Detailed circuit issues will not be discussed here (see, e.g., Ref. 6). After this brief discussion of the local foreground memories, we will now proceed with background memories of the sequential addressing type.

2.3 Sequentially Addressable Memories

Only two of the variety of possible types will be discussed. The memory matrix itself is the same for both and is also identical to the ones used for (multiport) RAMs.

2.3.1 First-In First-Out Structures

Such a FIFO is sometimes also referred to as a queue. In general (Fig. 6), it can be used to handle the communication between processing units P1 (source) and P2 (destination) which do the following:

1. Exhibit the same average data throughput

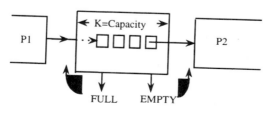

Figure 6 Communication between FIFO and two processors.

2. Require different production (P1) and consumption (P2) cycles

3. Treat the data in the same sequence

The required capacity K depends on the maximal skew between the consumption and production of the data. Optimization of K is possible if the "schedules" of P1 and P2 can be delayed relative to one another. The main principles of the organization of a dual-port FIFO, with one read RPTR and one write bus, are shown in Figure 7.

Note the circular shift registers (pointers) for both read and write address WPTR selection which contain a single 1 which is shifted to consecutive register locations, controlled by a SHR or SHW signal. The latter are based on external R or W commands. A small FSM is, in general, provided to supervise the FIFO operation. The result of a check whether the position of the two pointers is identical is usually also sent as a flag to this FSM. In this way, also the fact whether the FIFO is "full" or "empty" can be monitored and broadcasted to the "periphery."

2.3.2 First-In Last-Out Structures

Such a FILO is sometimes also referred to as a stack. In general, it can be used to handle the storage of data which has to be temporarily stored ("pushed" on the stack) and read out ("popped" off the stack) in the reverse order. It is regularly used in the design of recursive algorithms and especially in subroutine stacks.

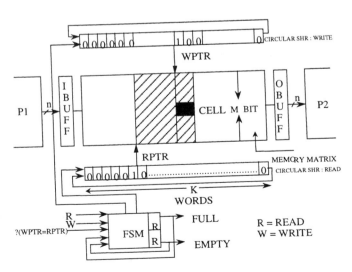

Figure 7 Internal FIFO organization.

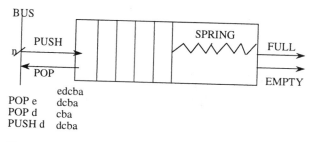

Figure 8 Dynamic FILO.

In principle, the stack can be made "dynamic" (Fig. 8) where the data are pushed and popped in such a way that all data move (as if a spring were present). This leads to a tremendous waste of power in a complementary metal-oxide semiconductor (CMOS) and should be used only in other technologies. A better solution in CMOS is to make the stack "static" as in Figure 9. Here, the only moves are made in a shift register (pointer) which can now move in two directions, as opposed to the unidirectional shift in the FIFO case.

3 CACHE MEMORY ORGANIZATION

The objectives of this section are as follows: (1) to discuss the fundamental issues about how cache memories operate; (2) to discuss the characteristic parameters

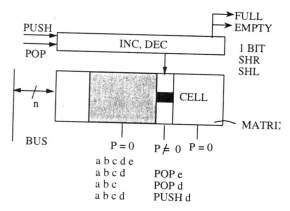

Figure 9 Static FILO.

of a cache memory and their impact on performance as well as power; (3) to briefly introduce the three types of cache miss; and (4) to discuss the differences between hardware- and software-controlled cache for the current state-of-the-art media and digital signal processors (DSPs).

3.1 Basic Cache Architecture

In this subsection, a brief overview of the different steps involved in the operation of a cache are presented. We will use a direct-mapped cache as shown in Figure 10 for explaining the operation of a cache, but the basic principles remain the same for other types of cache memory [7]. The following steps happen whenever there is a read/write from/to the cache: (1) address decoding; (2) selection based on index and/or block offset; (3) tag comparison; (4) data transfer. These steps are highlighted using an ellipse in Figures 10 and 11.

The first step (address decoding) performs the task of decoding the address, supplied by the CPU, into the block address and the block offset. The block address is further divided into tag and index address of the cache. Once the address decoding is complete, the index part of the block address is used to obtain the particular cache line demanded by the CPU. This line is chosen only if the valid bit is set and the tag portion of data cache and the block address match with each other. This involves the comparison of all the individual bits. Note

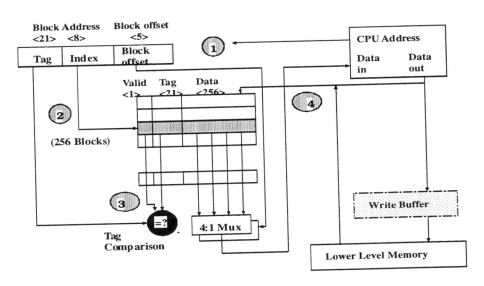

Figure 10   An 8-KB direct mapped cache with 32-byte blocks.

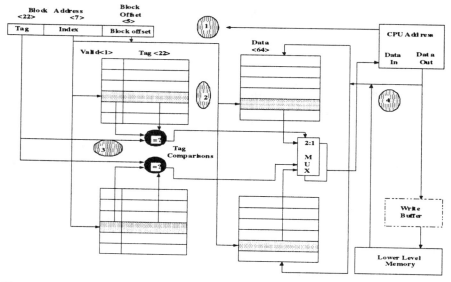

Figure 11 An 8-KB two-way associative cache with 32-byte blocks.

that for a direct-mapped cache, we have only one tag comparison. Once this process is done, the block offset is used to obtain the particular data element in the chosen cache line. This data element is now transferred to/from the CPU.

In Figure 10, we have used a direct-mapped cache, whereas if we had used an n-way set-associative cache, the following differences will be observed: (1) n tag comparisons instead of one and (2) less index bits and more tag bits. This is shown in Figure 11, which shows a two-way set associative cache. We will briefly discuss some of the issues that need to be considered during the design of cache memories in the next subsection.

3.2 Design Choices

In this subsection, we explain in brief some of the important decisions involved in the design of a cache. We shall discuss the choice of line size, degree of associativity, updating policy, and replacement policy.

1. Line size: The line size is the unit of data transfer between the cache and the main memory. The line size is also referred to as block size in the literature. As the line size increases from very small to very large, the miss ratio will initially decrease, because a miss will fetch more data at a time. Further increases will then cause the miss ratio

to increase, as the probability of using the newly fetched (additional) information becomes less than the probability of reusing the data that are replaced [8]. The optimal line size in the above cases is completely algorithm dependent. In general, very large line sizes are not preferred because this contributes to larger load latencies and increased cache pollution. This is true for both the general-purpose and embedded applications.

2. Mapping and associativity: The process of transferring data elements from main memory to the cache is termed "mapping." Associativity refers to the process of retrieving a number of cache lines and then determining if any of them is the target. The degree of associativity and the type of mapping having significant impact on the cache performance. Most caches are set associative [7], which means that an address is mapped into a set and then an associative search is made of that set (see Figs. 10 and 11). Empirically, and as one would expect, increasing the degree of associativity decreases the miss ratio. The highest miss ratios are observed for direct-mapped cache, two-way associativity is significantly better and four-way is slightly better still. Further increase in associativity only slowly decreases the misses. Nevertheless, cache with a larger associativity requires more tag comparisons and these comparators constitute a significant amount of total power consumption in the memories. Thus, for embedded applications where power consumption is an important consideration, associativities larger than four or eight are not commonly observed. Some architectural techniques for low-power cache are presented in Refs. 9 and 10.

3. Updating policy: The process of maintaining coherency between two consecutive memory levels is termed as updating policy. There are two basic approaches to updating the main memory: *write through* and *write back*. With write through, all the writes are immediately transmitted to the main memory (apart from writing to the cache); when using write back, (most) writes are written to the cache and are then copied back to the main memory as those lines are replaced. Initially, write through was the preferred updating policy because it is very easy to implement and solves the problems of coherence of data. However, it also generates a lot of traffic between various levels of memories. Hence, most current state-of-the-art media and DSP processors use a write-back policy. A trend toward giving control of the updating policy to the user (or compiler) is observed currently [11], which can be effectively exploited to reduce power, due to reduced write backs, by compile-time analysis.

4. Replacement policy: The replacement policy of a cache refers to the type of protocol used to replace a (partial or complete) line in the cache

on a cache miss. Typically a LRU (least recently used) type of policy is preferred, as it has acceptable results. Current state-of-the-art general-purpose as well as DSP processors have hardware-controlled replacement policies; that is, they have hardware counters to monitor the least recently used data in the cache. In general, we have observed that a policy based on compile-time analysis will always have significantly better results than fixed statistical decisions.

3.3 Classification of Cache Misses

A *cache miss* is said to have occurred whenever data requested by the CPU are not found in the cache. There are essentially three types of cache miss, namely compulsory, capacity, and conflict misses:

1. Compulsory misses: The first access to a block of data is not in the cache, so the block must be brought into the cache. These cache misses are termed compulsory cache misses.
2. Capacity misses: If the cache cannot contain all the blocks needed during the execution of a program, then some blocks need to be discarded and retrieved later. The cache misses due to discarding of these blocks are termed capacity misses. Alternatively, the total number of misses in a fully associative cache are termed capacity misses.
3. Conflict misses: For a set-associative cache, if too many blocks are mapped to the same set, then some of the blocks in the cache need to be discarded. These cache misses are termed conflict misses. The difference between the total number of cache misses in a direct-mapped or a set-associative cache and that of a fully associative cache are termed conflict misses.

For most real-life applications, the capacity and conflict misses are dominant. Hence, reducing these misses is vital to achieving better performance and reducing the power consumption. Figure 12 illustrates the detailed cache states for a fully associative cache. Note that a diagonal bar on an element indicates that the particular element was replaced by the next element, which is in the next column of the same row without a bar, due to the (hardware) cache-mapping policy. Hence, every diagonal bar represents a cache miss. The main memory layout is assumed to be single contiguous; namely array a[ ] is residing in locations 0 to 10 and array b[ ] in locations 11 to 21.

We observe from Figures 12 and 13 that the algorithm needs 32 data accesses. To complete these 32 data accesses, the fully associative cache requires 14 cache misses. Of these 14 cache misses, 12 are compulsory misses and the remaining 2 are capacity misses. In contrast, the direct-mapped cache requires 24 cache misses, as seen in Figure 13. This means that of 32 data accesses, 24 accesses are made to the off-chip memory and the remaining 8 are due to data

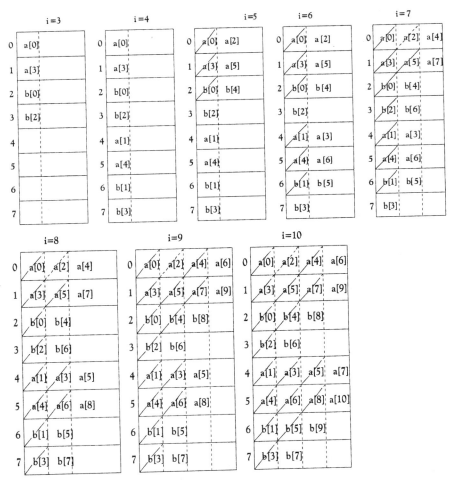

Figure 12 Initial algorithm and the corresponding cache states for a fully associative cache. For (i = 3; i < 11; i++), b[i − 1] = b[i − 3] + a[i] + a[i − 3].

reuse in the cache (on-chip). Thus, for the algorithm in our example, we have 10 conflict misses, 2 capacity misses, and 12 compulsory misses.

3.4 Hardware Versus Software Caches

Table 1 lists the major differences between hardware and software controlled caches for the current state-of-the-art multimedia and DSP processors. We will briefly discuss these differences:

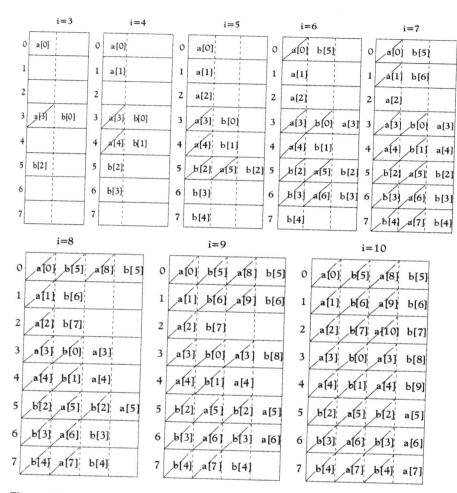

Figure 13 Initial algorithm and the corresponding cache states for a direct-mapped cache. For $(i = 3; i < 11; i++)$, $b[i - 1] = b[i - 3] + a[i] + a[i - 3]$.

1. The hardware-controlled caches rely on the hardware to do the data cache management. Hence, to perform this task, the hardware uses basic concepts of "cache lines" and "sets." For the software-controlled cache, the cache management is done by the compiler (or the user). The complexity of designing a software cache is much less and it requires only a basic concept of unit data transfer, which is "cache lines."

Table 1 Differences Between Hardware and Software Caches for Current State-of-the-Art Multimedia and DSP Processors

	Hardware-controlled cache	Software-controlled cache
Basic concepts	Lines and sets	Lines
Data transfer	Hardware	Partly software
Updating policy	Hardware	Software
Replacement policy	Hardware	NA

2. The hardware performs the data transfer based on the execution order of the algorithm at run time using fixed statistical measures, whereas for the software controlled cache, this task is performed either by the compiler or the user. This is currently possible using high-level compile-time directives like ''ALLOCATE( )'' and link time options like ''LOCK''—to lock certain data in part of the cache, through the compiler/linker [11].

3. The most important difference between hardware- and software-controlled cache is in the way the next higher level of memory is updated, namely the way coherence of data is maintained. For the hardware-controlled cache, the hardware writes data to the next higher level of memory either every time a write occurs or when the particular cache line is evicted, whereas for the software-controlled cache, the compiler decides when and whether or not to write back a particular data element [11]. This results in a large reduction in the number of data transfers between different levels of memory hierarchy, which also contributes to lower power and reduced bandwidth usage by the algorithm.

4. The hardware-controlled cache needs an extra bit for every cache line for determining the least recently used data, which will be replaced on a cache miss. For the software-controlled cache, because the compiler manages the data transfer, there is no need for additional bits or a particular replacement policy.

4 MAIN MEMORY ORGANIZATION

A large variety of possible types of RAM for use as main memories has been proposed in the literature and research on RAM technology is still very active, as demonstrated by the results in for example, the proceedings of the latest International Solid-State Circuits (ISSCC) and Custom Integrated Circuit (CICC) conferences. Summary articles are available in Refs. 4, 12, and 13. The general orga-

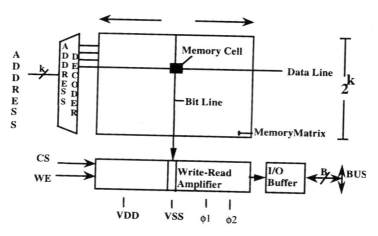

Figure 14   Basic floor plan for B-bit RAM with 2^k words.

nization will depend on the number of ports, but, usually, single-port structures are encountered in the large-density memories. This will also be the restriction here. Most other distinguishing characteristics are related to the circuit design (and the technological issues) of the RAM cell, the decoder, and the auxiliary R/W devices. In this section, only the general principles will be discussed. Detailed circuit issues fall outside the scope of this chapter, as mentioned earlier.

4.1   Floor-Plan Issues

For a B-bit organized RAM with a capacity of 2^k words, the floor plan in Figure 14 is the basis for everything. Note the presence of read and write amplifiers which are necessary to drive or sense the (very) long bit lines in the vertical direction. Note also the presence of write-enable (WE) (for controlling the R/W option) and a chip-select (CS) control signal, which is mainly used to save power. The different options and their effects are summarized in Table 2.

Table 2   Summary of Control Options for Figure 14

CS	WE	IN	OUT
0	X	X	Z
1	0	X	Valid
1	1	Valid	Z

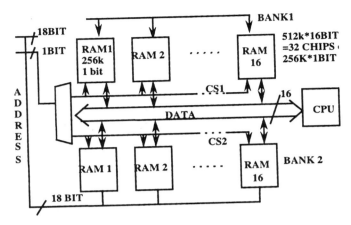

Figure 15 Example of memory bank organization.

The CS signal is also necessary to allow the use of memory banks (on separate chips) as needed for personal computers or workstations (Fig. 15). Note the fact that the CS1 and CS2 control signals can be considered as the most significant part of the 18-bit address. Indeed, in a way, the address space is split up vertically over two separate memory planes. Moreover, every RAM in a horizontal slice contributes only a single data bit.

For large-capacity RAMs, the basic floor plan of Figure 14 leads to a very slow realization because of the too long bit lines. For this purpose, the same principle as in large ROMs, namely postdecoding, is applied. This leads to the use of an X decoder and a Y decoder (Fig. 16) where the flexibility of the floor-plan shape is now used to end up with a near square (dimensions $x \times y$), which

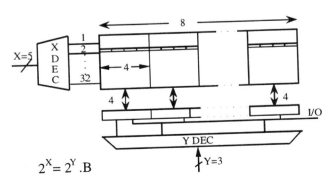

Figure 16 Example of postdecoding.

makes use of the chip area in the most optimal way and which reduces the access time and power (wire *th* related). In order to achieve this, the following equations can be applied:

$$x + y = k \quad \text{and} \quad x + y + \log_2 B$$

leading to $x = (\log_2 B + k)/2$ and $y = k - x$ for maximal "squareness."

This breakup of a large memory plane into several subplanes is very important also for low-power memories. In that case however, care should be taken to enable only the memory plane which contains data needed in a particular address cycle. If possible, the data used in successive cycles should also come from the same plane because activating a new plane takes up a significant amount of extra precharge power. An example floor plan in Figure 16 is drawn for $k = 8$ and $B = 4$.

The memory matrix can be split up in two or more parts too, to reduce the *th* of the word lines also by a factor 2 or more. This results in a typically heavily partitioned floor plan [3] as shown in a simplified form in Figure 17.

It should be noted also that the word *th* of data stored in the RAM is usually matched to the requirement of the application in the case of an on-chip RAM embedded in an ASIC. For RAM chips, this word organization normally has been standardized to a few choices only. Most large RAMs are 1-bit RAMs. However, with the push for more application-specific hardware, also 4-bit (nibble) and 8-

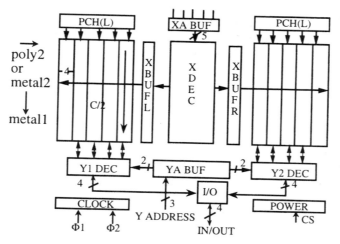

Figure 17 Partitioning of memory matrix combined with postdecoding.

bit (byte) RAMs are now commonly produced. In the near future, one can expect other formats to appear.

4.2 Synchronization and Access Times

An important aspect of RAMs are the access times, both for read and write. In principle, these should be balanced as much as possible, as the worst case determines the maximal clock rate from the point of view of the periphery. It should be noted that the RAM access itself can be either "asynchronous" or "clocked."

Most individual RAM chips are of the asynchronous type. The evolution of putting more and more RAM on-chip has lead to a state where stand-alone memories are nearly only DRAMs (dynamic RAMs). Within that category, an important subclass is formed by the so-called synchronous DRAMs or SDRAMs (see, e.g., Refs. 14 and 15). In that case, the bus protocol is fully synchronous, but the internal operation of the DRAM is still partly asynchronous. For the more conventional DRAMs which also have an asynchronous interface, the internal organisation involves special flags, which signal the completion of a read or write and thus the "readiness" for a new data and/or address word. In this case, a distinction has to be made between the address access delay t_{AA} and the chip (RAM) access delay t_{ACS}, as illustrated in Figure 18. Ideally, $t_{ACS} = t_{AA}$, but, in practice, t_{ACS} is the largest. Thus, special tricks have to be applied to approach this ideal.

For on-chip (embedded) RAMs as used in ASICs, typically a clocked RAM is preferred, as it is embedded in the rest of the (usually synchronous) architecture. These are nearly always SRAMs. A possible timing (clock) diagram for

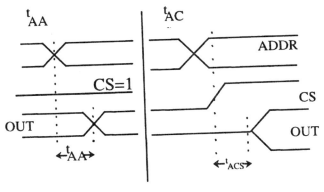

Figure 18 Distinction between t_{AA} and t_{ACS} for asynchronous RAM.

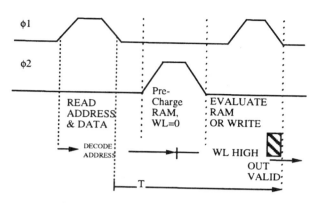

Figure 19   Timing diagram for clocked RAM.

such a RAM is illustrated in Figure 19, in which $\phi 1$ and $\phi 2$ are clock phases. The different pipeline stages in the reading/writing of a value from/to the synchronous RAM are indicated.

4.3   The External Data Access Bottleneck

In most cases, a processor requires one or more large external memories to store the long-term data (mostly of the DRAM type). For data-dominated applications, the total system power cost in the past was large due to the presence of these external memories on the board.

Because of the heavy push toward lower-power solutions in order to keep the package costs low, and recently also for mobile applications or due to reliability issues, the power consumption of such external DRAMs has been reduced significantly. Apart from circuit and internal organization techniques [16,17], also technology modifications such as the switch to SOI (silicon-on-insulator) [18] are considered. Because of all these principles to distribute the power consumption from a few "hot spots" to all parts of the architecture, the end result is indeed a very optimized design for power, where every part of the memory organization consumes a similar amount [17,19].

It is expected, however, that not much more can be gained because the "bag of tricks" now contains only the more complex solutions with a smaller return-on-investment. Note, however, that the combination of all of these approaches indicates a very advanced circuit technology, which still outperforms the current state of the art in data-path and logic circuits for low-power design (at least in industry). Hence, it can be expected that the relative power in the nonstorage parts can be more drastically reduced still in the future (on condition that similar invest-ments are done).

Combined with the advance in process technology, all of this has lead to a remarkable reduction of the DRAM related power: from several watts for the 16–32-MB generation to about 100 mW for 100-MHz operation in a 256-MB DRAM.

Hence, modern stand-alone DRAM chips, which are of the so-called syn-chronous (SDRAM) type, already offer low-power solutions, but this comes at a price. Internally, they contain banks and a small cache with a (very) wide width connected to the external high-speed bus (see Fig. 20) [15,20]. Thus, the low-power operation per bit is only feasible when they operate in burst mode with large data widths.

This is not directly compatible with the actual use of the data in the proces-sor data paths; therefore, without a buffer to the processors, most of the bits that are exchanged would be useless (and discarded). Obviously, the effective energy consumption per useful bit becomes very high in that case and also the effective bandwidth is quite low.

Therefore, a hierarchical and typically much more power-hungry intermedi-ate memory organization is needed to match the central DRAM to the data-ordering and bandwidth requirements of the processor data paths. This is also illustrated in Figure 21. The decrease of the power consumption in fast random-access memories is not yet as advanced as in DRAMs but that one is saturating, because many circuit and technology level tricks have been applied also in SRAMs. As a result, fast SRAMs keep on consuming on the order of watts for

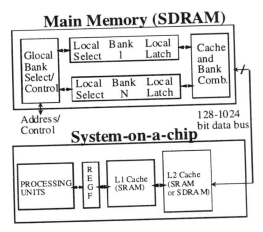

Figure 20   External data access bottleneck illustration with SDRAM.

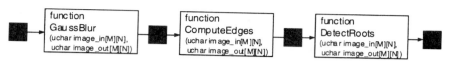

Figure 21 Initial cavity-detection algorithm.

high-speed operation around 500 MHz. Thus, the memory-related-system power bottleneck remains a very critical issue for data-dominated applications.

From the process technology point of view, this is not so surprising, especially for submicron technologies. The relative power cost of interconnections is increasing rapidly compared to the transistor-related (active circuit) components. Clearly, local data paths and controllers themselves contribute little to this overall interconnect compared to the major data/instruction buses and the internal connections in the large memories. Hence, if all other parameters remain constant, the energy consumption (and also the delay or area) in the storage and transfer organization will become even more dominant in the future, especially for deep submicron technologies. The remaining basic limitation lies in transporting the data and the control (like addresses and internal signals) over large on-chip distances and in storing them.

One last technological recourse to try to alleviate the energy-delay bottleneck is to embed the memories as much as possible on-chip. This has been the focus of several recent activities e.g., in the Mitsubishi announcement of an SIMD processor with a large distributed DRAM in 1996 [21] (followed by the offering of the "embedded DRAM" technology by several other vendors) and the IRAM initiative of Dave Patterson's group at the University of California, Berkeley [22]. The results show that the option of embedding logic on a DRAM process leads to a reduced power cost and an increased bandwidth between the central DRAM and the rest of the system. This is indeed true for applications where the increased processing cost is allowed [23]. However, it is a one-time drop, after which the widening energy–delay gap between the storage and the logic will keep on progressing, due to the unavoidable evolution of the relative interconnect contributions (see above). Thus, in the longer term, the bottleneck should be broken also by other means.

In Section 5–7, it will be shown that this is feasible, with quite spectacular effects at the level of the system design methodology. The price paid there will be increased design complexity, which can, however, be offset with appropriate design methodology support tools.

In addition to the mainstream evolution of these SRAMs, DRAMs, and SDRAMs, also more customized large-capacity high-bandwidth memories are proposed, intended for more specialized purposes. Examples are video RAMs, very wide DRAMs, and SRAMs with more than two ports (see, e.g., the eight-port SRAM in Ref. 24) [4].

5 CODE REWRITING TECHNIQUES FOR ACCESS LOCALITY

Code rewriting techniques, consisting of loop and data flow transformations, are an essential part of modern optimizing and parallelizing compilers. They are mainly used to enhance the temporal and spatial locality for cache performance and to expose the inherent parallelism of the algorithm to the outer (for asynchronous parallelism) or inner (for synchronous parallelism) loop nests [25–27]. Other application areas are communication-free data allocation techniques [28] and optimizing communications in general [29].

It is thus no surprise that these code rewriting techniques are also at the heart of our DTSE methodology. As the first step (after the preprocessing and pruning) in the script, they are able to significantly reduce the required amount of storage and transfers. As such however, they only increase the locality and regularity of the code. This enables later steps in the script [notably the data reuse, memory (hierarchy) assignment and in-place mapping steps] to arrive at the desired reduction of storage and transfers.

Crucial in our methodology is that these transformations have to be applied globally (i.e. with the entire algorithm as scope). This is in contrast with most existing loop transformation research, where the scope is limited to one procedure or even one loop nest. This can enhance the locality (and parallelization possibilities) within that loop nest, but it does not change the global data flow and associated buffer space needed between the loop nests or procedures. In this section, we will also illustrate our preprocessing and pruning step, which is essential to apply global transformations.

In Section 5.1, we will first give a very simple example to show how loop transformations can significantly reduce the data storage and transfer requirements of an algorithm. Next, we will demonstrate our approach by applying it to a cavity-detection application for medical imaging. This application is introduced in Section 5.2 and the code rewriting techniques are applied in Section 5.3. Finally (Sec. 5.4), we will also give a brief overview of how we want to perform global loop transformations automatically in the DTSE context.

5.1 Simple Example

This example consists of two loops: The first loop produces an array A[ ] and the second loop reads A[ ] to produce an array B[ ]. Only the B[ ] values have to be kept in memory afterward:

```
for (i = 1; i <= N; ++i) {
    A[i] = . . .;
}
for (i = 1; i < = N; ++i) {
    B[i] = f(A[i]);
}
```

Should this algorithm be implemented directly, it would result in high storage and bandwidth requirements (assuming that N is large), as all A[ ] signals have to be written to an off-chip background memory in the first loop and read back in the second loop. Rewriting the code using a loop merging transformation gives the following:

```
for (i = 1; i < = N; ++i) {
    A[i] = . . .;
    B[i] = f(A[i]);
}
```

In this transformed version, the A[ ] signals can be stored in registers because they are immediately consumed after they have been produced and because they are not needed afterward. In the overall algorithm, this significantly reduces storage and bandwidth requirements.

5.2 The Cavity-Detection Demonstrator

The cavity-detection algorithm is a medical image processing application which extracts contours from images to help physicians detect brain tumors. The initial algorithm consists of a number of functions, each of which has an image frame as input and one as output, as shown in Figure 21. In the first function, a horizontal and vertical gauss-blurring step is performed, in which each pixel is replaced by a weighted average of itself and its neighbours. In the second function [Computer-Edges], the difference with all eight neighbors is computed for each pixel, and this pixel is replaced by the maximum of those differences. In the last function [DetectRoots()], the image is first reversed. To this end, the maximum value of the image is computed, and each pixel is replaced by the difference between this maximum value and itself. Next, we look, for each pixel, whether a neighbor pixel is larger than itself. If this is the case, the output pixel is false, otherwise it is true. The complete cavity-detection algorithm contains some more functions, but these have been left out for simplicity. The initial code looks as follows:

```
void GaussBlur (unsigned char image_in[M][N], unsigned char gxy[M][N]) {
    unsigned char gx[M][N];
```

```
      for (y = 0; y < M; ++y)
        for (x = 0; x < N; ++x)
          gx[y][x] = . . . //Apply horizontal gaussblurring
      for (y = 0; y < M; ++y)
        for (x = 0; x <N; ++ x)
          gxy[y][x] = . . . // Apply vertical gaussblurring
}
void ComputeEdges (unsigned char gxy[M][N], unsigned char ce[M][N]) {
    for (y = 0; y < M; ++y)
      for (x = 0; x < N; ++x)
        ce[y][x] = . . . // Replace pixel with the maximum difference
        with its neighbors
}
void Reverse (unsigned char ce[M][N], unsigned ce_rev[M][N]) {
    for (y = 0; y < M; ++y)
      for (x = 0; x < N; ++x)
        maxval = . . . // Compute maximum value of the image
      //Subtract every pixelvalue from this maximum value
    for (y = 0; y < M; ++y)
      for (x = 0; x < N; ++x)
        ce_rev[y][x] = maxval − ce[y][x];
}
void DetectRoots (unsigned char ce[M][N], unsigned char
image_out[M][N]) {
  unsigned char ce_rev[M][N];
  Reverse (ce, ce_rev); // Reverse image
  for (y = 0; y < M; ++y)
    for (x = 0; x < N; ++x)
      image_out[y][x] = . . . // Is true if no neighbors are bigger than
      current pixel
}
void main () {
  unsigned char image_in[M][N], gxy[M][N], ce[M][N], image_out[M][N];
  //...(read image)
  GaussBlur(image_In, gxy);
  ComputeEdges(gxy, ce);
  DetectRoots (ce, image_out);
```

5.3   Code Rewriting for the Cavity-Detection Demonstrator

For the initial cavity-detection algorithm, as given in Figure 21, the data transfer and storage requirements are very high. The main reason is that each of the functions reads an image from off-chip memory and writes the result back to this memory. After applying our DTSE methodology, these off-chip memories and transfers will be heavily reduced, resulting in much less off-chip data storage and transfers. Note that all steps will be performed in an application-independent systematic way.

5.3.1 Preprocessing and Pruning

First of all, the code is rewritten in a three-level hierarchy. The top level (level 1) contains system-level functions between which no optimizations are possible. In level 2, all relevant data-dominated computations are combined into one single procedure, which is more easily analyzable than a set of procedures/functions. Level 3 contains all low-level (e.g., mathematical) functions, which are not relevant for the data flow. Thus, all further optimizations are applied to the level 2 function. This is a key feature of our approach because it allows the exposure of the available freedom for the actual exploration steps. The code shown further on is always extracted from this level 2 description.

Next, the data flow is analyzed and all pointers are substituted by indexed arrays, and the code is transformed into single-assignment code such that the flow dependencies become fully explicit. This will allow for more aggressive data flow and loop transformations. Furthermore, it will also lead to more freedom for our data-reuse and in-place mapping stages later. This will allow the further compaction of the data in memory, in a more global and more efficient way than in the initial algorithm code.

5.3.2 Global Data Flow Transformations

In the initial algorithm, there is a function Reverse which computes the maximum value of the whole image. This computation is a real bottleneck for DTSE. From the point of view of computations, it is almost negligible, but from the point of view of transfers, it is crucial, as the whole image has to be written to off-chip memory before this computation and then read back afterward. However, in this case, this computation can be removed by a data flow transformation.

Indeed, the function Reverse( ) is a direct translation from an original system-level description of the algorithm, where specific functions have been reused. It can be avoided by adapting the next step of the algorithm [DetectRoots( )] by means of a data flow transformation. Instead of `image_out[y][x] = if (p > {q})` . . ., where p and q are pixel elements produced by Reverse( ), we can write `image_out[y][x] = if (-p < {-q})` . . . or `image_out[y][x] = if (c - p < {c - q})` . . ., where `c = maxval` is a constant. Thus, instead of performing the Reverse( ) function and implementing the original DetectRoots( ), we will omit the Reverse( ) function and implement the following:

```
void cav_detect (unsigned char image_in[M][N], unsigned char image_out
[M][N]) {
    for (y = 0; y < M; ++y)
        for (x = 0; x < N; ++x)
            gx[y][x] = . . . // Apply horizontal gaussblurring
```

```
for (y = 0; y < M; ++y)
    for (x = 0; x < N; ++x)
        gxy[y][x] = . . . // Apply vertical gaussblurring
for (y = 0; y < M; ++y)
    for (x = 0; x < N; ++x)
        ce[y][x] = . . . // Replace pixel with the maximum difference
        with its neighbors
for (y = 0; y < M; ++y)
    for (x = 0; x < N; ++x)
        image_out[y][x] = . . . // Is true if no neighbors are smaller
        than current pixel
}
```

Next, another data flow transformation can be performed to reduce the initializations. In the initial version, these are always done for the entire image frame. This is not needed; only the borders have to be initialized, which saves a lot of costly memory accesses. In principle, designers are aware of this, but we have found that, in practice, the original code usually still contains a large amount of redundant accesses. By systematically analyzing the code for this (which is heavily enabled by the preprocessing phase), we can identify all redundancy in a controlled way.

5.3.3 Global Loop Transformations

The loop transformations which we apply in our methodology are relatively conventional as such, but we apply them much more globally (over all relevant loop nests together) than conventionally done, which is crucial to optimize the global data transfers. Thus, the steering guidelines for this clearly differ from the traditional compiler approach.

In our example, a global y-loop folding and merging transformation is first applied. The resulting computational flow is depicted in Figure 22; it is a line-based pipelining scheme. This is possible because, after the data flow transformations, all computations in the algorithm are neighborhood computations. The code

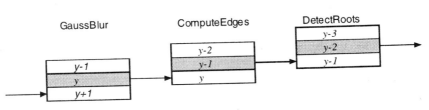

Figure 22 Cavity-detection algorithm after y-loop transformation.

now looks as follows (after leaving out some conditions on y to reduce the complexity):

```
void cav_detect (unsigned char in_image[M][N], unsigned char
out_image[M][N]) {
    for (y = 0; y < M + 3; ++y)
        for (x = 0; x < N; ++x)
            gx[y][x] = ...// Apply horizontal gaussblurring
        for (x = 0; x < N; ++x)
            gxy[y - 1][x] = . . . // Apply vertical gaussblurring
        for (x = 0; x < N; ++x)
            ce[y - 2][x] = . . . // Replace pixel with max. difference
            with its neighbors
        for (x = 0; x < N; ++x)
            image_out[y - 3] = . . . //Is true if no neighbors are smaller
            than this pixel
}
```

A global x-loop folding and merging transformation is applied too. This further increases the locality of the code and thus the possibilities for data reuse. As a result, the computations are now performed according to a fine-grain (pixel-based) pipelining scheme (see Fig. 23). The code now looks as follows (also here, conditions on x and y have been left out of the code):

```
void cav_detect (unsigned char in_image[M][N], unsigned char
out_image[M][N]) {
    for (y = 0; y < M + 3; ++y)
        for (x = 0; x < N + 2; ++x) {
            gx[y][x] = . . . // Apply horizontal gaussblurring
            gxy[y - 1][x] = . . . // Apply vertical gaussblurring
            ce[y - 2][x - 1] = . . . // Replace pixel with max. difference
            with its neighbors
            image_out[y - 3][x - 2] = . . . // Is true if no neighbors are
            smaller than this pixel
    }
}
```

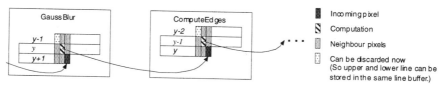

Figure 23 Cavity-detection algorithm after x-loop transformation.

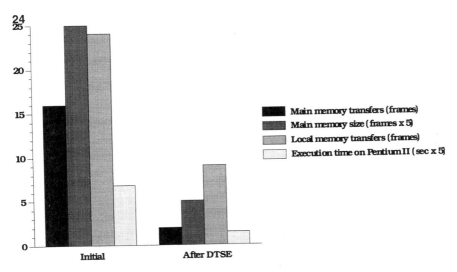

Figure 24   Results for cavity-detection application.

The result of applying these transformations is a greatly improved locality, which will be exploited by the data reuse and in-place mapping steps to reduce the storage and bandwidth requirements of the application. It is clear that only three line buffers per function are needed for the intermediate memory between the different steps of the algorithm (as opposed to the initial version, which needed frame memories between the functions). In-place mapping can further reduce this to two line buffers per function.

The final results for the cavity-detection application are given in Figure 24. The figure shows that the required main memory size has been reduced by a factor of 8. This is especially important for embedded applications, where the chip area is an important cost factor. Moreover, the accesses to both main memory and local memory have been heavily reduced, which will result in a significantly lower power consumption. Because of the increased locality, also the number of cache misses (e.g., on a Pentium II processor) is much lower, resulting in a performance speedup by a factor of 4.

5.4   A Methodology for Automating the Loop Transformations

To automate the loop transformations, we make use of a methodology based on the polytope model [30,31]. In this model, each n-level loop nest is represented

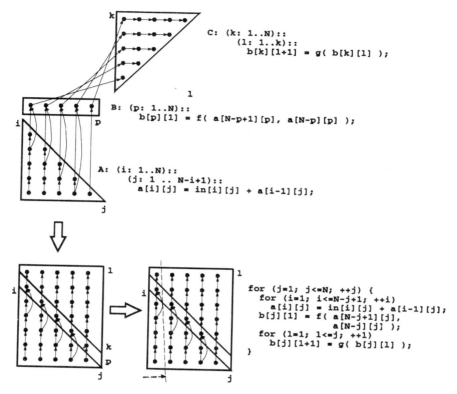

Figure 25 Example of automatable loop transformation methodology.

geometrically by an *n*-dimensional polytope. An example is given at the top of Figure 25, where, for example, the loop nest with label A is two dimensional and has a triangular polytope representation, because the inner loop bound is dependent on the value of the outer-loop index. The arrows in Figure 25 represent the data dependencies; they are drawn in the direction of the data flow. The order in which the iterations are executed can be represented by an ordering vector which traverses the polytope.

To perform global loop transformations, we have developed a two-phase approach. In the first phase, all polytopes are placed in one common iteration space. During this phase, the polytopes are merely considered as geometrical objects, without execution semantics. In the second phase, a global ordering vector is defined in this global iteration space. In Figure 25, an example of this methodology is given. At the top, the initial specification of a simple algorithm is shown; at the bottom left, the polytopes of this algorithm are placed in the common iteration space in an optimal way, and at the bottom right, an optimal ordering vector is defined and the corresponding code is derived.

Most existing loop transformation strategies work directly on the code. Moreover, they typically work on single-loop nests, thereby omitting the global transformations which are crucial for storage and transfers. Many of these techniques also consider the body of each loop nest as one union [32], whereas we have a polytope for each statement, which allows more aggressive transformations. An exception are ''affine-by-statement'' techniques [33] which transform each statement separately, but our two-phase approach still allows a more global view of the data transfer and storage issues.

6 TASK VERSUS DATA-PARALLELISM EXPLOITATION

Parallelization is a standard way to improve the performance of a system when a single processor cannot do the job. However, it is well known that this is not obvious, because many possibilities exist to parallelize a system, which can severely differ in performance. That is also true for the impact on data storage and transfers.

Most of the research effort in this area addresses the problem of parallelization and processor partitioning [25,34,35]. These approaches do not take into account the background-storage-related cost when applied on data-dominated applications. Only speed is optimized and not the power or memory size. The data communication between processors is usually taken into account in most recent methods [36], but they use an abstract model (i.e., a virtual processor grid, which has no relation with the final number of processors and memories). A first approach for more global memory optimization in a parallel processor context was described in Ref. 37, in which we showed that an extensive loop reorganization has to be applied *before* the parallelization steps. A methodology to combine parallelization with code rewriting for DTSE was presented in Ref. 38.

In this section, we will focus on the parallelization itself. Two main alternatives are usually task and data parallelism. In task parallelism, the different subsystems of an application are assigned to different processors. In data parallelism, each processor executes the whole algorithm, but only to a part of the data. Also, hybrid task-data parallel alternatives are possible though. When data transfer and storage optimization is an issue, even more attention has to be paid to the way in which the algorithm is parallelized. In this section, we will illustrate this on two examples, namely the cavity-detection algorithm and an algorithm for video compression (QSDPCM).

6.1 Illustration on Cavity-Detection Application

We will look at some ways to parallelize the cavity-detection algorithm, assuming that a speedup of about 3 is required. In practice, most image processing algorithms do not require large speedups, so this is more realistic than massive speedup. In the following subsections, two versions (initial and globally optimized) of the algorithm are parallelized in two ways:task and data parallelism.

6.1.1 Initial Algorithm

Applying task parallelization to the initial algorithm leads to a coarse-grain pipelining solution (at the level of the image frames). This can work well for load balancing on a three-processor system, but it is clearly an unacceptable method if we have efficient memory management in mind.

Data parallelism is a better choice. Neglecting some border effects, the cavity-detection algorithm lends itself very well to this kind of parallelism, as there are only neighbor-to-neighbor dependencies. Thus, each processor can work more or less independently of the others, except at the boundaries, where some idle synchronization and transfer cycles will occur. Each processor will still need two frame buffers, but now these buffers are only a third of a frame. Thus, the data-parallel solution will, in fact, require the same amount of buffers as the monoprocessor case.

6.1.2 Globally Optimized Algorithm

Applying task parallelization consists of assigning each of the steps of the algorithm to a different processor, but now we arrive at a fine-grain pipelining solution. Processor 1 has a buffer of two lines ($y - 1$ and y). Line $y + 1$ enters the processor as a scalar stream; synchronously, the GaussBlur step can be performed on line y, the result of which can be sent to the second processor as a scalar stream. This one can concurrently (and synchronously) apply the ComputeEdges step to line $y - 1$ and so on. In this way, we only need a buffer of two lines per processor, or six in total! This is the same amount as we needed for the monoprocessor case. Therefore, we have achieved what we were looking for: improved performance without sacrificing storage and transfer overhead (which would translate in area and power overhead). Because the line buffer accesses are in FIFO order, cheap FIFO buffers can be selected to implement them.

When we use data parallelism with the globally optimized version, we will need 6 line buffers per processor, or 18 in total. The *th* of the buffers will depend on the way in which we partition the image. If we use a rowwise partitioning (e.g., the first processor processes the upper third of the image, etc.), the 18 buffers are of the same size as in the monoprocessor case. A columnwise partitioning yields better results: We still need 18 buffers, but their *th* is only a third

Table 3   Results for Cavity-Detection Application

Version + parallelism	Frame mem.	Frame transfers	Line buffers
Initial, data	2	30	0
Initial, task	6	30	0
Transformed, data	0	0	6 SRAM
Transformed, task	0	0	6 FIFO

of a line (thus equivalent to 6 line buffers of full th). Because the accesses are not FIFO compatible in this case, the buffers will have to be organized as SRAMs, which are more expensive than FIFOs.

The results are summarized in Table 3. It is clear that the task-parallel version is the optimal solution here. Note that the load balancing is less ideal than for the data-parallel version, but it is important to trade off performance and DTSE; for example, if we can avoid a buffer of 32 Kbits by using an extra processor, this can be advantageous even if this processor is idle 90% of the time (which also means that we have a very bad load balance), because the cost of this extra processor in terms of area and power is less than the cost of a 32-Kbit on-chip memory.

6.2 QSDPCM Video Processing Application

The QSDPCM (Quadtree Structured Difference Pulse Code Modulation) technique is an interframe compression technique for video images. It involves a motion estimation step and a quadtree-based encoding of the motion-compensated frame-to-frame difference signal. A global view of the algorithm is given in Figure 26. We will not explain the functionality of the different subsystems in detail here.

The partitioning of the QSDPCM application is based on the computational load (i.e., on the number of operations to be executed on each processor). To keep the load balanced, approximately the same amount of computation must be assigned to each of the processors. The total number of operations for the coding of one frame is 12,988K. We will assign about 1000K operations to each processor; therefore, we need 13 processors. The total size of the array signals in the application is 1011K words. The number of accesses to these signals is 9800K per frame. The array signals can be divided in two categories:

- Category A (532K): those which are either inputs or outputs of the algorithm on a per frame basis (coding of one frame)
- Category B (479K): those which are intermediate results during the processing of one frame

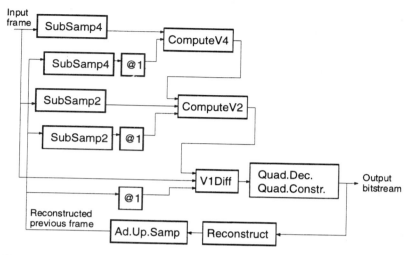

Figure 26 QSDPCM application.

The category A signals can be reduced to 206K by aggressive in-place mapping, but they are always needed, independent of the partitioning. Therefore, our comparisons further on will consider the category B signals.

We will compare (wrt DTSE) different parallelization alternatives for this application: a pure data-level partitioning, a modified task-level partitioning, a pure task-level partitioning, and two hybrid versions (all of these have been obtained by hand). Except for the first partitioning, all alternatives have been evaluated for an initial and a globally transformed version of the code.

6.2.1 Pure Data-Level Partitioning

In data-level partitioning, each of the 13 processors will perform the whole algorithm on its own part of the current frame (so each processor will work on approximately 46 blocks of 8×8 pixels). The basic advantage of this approach is that it is simple to program. It also requires reduced communication between the processors. On the other hand, each processor has to run the full code for all the tasks.

The memory space required for this partitioning is 541K words (42K per processor), whereas the number of accesses for the processing of 1 frame is 12,860K. The main reason for these huge memory space requirements is that each processor operates on a significant part of the incoming frame and buffers are required to store the intermediate results between two submodules of the QSDPCM algorithm. Because of their large size (541K), these buffers cannot be stored on-chip. The increase of the number of accesses (from 9900K in the non-parallel solution to 12860K) is due to the overlapping motion-estimation regions of the blocks in the boundaries of neighboring frame areas.

6.2.2 Modified Task-Level Partitioning

In task-level partitioning, the different functions (or parts of them) of the QSDPCM algorithm are assigned in parts of about 1 million operations over the 13 processors. This partitioning is more complex than data-level partitioning, and for this reason, it requires more design time. It also requires increased communication (through double buffers) between processors. However, each processor runs only a small part of the code of the QSDPCM application. Here, a modified task-level solution is discussed (i.e., one which is, in fact, not a pure task-level partitioning, as it already includes some data parallelism). A pure task-level partitioning will be discussed in the next subsection.

Figure 27 shows the modified task-level partitioning. The required memory space for the category B signals is 287K words, which is significantly smaller than for the data-level partitioning (541K). The main reason for this is that all of the buffers storing the intermediate signals are only present between two processors. In data-level partitioning, these buffers are present in each processor. The number of memory accesses required for the processing of one frame in the task-level case (13,640K) is increased in comparison to the data-level case

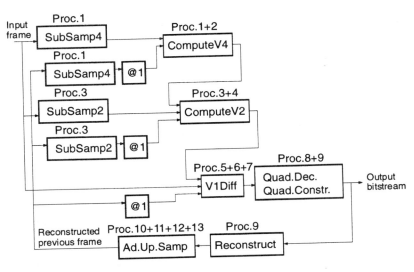

Figure 27 Modified task-level partitioning.

(12,860K). This is a result of the increased communication between processors (through double buffers).

6.2.3   Pure Task-Level Partitioning

There are two points in the previous partitioning in which data-level partitioning is performed for specific groups of processors (see Fig. 27). Using task-level partitioning at these points also requires 210K words for the storage of the difference blocks, as well as 2737K accesses to these words. This is an overhead in comparison to the previous solution. Similar conclusions hold for processors 10, 11, 12, and 13. Thus, it is clear that a pure task-level partitioning of the QSDPCM algorithm is very inefficient in terms of area and power.

6.2.4   Modified Task-Level Partitioning Based on a Reorganized Description

Now, the modified task-level partitioning is performed again, but after applying extensive loop transformations in the initial description (as described in Sec. 6). The aim of the loop reorganization is to reduce the memory required to store the intermediate arrays (signals of category B) and the number of off-chip memory accesses, because on-chip storage of these signals will become possible after the size reduction. As a result of the loop transformations, the intermediate array signals require only 1190 words to be stored. This means that on-chip storage is indeed possible.

6.2.5   Pure Task-Level Partitioning Based on the Reorganized Description

The pure task-level partitioning has been analyzed for the loop reorganized description too. The result is that it imposes an overhead of 1024 words and 2737K accesses to these words. Thus, it is clear that the overhead is significant.

6.2.6   Hybrid Partitioning 1, Based on the Initial Description

This hybrid task–data-level partitioning is based on a combination of the pure task and data-level partitionings. The functions of the QSDPCM application are divided into groups which are executed by different groups of processors. However, within each group of processors, each processor performs all of the functions of the group to a different part of the data. The proposed hybrid partitioning is described in Figure 28. As far as the intermediate array signals are concerned, the memory size required for their storage (245K) is smaller than the size required by the modified task-level partitioning (287K). In the same way, the number of accesses to these signals is reduced.

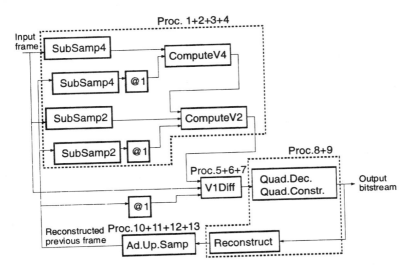

Figure 28 Hybrid partitioning 1.

6.2.7 Hybrid Partitioning 1, Based on the Loop Reorganized Description

For this version, the memory size required for the storage of the intermediate array signals is 1872 words. This memory size is higher in comparison to the corresponding size required by the task-level partitioning (1190 words). Although some buffers between procedures SubSamp4 and V4 as well as between Sub-Samp2 and V2 were eliminated, the gain was offset by replicates of array signals, with the same functionality in all the processors performing the same task on different data.

6.2.8 Hybrid Partitioning 2, Based on the Initial Description

The second hybrid partitioning alternative is oriented more toward the modified task-level partitioning. The only difference is in the assignment of tasks to processors 1 and 2. The memory size required by the intermediate array signals is now 282K words and is slightly reduced in comparison to the task-level partitioning. The number of accesses to these signals is also reduced by 10K in comparison to the task-level partitioning as a result of the reduced communication between processors 1 and 2. This second hybrid partitioning requires more words for the storage of the intermediate array signals and more accesses to these signals in comparison to the first hybrid partitioning.

6.2.9 Hybrid Partitioning 2, Based on the Loop Reorganized Description

In this case, the number of words required to store the intermediate signals (1334 words) is slightly larger than for the corresponding task-level partitioning (1190 words). However, this memory size is now smaller than in the first hybrid partitioning. The number of memory accesses to these signals is smaller than in task-level partitioning but larger than in the first hybrid partitioning.

6.3 Conclusions

As far as the memory size required for the storage of the intermediate array signals is concerned, the results of the partitionings based on the initial description prove that this size is reduced when the partitioning becomes more data oriented. This size is smaller for the first hybrid partitioning (245K), which is more data oriented than the second hybrid partitioning (282K) and the task-level partitioning (287 K).

For the reorganized description; the results indicate the opposite. In terms of the number of memory accesses to the intermediate signals, the situation is simpler: This number always decreases as the partitioning becomes more data oriented.

Table 4 shows an overview of the achieved results. The estimated area and power figures were obtained using a model of Motorola (this model is proprietary so we can only give relative values). From Table 4, it is clear that the rankings for the different alternatives (initial and reorganized) are clearly distinct. For the reorganized description, the task-level-oriented hybrids are better. This is true because this kind of partitioning keeps the balance between double buffers (present in task-level partitioning) and replicates of array signals with the same func-

Table 4 Results for QSDPCM

Version	Partitioning	Area	Power
Initial	Pure data	1	1
	Pure task	0.92	1.33
	Modified task	0.53	0.64
	Hybrid 1	0.45	0.51
	Hybrid 2	0.52	0.63
Reorganized	Pure task	0.0041	0.0080
	Modified task	0.0022	0.0040
	Hybrid 1	0.0030	0.0050
	Hybrid 2	0.0024	0.0045

tionality in different processors (present in data-level partitioning). However, we believe that the optimal partitioning depends highly on the number of submodules of the application and on the number of processors.

7 DATA LAYOUT REORGANIZATION FOR REDUCED CACHE MISSES

In section 3, we introduced the three types of cache miss and identified that conflict misses are one of the major hurdles to achieving better cache utilization. In the past, source-level program transformations to modify execution order to enhance cache utilization by improving the data locality have been proposed [39–41]. Storage-order optimizations are also helpful in reducing cache misses [42,43]. However, existing approaches do not eliminate the majority of conflict misses. In addition [39,42], very little has been done to measure the impact of data layout optimization on the cache performance. Thus, advanced data layout techniques need to be identified to eliminate conflict misses and improve the cache performance.

In this section, we discuss the memory data layout organization (MDO) technique. This technique allows an application designer to remove most of the conflict cache misses. Apart from this, MDO also helps in reducing the required bandwidth between different levels of memory hierarchy due to increased spatial locality.

First, we will briefly introduce the basic principle behind memory data layout organization with an example. This is followed by the problem formulation and a brief discussion of the solution to this problem. Experimental results using a source-to-source compiler for performing data layout optimization and related discussions are presented to conclude this section.

7.1 Example Illustration

Figure 29 illustrates the basic principle behind MDO. The initial algorithm needs three arrays to execute the complete program. Note that the initial memory data layout is single contiguous irrespective of the array and cache sizes. The initial algorithm can have $3N$ (cross-) conflict cache misses for a direct-mapped cache, in the worst case [i.e., when each of the arrays are placed at an (base) address, which is a multiple of the cache size]. Thus, to eliminate all of the conflict cache misses, it is necessary that none of the three arrays get mapped to the same cache locations, in this example.

The MDO optimized algorithm will have no conflict cache misses. This is because in the MDO optimized algorithm, the arrays always get mapped to fixed nonoverlapping locations in the cache. This happens because of the way the data

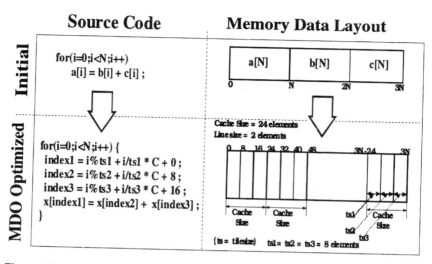

Figure 29 Example illustration of MDO optimization on a simple case. Note the source code generated and the modified data layout.

are stored in the main memory, as shown in Figure 29. To obtain this modified data layout, the following steps are carried out: (1) The initial arrays are split into subarrays of equal size. The size of each subarray is called *tile-size*. (2) Merge different arrays so that the sum of their tile-size's equals cache size. Now, store the merged array(s) recursively until all of the arrays concerned are mapped completely in the main memory. Thus, we now have a new array which comprises all the arrays, but the constituent arrays are stored in such a way that they get mapped into cache so as to remove conflict misses and increase spatial locality. This new array is represented by x[ ] in Figure 29.

In Figure 29, two important observations need to be made: (1) There is a recursive allocation of different array data, with each recursion equal to the cache size and (2) the generated addressing, which is used to impose the modified data layout on the linker.

7.2 The General Problem

In this section, we will first present a complete problem definition and then discuss the potential solutions. The general memory data layout organization problem for efficient cache utilization (DOECU) can be stated as, "For a given program with *m* loop nests and *n* variables (arrays), obtain a data layout which has the least possible conflict cache misses." This problem has two subproblems.

First, the tile size evaluation problem and, second, the array merging/clustering problem.

7.2.1 Tile-Size Evaluation Problem

Let X_i be the tile size of the array i and C be the cache size. For a given program, we need to solve the m equations in Eq. (1) to obtain the needed (optimal) tile sizes. This is required for two reasons. First, an array can have different effective size in different loop nests. We define effective size as "the number of elements of an array accessed in a loop nest." This number can thus represent either the complete array size or a partial size and is represented as *effsize*. The second reason is that different loop nests have a different number of arrays which are simultaneously alive.

$$
\begin{aligned}
L_1 &= x_1 + x_2 + x_3 + \cdots + x_n \leq C \\
L_2 &= x_1^1 + x_2^1 + x_3^1 + \cdots + x_n^1 \leq C \\
&\vdots \\
L_m &= x_1^{(m-1)} + x_2^{(m-1)} + x_3^{(m-1)} + \cdots + x_n^{(m-1)} \leq C
\end{aligned}
\tag{1}
$$

These equations need to be solved so as to minimize the number of conflict misses. The conflict misses can be estimated using techniques such as cache miss equations [43]. In this chapter, we assume that all of the arrays which are simultaneously alive have an equal probability to conflict (in the cache). The optimal solution to this problem comprises solving the ILP problem [44,45], which requires large CPU time. Hence, we have developed heuristics which provide good results in a reasonable CPU time. A more detailed discussion of this topic can be found in Refs. 46 and 47.

7.2.2 Array Merging/Clustering Problem

We now further formulate the general problem using the loop weights. The weight in this context is the probability of conflict misses calculated based on the simultaneous existence of arrays for a particular loop nest (i.e., sum of effective sizes of all the arrays) as given by

$$
L_{wk} = \sum_{i=1}^{n} effsize_i
\tag{2}
$$

Hence, now the problem to be solved is which variables are to be clustered or merged and in what order (i.e., from which loop nest onward) so as minimize the cost function. Note that we have to formulate the array merging problem this way, because we have many tile sizes for each array* and there are a different number of arrays alive in different loop nests. Thus, using the above loop weights, we can identify loop nests which can potentially have more conflict misses and focus on clustering arrays in these loop nests.

7.3 The Pragmatic Solution

We now discuss some pragmatic solutions for the above problem. These solutions comprise heuristics, which are less complex and faster from the point of view of automation. First, we briefly discuss how the two stages of the problem are solved:

1. The first step involves evaluation of the effective sizes for each array instances in the program. Next, we perform a proportionate allocation based on the effective size of every array in every loop nest. This means that arrays with larger effective sizes get larger tile sizes and vice versa. Thus, the remaining problem is the merging of different arrays.

2. The second step involves the merging/clustering of different arrays with their tile sizes. To achieve this, we first arrange all the loop nests (in our internal model) in ascending order of their loop weights, as calculated earlier. Next, we start merging arrays from the loop nest with highest loop weight and go on until the last remaining array has been merged. Note that once the total tile size is equal to the cache size, we start a second cluster and so on. This is done in a relatively greedy way, because we do not explore for the best possible solution extensively.

We have automated two heuristics in a prototype tool, which is a source-to-source (*C*-to-*C*) precompiler step. The basic principle of these two heuristics are as follows:

1. **DOECU I:** In the first heuristic, the tile size is evaluated individually for each loop nest which means that the proportionate allocation is performed based on the effective sizes of each array in the particular loop nest itself. Thus, we have many alternatives† for choosing the tile size for an array. In the next step, we start merging the arrays from the loop nest with the highest weight, as calculated earlier, and move to the loop nest with the next highest weight and so on until all of the arrays are merged. In summary, we evaluate the tile sizes locally but perform the merging globally based on loop weights.

---

* In the worst case, one tile size for every loop nest in which the array is alive.
† In the worst case, we could have a different tile size for every array in every loop nest for the given program.

2. **DOECU II**: In the second heuristic, the tile sizes are evaluated by a more global method. Here, we first accumulate the effective sizes for every array over all of the loop nests. Next, we perform the proportionate allocation for every loop nest based on the accumulated effective sizes. This results in smaller difference between tile size evaluated for an array in one loop nest compared to the one in another loop nest. This is necessary because suboptimal tile sizes can result in larger self-conflict misses. The merging of different arrays is done in a similar way to that in the first heuristic.

7.4 Experimental Results

This subsection presents the experimental results of applying MDO, using the prototype DOECU tool, on three different real-life test vehicles, namely a cavity-detection algorithm used in medical imaging, a voice-coder algorithm which is widely used in speech processing, and a motion-estimation algorithm used commonly in video processing applications. Cavity-detection algorithm is explained in Section 5.2. We will not explain the other algorithms here, but it is important to note that they are data dominated and comprise 2–10 pages of C code.

The initial C source code is transformed using the prototype DOECU tool, which also generates back the transformed C code. These two C codes, initial and MDO optimized, are then compiled and executed on the Origin 2000 machine and the performance monitoring tool "perfex" is used to read the hardware counters on the MIPS R10000 processor.

Tables 5–7 give the obtained results for the different measures for all the three applications. Note that Table 7 has same result for both the heuristics be-

Table 5 Experimental Results on Cavity-Detection Algorithm Using MIPS R10000 Processor

	Initial	DOECU-I	DOECU-II
Avg. mem. acc. time	0.48218	0.2031	0.187943
L1 cache line reuse	423.2192	481.1721	471.0985
L2 cache line reuse	4.960771	16.65545	23.19886
L1 cache hit rate	0.997643	0.997926	0.997882
L2 cache hit rate	0.832236	0.94336	0.958676
L1–L2 BW (Mb/sec)	13.58003	4.828789	4.697513
L2–mem. BW (Mb/sec)	8.781437	1.017692	0.776886
L1–L2 data transfer (Mb)	6.94	4.02	3.7
L2–mem. data transfer (Mb)	4.48	0.84	0.61

Table 6 Experimental Results on Voice Coder Algorithm Using MIPS R10000 Processor

	Initial	DOECU-I	DOECU-II
Avg. mem. acc. time	0.458275	0.293109	0.244632
L1 cache line reuse	37.30549	72.85424	50.88378
L2 cache line reuse	48.51464	253.4508	564.5843
L1 cache hit rate	0.973894	0.98646	0.980726
L2 cache hit rate	0.979804	0.99607	0.998232
L1–L2 BW (Mb/sec)	115.4314	43.47385	49.82194
L2–mem. BW (Mb/sec)	10.13004	0.707163	0.31599
L1–L2 data transfer (Mb)	17.03	10.18	9.77
L2-mem. data transfer (Mb)	1.52	0.16	0.06

cause the motion-estimation algorithm has only one (large) loop nest with a depth of six, namely six nested loops with one body.

The main observations from all the three tables are as follows. MDO optimized code has a larger spatial reuse of data both in the L1 and L2 cache. This increase in spatial reuse is due to the recursive allocation of simultaneously alive data for a particular cache size. This is observed from the L1 and L2 cache line reuse values. The L1 and L2 cache hit rates are consistently greater too, which indicates that the tile size evaluated by the tool were nearly optimal because for suboptimal tile sizes, there will more self-conflict cache misses.

Because the spatial reuse of data is increased, the memory access time is reduced by an average factor 2 all of the time. Similarly, the bandwidth used between the L1 and L2 caches is reduced by 40% to a factor of 2.5 and the

Table 7 Experimental Results on Motion-Estimation Algorithm Using MIPS R10000 Processor

	Initial	DOECU-I/II
Avg. mem. acc. time	0.782636	0.28985
L1 cache line reuse	9,132.917	13,106.61
L2 cache line reuse	13.5	24.22857
L1 cache hit rate	0.999891	0.999924
L2 cache hit rate	0.931034	0.960362
L1–L2 BW (Mb/sec)	0.991855	0.299435
L2–mem. BW (Mb/sec)	0.31127	0.113689
L1–L2 data transfer (Mb)	0.62	0.22
L2–mem. data transfer (Mb)	0.2	0.08

bandwidth between the L2 cache and the main memory is reduced by a factor of 2–20. This indicates that although the initial algorithm had larger hit rates, the hardware was still performing many redundant data transfers between different levels of the memory hierarchy. These redundant transfers are removed by the modified data layout and heavily decreased the system bus loading. This has a large impact on the global system performance, because most multimedia applications are required to operate with peripheral devices connected using the off-chip bus.

Because we generate complex addressing, we also perform address optimizations [48] to remove the addressing overhead. Our studies have shown that we are able to not only remove the complete overhead in addressing but also gain by up to 20% in the final execution time, on MIPS R10000 and HP PA-8000 processors, compared to the initial algorithm, apart from obtaining the large gains in the cache and memory hierarchy.

REFERENCES

1. G Lawton. Storage technology takes the center stage. IEEE Computer Mag 32(11): 10–13, 1999.
2. R Evans, P Franzon. Energy consumption modeling and optimization for SRAMs. IEEE J Solid-State Circuits 30(5):571–579, 1995.
3. K Itoh, Y Nakagome, S Kimura, T Watanabe. Limitations and challenges of multi-gigabit DRAM chip design. IEEE J Solid-State Circuits 26(10), 1997.
4. B Prince. Memory in the fast lane. IEEE Spectrum 38–41, 1994.
5. R Jolly. A 9ns 1.4GB/s 17-ported CMOS register file. IEEE J Solid-State Circuits 26(10):1407–1412, 1991.
6. N Weste, K Esharaghian. Principles of CMOS VLSI Design. 2nd ed. Reading, MA: Addison-Wesley, 1993.
7. D Patterson, J Hennessey. Computer Architecture: A quantitative Approach. San Francisco: Morgan Kaufmann, 1996.
8. AJ Smith. Line size choice for CPU cache memories. IEEE Trans Computers 36(9), 1987.
9. CL Su, A Despain. Cache design tradeoffs for power and performance optimization: a case study. Proc. Int. Conf. on Low Power Electronics and Design (ICLPED), 1995, pp 63–68.
10. U Ko, PT Balsara, A Nanda. Energy optimization of multi-level processor cache architectures. Proc. Int. Conf. on Low Power Electronics and Design (ICLPED), 1995, pp 63–68.
11. Philips. TriMedia TM1000 Data Book. Sunnyvale, CA: Philips Semiconductors, 1997.
12. R Comerford, G Watson. Memory catches up. IEEE Spectrum 34–57, 1992.
13. Y Oshima, B Sheu, S Jen. High speed memory architectures for multimedia applications. IEEE Circuits Devices Mag 8–13, 1997.

14. Eto S, et al. A 1-GB SDRAM with ground-level precharged bit-line and nonboosted 2.1V word-line. IEEE J Solid-state Circuits 33:1697–1702, 1998.

15. Kirihata T, et al. A 220 mm², four- and eightbank, 256 Mb SDRAM with single-sided stiched WL architecture. IEEE J Solid-State Circuits 33:1711–1719, 1998.

16. T Yamada (Sony). Digital storage media in the digital highway era. Proc. IEEE Int. Solid-State Circuits Conf. (ISSCC), 1995, pp 16–20.

17. K Itoh. Low voltage memory design. IEEE Int. Symp. on Low Power Design (ISLPD), Tutorial on Low Voltage Technologies and Circuits, 1997.

18. S Kuge, F Morishita, T Suruda, S Tomishima, N Tsukude, T Yamagata, K Arimoto. SOI–DRAM circuit technologies for low power high speed multigiga scale memories. IEEE J Solid-State Circuits 31:586–596, 1996.

19. T Seki, E Itoh, C Furukawa, I Maeno, T Ozawa, H Sano, N Suzuki. A 6-ns 1-Mb CMOS SRAM with latched sense amplifier. IEEE J Solid-State Circuits 28(4):478–483, 1993.

20. Kim C, et al. A 64 Mbit, 640 MB/s bidirectional data-strobed double data rate SDRAM with a 40mW DLL for a 256 MB memory system. IEEE J Solid-State Circuits 33:1703–1710, 1998.

21. T Tsuruda, M Kobayashi, T Tsukude, T Yamagata, K Arimoto. High-speed, high bandwidth design methodologies for on-chip DRAM core multimedia system LSIs. Proc. IEEE Custom Integrated Circuits Conf. (CICC), 1996, pp 265–268.

22. D Patterson, T Anderson, N Cardwell, R Fromm, K Keeton, C Kozyrakis, R Thomas, K Yelick. Intelligent RAM (IRAM): Chips that remember and compute. Proc. IEEE Int. Solidstate Circuits Conf. (ISSCC), 1997, pp 224–225.

23. N Wehn, S Hein. Embedded DRAM architectural trade-offs. Proc. 1st ACM/IEEE Design and Test in Europe Conf. 1998, pp 704–708.

24. T Takayanagi et al. 350 MHz time-multiplexed 8-port SRAM and word-size variable multiplier for multimedia DSP. Proc. IEEE Int. Solid-State Circuits Conf. (ISSCC), 1996, pp 150–151.

25. S Amarasinghe, J Anderson, M Lam, C Tseng. The SUIF compiler for scalable parallel machines. Proc. 7th SIAM Conf. on Parallel Processing for Scientific Computing, 1995.

26. M Wolf. Improving data locality and parallelism in nested loops. PhD thesis, Stanford University, 1992.

27. U Banarjee, R Eigenmann, A Nicolau, D Padua. Automatic program parallelisation. Proc IEEE 81, 1993.

28. T-S Shen, J-P Sheu. Communication free data allocation techniques for parallelizing compilers on multicomputers. IEEE Trans Parallel Distrib Syst 5(9):924–938, 1994.

29. M Gupta, E Schonberg, H Srinivasan. A unified framework for optimizing communication in data-parallel programs. IEEE Trans Parallel Distrib Syst 7(7):689–704, 1996.

30. C Lengauer. Loop parallelization in the polytope model. Proc. 4th Int. Conf. on Concurrency Theory (CONCUR), 1993.

31. M van Swaaij, F Franssen, F Catthoor, H de Man. Modelling data and control flow for high-level memory management. Proc. 3rd ACM/IEEE European Design Automation Conf. (EDAC), 1992.

32. A Darte, Y Robert. Scheduling uniform loop nests. Internal Report 92-10, ENSL/ IMAG, Lyon, France, 1992.

33. A Darte, Y Robert. Affine-bystatement sheduling of uniform loop nests over parametric domains. Internal Report 92-16, ENSL/IMAG, Lyon, France, 1992.

34. M Neeracher, R Ruhl. Automatic parallelisation of LINPACK routines on distributed memory parallel processors. Proc. IEEE Int. Parallel Proc. Symp. (IPPS), 1993.

35. C Polychronopoulos. Compiler optimizations for enhancing parallelism and their impact on the architecture design. IEEE Trans Computer 37(8):991–1004, 1988.

36. A Agarwal, D Krantz, V Nataranjan. Automatic partitioning of parallel loops and data arrays for distributed sharedmemory multiprocessors. IEEE Trans Parallel Distrib Syst 6(9):943–962, 1995.

37. K Danckaert, F Cathhoor, H de Man. System-level memory management for weakly parallel image processing. In Proc. Euro-Par Conf. Lecture Notes in Computer Science Vol. 1124. Berlin: Springer Verlag, 1996.

38. K Danckaert, F Cathhoor, H de Man. A loop transformation approach for combined parallelization and data transfer and storage optimization. Proc. Conf. on Parallel and Distributed Processing Techniques and Applications, 2000, Volume V, pp 2591–2597.

39. M Kandemir, J Ramanujam, A Choudhary. Improving cache locality by a combination of loop and data transformations. IEEE Trans Computers 48(2):159–167, 1999.

40. M Lam, E Rothberg, M Wolf. The cache performance and optimization of blocked algorithms. Proc. Int. Conf. on Architectural Support for Programming Languages and Operating Systems, 1991, pp 63–74.

41. D Kulkarni, M Stumm. Linear loop transformations in optimizing compilers for parallel machines. Austral Computer J 41–50, 1995.

42. PR Panda, ND Dutt, A Nicolau. Memory data organization of improved cache performance in embedded processor applications. Proc. Int. Symp. on System Synthesis, 1996, pp 90–95.

43. E De Greef. Storage size reduction for multimedia applications. PhD thesis, Department of Electrical Engineering, Katholieke Universiteit, Leuven, Belgium, 1998.

43. S Ghosch, M Martonosi, S Malik. Cache miss equations: A compiler framework for analyzing and tuning memory behaviour. ACM Trans Program Lang Syst 21(4): 702–746, 1999.

44. CL Lawson, RJ Hanson. Solving Least-Square Problems. Classics in Applied Mathematics. Philadelphia: SIAM, 1995.

45. GL Nemhauser, LA Wolsey. Integer and Combinatorial Optimizations. New York: Wiley, 1988.

46. C Kulkarni. Cache conscious data layout organization for embedded multimedia applications. Internal Report, IMEC-DESICS, Leuven, Belgium, 2000.

47. C Kulkarni, F Cathhoor, H de Man. Advanced data layout optimization for multimedia applications. Proc workshop on Parallel and Distributed Computing in Image, Video and Multimedia Processing (PDIVM) of IPDPS 2000. Lecture Notes in Computer Science Vol. 1800, Berlin: Springer-Verlag, 2000, pp 186–193.

48. S Gupta, M Miranda, F Cathhoor, R Gupta. Analysis of high-level address code transformations for programmable processors. Proc. 3rd ACM/IEEE Design and Test in Europe Conf., 2000.

Index